# Bioethics and Human Rights

A READER FOR HEALTH PROFESSIONALS

# Bioethics and Human Rights

EDITED BY

## Elsie L. Bandman, R.N., Ed.D.

Associate Professor of Nursing, Hunter College–Bellevue School of
Nursing, Hunter College of the City University of New York

## Bertram Bandman, Ph.D.

Professor of Philosophy, Brooklyn Center, Long Island
University, Brooklyn

LITTLE, BROWN AND COMPANY, BOSTON

Medical ethics.
Bioethics.
Civil rights.

Library of Congress Catalog Card No. 78-57419

ISBN 0-316-07998-7

Printed in the United States of America

TO OUR STAUNCHEST AND MOST LOYAL SUPPORTERS:

*Nancy Bandman*
*Alfred Rifkin*
*Dorothy Nayer*

> *A human right is something of which no one may be deprived without a grave affront to justice.*
> *— Maurice Cranston*

> *. . . rights are political trumps held by individuals.*
> *— Ronald Dworkin*

# Contents

# Preface

This book is written for the health professional, who, as the person with the most frequent and intimate contact with the health consumer, is in a position of influence and trust. In it papers from varying perspectives are presented, including the topics of human rights, genetics, abortion, euthanasia, the rights of especially vulnerable groups, the medical structure, the problem of alienation, and the rights within and to health care. The book is the product of interdisciplinary thought and values.

As presented, the function of bioethics is to explicate questions of moral rights. In this work the relevance of moral rights to problems and issues that occur in a health care context is demonstrated, and ethical principles are related to specific problematic situations.

It is now commonplace for a health professional to be confronted by a client on matters affecting the course of his or her illness or the morality of his or her choice about treatment. It becomes incumbent, therefore, that the health professional assume responsibility for responding to the client in ways that support the client's moral deliberation and human rights.

An assumption made throughout the book is that the right to the enjoyment of good health is a fundamental necessity of life as vital as food, shelter, and safety. Health care is one way of achieving good health. Good health is defined as encompassing all the structures, processes, individuals, and resources that affect the health and well-being of persons, groups, and families. It is viewed as placing the person in the best possible condition for coping with his or her life through care, cure, and compassion.

The client is seen at the center of this vortex, as the moral source of authority for decisions about personal health and well-being. The role of the health professional is perceived ideally as a relationship of mutuality and complementarity to the client. In a supportive environment of justice, the health professional protects the client's rights and facilitates the client's efforts to arrive at his or her own conclusions about health care.

The new definition of relations between client and provider, which stresses equality of rights, is an outgrowth of the wave of "consumerism" and "rights talk," reflecting public awareness of health affairs. The consumer seeks partnership in decisions of policy and expects accountability for the quality of care given and monies spent. The public clamors for accessible health care that is given in a manner that enhances the self-respect and dignity of the recipient and provider alike. The health care professional is now in a position to participate in worthwhile change.

Further opportunity for the health professional to increase his or her moral responsibility to the client is evident in the emerging patterns of medical care. The traditional relationship of the physician to the patient is giving way to group practice, "team medicine," and to the use of assistants of various kinds. The highly trained physician may be too busy to provide the patient with

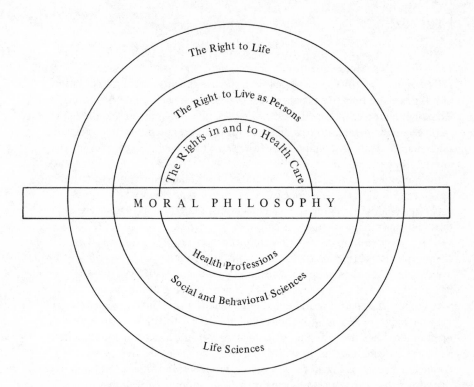

The Right to Life

The Right to Live as Persons

The Rights in and to Health Care

MORAL PHILOSOPHY

Health Professions

Social and Behavioral Sciences

Life Sciences

*The relation of moral-philosophical foundations to the rights to life, to personhood, and to health care, which correspond to the topics of this book.*

explanations of medical procedures or to work through language or cultural barriers to understanding.

Increasing the use of technological measures for diagnosis and treatment of patients is another means for loosening the attachment between physician and patient. The patient's contact with many complicated medical procedures and with diverse health professionals intimately involved in the diagnoses, treatment, and care of the patient inhibits the exclusive relation between patient and physician. As the sources of information and support enlarge, the client's attachment to the physician lessens, and the health professional (other than the physician) can be an effective part of the patient's deliberative process. This book assists that endeavor by enlarging the professional's knowledge about alternative moral decisions and underlying assumptions. Accordingly, the issues of health and illness are no longer the prerogative of a few to decide. The growth of knowledge and the increased sensitivity to human values enable health professionals to speak on behalf of patient autonomy and self-determination.

This book aims to increase such knowledge and sensitivity. Reference to a large body of thought and deliberation on each side of an issue are integral parts of the presentation. More than a dozen different disciplines and professions are represented. Different views are expressed about the meaning of rights. Views on rights range from Abraham Melden, who argues that rights are central to moral discourse, to James McClellan, who holds, on the contrary, that rights are "a form of idolatry."

*Bioethics and Human Rights* is divided into four topics, beginning with a discussion of the moral-philosophical foundations of rights in Topic I that permeates the other three topics. The purpose is to help the reader evaluate the relation between human rights and the fundamental biological sciences, the social sciences, and their professional applications (see figure). Topic II on the right to life examines the problems inherent in genetics, abortion, and euthanasia and their implications for public policy. Topic III, the right to live as persons, is concerned with the needs of special groups and policy recommendations based on assumptions from political science, economics, and social psychology. Issues of rights of groups that are considered special because of their vulnerability, such as the newborn, aged, mentally ill, and prisoners, are explored.

Topic IV, on the right to health care, applies the work of moral philosophy as well as the findings of the life sciences and the social sciences to an examination of policies and problems in health care institutions. Issues of the right to life and the quality of life are examined in the dynamics of distributing health care services.

A teaching device that may be helpful in illustrating some problems associated with formulating rational and just policies on abortion, euthanasia, and the distribution of scarce health care resources consists in drawing two concentric circles on the board. If the inner, smaller circle represents available, finite resources at any time and is kept small in relation to the outer, larger circle (which represents total needs and claims), the inner circle will represent an adequate distribution of health care and related goods and services. But if the inner circle provides only for a small number of people, many people on earth will be outside the inner circle, with virtually no health care goods and services distributed to those outside the inner circle. The larger the inner circle, the fewer beings will be excluded. But the larger the inner circle, the less there will be for each individual within the inner circle. If the inner circle approaches the outer circle, it will be like distributing a slice of bread to 50 people. But to return to making the inner circle smaller, to give one slice to 1 person would mean 49 people would be excluded. The resulting dilemma for proabortionists and antiabortionists alike, along with euthanasia advocates and opponents as well as proponents and opponents of health care rights, is: How large can we justly make the inner circle? The essays in this book can be viewed as alternative attempts at taking defensible positions on how small or large to make the inner circle; whether, for example, to include as rightholders fetuses, prisoners, the mentally ill, the old, the poor, the retarded, and infants.

It is hoped that the reader will find this anthology sufficiently provocative to generate well-considered solutions in place of stereotypical thinking. The issues posed are not easily amenable to resolution but may be ameliorated by individuals who develop moral sensitivity and principle.

We acknowledge our indebtedness to all our authors, who were steadfast and supportive throughout the development of this book. We appreciate the help of the Society for Philosophy and Public Affairs for the wholehearted participation of many of its members in this work. We thank the students, faculty, and administration of Long Island University, particularly the Dean of Instructional Resources, Gary Marotta; Deans Felice Lewis and Ray Longtin; Professors Robert Spector and Elinor West; and Julius Prado and Ms. Cyd Smith for help and encouragement. At an earlier phase, the New York Council for the Humanities and the Andrew W. Mellon Foundation supported a three-day interdisciplinary conference, which contributed to the conception of this book. We thank Ronald Florence, Director of the New York Council for the Humanities, for his encouragement of this project.

We also thank Christopher R. Campbell, former Nursing Editor of Little, Brown, for his faith and confidence in the project and his important advice throughout; Julie Stillman, Nursing Editor of Little, Brown, for her continuous, valuable help; and Jane MacNeil, Copyediting Supervisor of Little, Brown, for the excellence of the final editing work. Finally, we are profoundly grateful and owe much more than we can say to Dorothy Nayer for her invaluable editorial assistance and for whom this project was a labor of love.

E. L. B.
B. B.

New York City

# Contributing Authors

RAZIEL ABELSON, Ph.D.
*Professor of Philosophy, New York University, New York*

NATALIE ABRAMS, Ph.D.
*Assistant Professor of Philosophy, New York University Medical Center, New York*

ROBERT BAKER, Ph.D.
*Assistant Professor of Philosophy, Union College, Schenectady, New York*

BERTRAM BANDMAN, Ph.D.
*Professor of Philosophy, Brooklyn Center, Long Island University, Brooklyn*

ELSIE L. BANDMAN, R.N., Ed.D.
*Associate Professor of Nursing, Hunter College–Bellevue School of Nursing, Hunter College of the City University of New York*

BERNARD H. BAUMRIN, Ph.D., J.D.
*Professor of Philosophy, Lehman College and the Graduate School of the City University of New York*

BERNARD J. BERGEN, Ph.D.
*Professor of Psychiatry, Department of Psychiatry, Dartmouth Medical School, Hanover, New Hampshire*

JEFFREY BLUSTEIN, Ph.D.
*Assistant Professor of Philosophy, Barnard College, New York*

PHILIP W. BRICKNER, M.D.
*Assistant Professor of Clinical Medicine, New York University School of Medicine; Director, Department of Community Medicine, St. Vincent's Hospital, New York*

JAMES F. CHILDRESS, Ph.D.
*Joseph P. Kennedy, Senior, Professor of Christian Ethics, Kennedy Institute, Center for Bioethics, Georgetown University, Washington, D.C.*

JUAN COBARRUBIAS, Ph.D.
*Assistant Professor of Philosophy, William Patterson College, Wayne, New Jersey*

MARIA COBARRUBIAS, M.A.
*Adjunct Assistant Professor of Philosophy, Hunter College, City University of New York*

ARTHUR J. DYCK, Ph.D.
*Mary B. Saltonstall Professor of Population Ethics, School of Public Health, Member of the Faculty of the Divinity School, and Codirector, Kennedy Interfaculty Program in Medical Ethics, Harvard University*

LEE EHRMAN, Ph.D.
*Professor of Biology, Division of Natural Science, State University of New York, Purchase*

GERTRUDE EZORSKY, Ph.D.
*Professor of Philosophy, Brooklyn College and the Graduate School of the City University of New York*

JOEL FEINBERG, Ph.D.
*Professor of Philosophy, University of Arizona, Tucson*

SALLY GADOW, Ph.D.
*Assistant Professor of Philosophy, Johns Hopkins University, School of Health Services, Baltimore*

WILLARD GAYLIN, M.D.
*President, The Hastings Center, Institute of Society, Ethics and the Life Sciences, Hastings-on-Hudson, New York*

JUNE RESNICK GERMAN, J.D.
*Supervising Attorney, Mental Health Information Service, Appellate Division, First Judicial Department, New York*

JUNE P. GIKUURI, R.N., M.S.
*Director of Patient Relations/Community Board, Kings County Hospital Center, Brooklyn*

MARTIN P. GOLDING, Ph.D.
*Professor of Philosophy, Duke University, Durham, North Carolina*

VIRGINIA HELD, Ph.D.
*Associate Professor of Philosophy, Hunter College, City University of New York*

ADELE D. HOFMANN, M.D.
*Associate Professor of Clinical Pediatrics, New York University School of Medicine; Director, Adolescent Medical Unit, New York University Medical Center, Bellevue Hospital Center*

MARVIN KOHL, Ph.D.
*Professor of Philosophy, State University College at Fredonia, Fredonia, New York*

MARC LAPPÉ, Ph.D.
*Chief, Office of Health, Law, and Value, State Department of Health, Sacramento, California*

BURTON M. LEISER, Ph.D.
*Professor and Chairman, Department of Philosophy, Drake University, Des Moines, Iowa*

HELEN BLOCK LEWIS, Ph.D.
*Adjunct Professor of Psychology, Yale University, New Haven*

FLORENCE LIEBERMAN, D.S.W.
*Associate Professor of Social Work, Hunter College, City University of New York*

JACOB J. LINDENTHAL, Ph.D.
*Chief of Behavioral Sciences, Department of Psychiatry, College of Medicine and Dentistry of New Jersey, Newark*

MALCOLM MacKAY, J.D.
*Senior Vice President, New York Life Insurance Company, New York*

JOSEPH MARGOLIS, Ph.D.
*Professor of Philosophy, Temple University, Philadelphia*

GARY MAROTTA, Ph.D.
*Dean of Instructional Resources, Brooklyn Center, Long Island University, Brooklyn*

ANGELA BARRON McBRIDE, M.S.N.
*Doctoral candidate in Developmental Psychology, Purdue University, Lafayette, Indiana*

WILLIAM LEON McBRIDE, Ph.D.
*Professor of Philosophy, Purdue University, Lafayette, Indiana*

MICHAEL R. McGARVEY, M.D.
*New York State Office of Health Systems Management; Deputy Director, Health Facilities Standards and Control and Chief Medical Officer, New York*

HARRY R. MOODY, Ph.D.
*Executive Secretary, Brookdale Center on Aging, Hunter College, City University of New York*

SIDNEY MORGENBESSER, Ph.D.
*Professor of Philosophy, Columbia University, New York*

CATHERINE P. MURPHY, R.N., Ed.D.
*Assistant Professor, Graduate Medical-Surgical Nursing Program, Boston University School of Nursing, Boston*

ERNEST NAGEL, Ph.D.
*University Professor Emeritus of Philosophy, Columbia University, New York*

LOIS LYON NEUMANN, M.D.
*Associate Professor of Pediatrics, New York University School of Medicine, New York; Director of Neonatology, New York University-Bellevue Medical Center, New York*

LISA H. NEWTON, Ph.D.
*Associate Professor of Philosophy, Fairfield University, Fairfield, Connecticut*

R. JOSEPH NOVOGROD, Ph.D.
*Professor of Political Science and Director, Criminal Justice Programs, Long Island University, Brooklyn Center, Brooklyn*

HILDEGARD E. PEPLAU, R.N., Ed.D.
*Professor Emeritus, Rutgers College of Nursing, Rutgers University, Brunswick, New Jersey*

WILLIAM RUDDICK, Ph.D.
*Associate Professor of Philosophy, New York University, New York*

DEBORAH MICHELLE SANDERS, J.D.
*Lawyer in private practice, Berkeley, California*

DANIEL H. SCHWARTZ, Ph.D.
*Associate Director, Montefiore Hospital and Medical Center, Bronx*

VICTOR W. SIDEL, M.D.
*Professor of Community Health and Chairman, Department of Social Medicine, Montefiore Hospital and Medical Center, Albert Einstein College of Medicine of Yeshiva University, Bronx*

MATTHEW IES SPETTER, Ph.D.
*Associate Professor in Social Psychology, Terris Peace Studies Institute of Manhattan College, New York; First President and Founder, Riverdale Mental Health Clinic, Bronx*

CLAUDEWELL S. THOMAS, M.D., M.P.H.
*Professor and Chairman, Department of Psychiatry and Mental Health Science, College of Medicine and Dentistry of New Jersey, New Jersey Medical School, Newark*

SUSAN H. WEBB, M.A.
*Member of Vermont State General Assembly, Montpelier*

PETER WILLIAMS, Ph.D., J.D.
*Associate Professor of Philosophy, Division of Social Sciences and Humanities, Health Sciences Center, State University of New York at Stony Brook*

EMMETT WILSON, Jr., M.D.
*Assistant Clinical Professor of Psychiatry, Payne Whitney Children's Clinic, Cornell University Medical College, New York*

# Bioethics and Human Rights

# Foundations of Human Rights in Health Care

## THE PURPOSE OF THE BOOK: ITS SCOPE AND SIGNIFICANCE TO HEALTH PROFESSIONALS AND HEALTH CONSUMERS

The purpose of this book is to examine the nature and significance of human rights in a biomedical context. The issues addressed include the right to life, the right to live as persons, and the right to health care. The goal of this book is to question basic assumptions about life, death, and health issues without, however, providing final answers. In the book an attempt is made to provide a variety of reasoned points of view to help the reader arrive at justified beliefs.

The book is topically organized and consists of a focused exchange of views, beliefs, evidence, reasons, and arguments. It is organized into four topics and is designed to sharpen considerations that affect the formulation of institutional and public health policies. A general consideration of the nature and significance of rights is discussed in Topic I. Topic II is about the right to life and is subdivided into genetics, abortion, and euthanasia. It is concerned with the right to plan future generations, the right to one's body, the right of an unborn being to come into the world, and the right to begin and end life.

Put another way, the bioethical problem that requires policy clarification is that of determining where the boundaries of the concept of a human being, or more narrowly, that of a person, are to be drawn. This distinction would enable one to say that a given act of abortion or euthanasia was or was not an unjustified act of killing.

The questions of genetics involve the right to experiment with genes and the right to determine the criteria of selectivity and diversity of future human beings. When determining if abortion is an act of unjustified killing, the right of a woman to control her body in relation to the right of a fetus to be born must be considered together with the question of the stage of pregnancy in which the woman is. These problems are inherent in the topic of abortion. At the other end of the spectrum of life are the problems that center around such issues as killing a person in dire pain who requests and wishes to have his life terminated, for example, a person who is trapped under a burning truck with his body ablaze. Another problem of euthanasia is whether or not a right exists to kill a comatose human being whose explicit consent is unobtainable.

Undoubtedly there are other types of problems concerning the killing of a mature person. The only way to save several lives may be at the loss of one life, or the decision to save one life may endanger the lives of several others. A significant public concern is that the killing of some types of persons may be thought to justify the extermination of people judged undesirable by others, such as the mentally ill or the old or senile.

In Topic III the right to live as persons and the responsibilities for changing behavior are explored. Once a set of persons occupies a place within the circle of life, the problem of the right to live as persons arises. The limits to this right are found in the right a person has to be free to live well, without unjustified interferences of his or her freedom. Conversely, being free — and one can only be free if unmolested and free from unjustified harm, injury, or threat of violence — is the presence of rules and norms for changing undesirable behavior. To induce changed behavior, reward and punishment are commonly used. In this connection, a remarkable insight of Aristotle's [4, Book 10, p. 273] long predates Bentham's famous remark that "all mankind has two masters, pain and pleasure" [10, p. 1]. According to Aristotle, all behavior with which to steer human action stems from two rudders, pain and pleasure [4]. But a question arises as to what extent behavior modification violates the right to live as a person.

Few people would argue against the need to control some forms of behavior. To control behavior a number of behavior modifiers are used, ranging from parental and ministerial preachments, teachers' ways, and orthodox psychoanalysis to drugs of all kinds, electroshock therapy, and psychosurgery. In the face of bewildering devices for changing behavior, a question of public policy that needs clarification is the determination of what rights the mentally ill, for example, have as well as what the scope, limits, and justification of others are to limit their rights. Do the mentally ill have the right to refuse medical and chemotherapeutic treatments as well as electroconvulsive therapy? Do they have a right to leave purely custodial institutions that fail to provide effective therapy? Should deviant behavior be classified as "mental illness"? These issues, in turn, depend on a further clarification of public policy concerning the question of what justifies a norm of standard behavior and a corresponding authorization to alter nonstandard behavior in accordance with the norm.

Three particular groups are especially prone to behavior modification: prisoners, children, and the aging. The members of these groups are especially vulnerable to having their rights abused on the assumption, which is not always granted, that they have rights. When minority groups and poor people are added to these groups, they, too, become steady victims of unwarranted behavior control as if they had no rights.

The right to life and the right to live as a person are insufficient characterizations of a full human life, which should also include the right to live well in health. This issue is considered in Topic IV. Since living as a person involves health care, a question arises as to whether or not health care is a right or a privilege [17]. If health care is a privilege, those persons who can afford it will be the only ones to receive it. If, however, health care is a right, and thus free to everyone, the resulting extension of rights may impose an excessive burden upon those persons who are correspondingly obliged to provide for and to finance such care [17]. In the language of rights, this requirement that rights, if recognized, imply corresponding obligations, as noted by Jeremy Bentham [10, pp. 224–225], means that the right to health care implies a correlative obligation on someone's part to provide it. In a world in which "there are no free lunches," it is increasingly difficult to persuade those persons capable of assuming the duty of providing health care to do so [17]. For the right to health care to be effective depends on the formation of public policies aimed at carving out an appropriate tax support for health care. Some writers argue, however, that the right to health care so extends the concept of rights that rights no longer imply corresponding obligations. Rights become, in effect, *thinned out*.

Beyond the problems of hospital care, further problems of rights to health care are discussed in terms of national and international economic and ecological considerations, with comparisons of their effects in the health care delivery systems of China, Sweden, Britain, and America. The concept of rights vested in the hospital-ized client and the correlative duties of the care provider [2] are said by some to result in adversarial relations rather than relations of trust [3, 8]. Other writers view the delineation of patients' rights as a simple restoration of human rights in a biomedical context [2, 6].

Human rights are considered significant to the care provider, the client, and the consumer in a context of health care by providing choices among alternatives of care [2], based on reasoned beliefs about individual freedom, correlative obligations, and the importance and limits of these obligations. The health consumer, as the recipient, the intended beneficiary, and the utilizer of health care services, is perhaps most vulnerable to the uncertainties and complexities of the health care network. The process of health care itself requires a submission of self to a variety of desig-nated persons, who are usually strangers. Each of these persons in turn performs an aspect of the health care, often without knowledge of its impact on the individual. The client thereby is forced into the alternative either of placing unconditional trust in each health worker or becoming informed about these matters that vitally affect his health and welfare. Even for the informed client, this knowledge may create the illusion of autonomy, since a layperson's knowledge of technical and scientific data remains fragmented and incomplete against the bewildering array of current data. Therefore, the health professional's protection of the person's right to know what is happening to his or her body, which includes knowledge of the risks and alternatives of the treatment, is a significant contribution to the physical, mental, and social well-being of the person [1, 2, 6, 8].

The health professional is in the position to enable the client to become an active, responsible and responsive participant in the care provided [2]. Otherwise, an anxious person, new to the patient role and unfamiliar with technical terms, may be simply unable to grasp the implications and consequences of the treatment plan and may inadvertently nullify it. Even if the treatment plan is understood, the client may be ignorant of his or her human rights and the extent of self-determination that this knowledge implies in terms of accepting or refusing treatment, for example. The Patient's Bill of Rights [2], which is not a legal document but a manifesto, is a publicly articulated statement of intent that becomes a standard of care. The bill begins as a claim to a right that appeals for incorporation into the social and legal systems. The knowledge of this bill, where operative, deepens and extends the awareness of the significance of human rights to both care provider and to the recip-ient. The health professional who has a sense of the importance of human rights to self and others sees new opportunities for additional functions. The health worker may extend the professional role to a position of advocacy, appealing on behalf of the cause of another or of defending or maintaining a proposal or cause [3]. This role may be accomplished by a nurse presenting the family's view as an alternative approach to a problematic situation, by a social worker negotiating another system on behalf of the client, or by all of the professionals in an agency protesting bud-getary cuts and policies antithetical to the agency's mission.

Does this position imply that the health professional must take a stand as moral agent in favor of or against policies affecting procedures of health care and human welfare services? It is believed to be virtually impossible for a competent practicing professional to operate without beliefs and values that shape behavior and determine choice. Based on the preceding assumption, in this book an attempt is made to help explicate those beliefs and values surrounding health issues, so that a more reasoned basis for decision based on philosophically defensible positions regarding human rights and bioethical issues is provided.

Strong currents of change are forthcoming in the structure, functions, and financing of health care delivery systems. Many of the predicted changes appear to be motivated primarily by economic and political factors. Further consideration reveals changes based primarily on underlying assumptions of values. The issue of everyone's right to health care, for example, presupposes value assumptions regarding priorities among the allocations of scarce, finite resources, the ultimate worth of the individual, and the duties and obligations of society to him or her. Complicating the issue are considerations of the person's rights and responsibility for his or her own body and behavior that adversely affects the body (smoking, obesity, alcoholism), and finally, whether or not society is then obligated to care for the cancer induced by smoking, the cardiovascular disease aggravated by obesity, and the deterioration of the alcoholic and of his or her dependent family. To continue this line of reasoning is to ask then who shall decide and how will this decision concerning ultimate authority be made? The next question might be what are the rights of special groups, such as health professionals, paraprofessionals, or broadly based groups such as the community and the consumer, to participate in decisions affecting policies and governance. How should authority and power be distributed? Will it be allocated traditionally in hierarchical arrangements with the administrative or clinical director or both at the head, with power, directives, and resources planning from the top in a downward direction? What are the rights of patients or clients in this structure and how can they be manifested? What are the rights and corresponding duties and obligations of health professionals in the system to the system? What responsibilities and roles does the acknowledgment of human rights imply for the daily practice of the diverse professions engaged in health care? Must the patient surrender himself or herself to the physician in a state of complete trust? What are the obligations of the physicians to his or her patient? Are they absolute or relative? What about the position of other health professionals toward the client for telling the truth? Should the focus of current health and welfare systems be on the repair of disabled persons, or should the major attention instead be on preventive services and on promotion of health, parenting, and benign life-styles? How is the obligation to people beyond the familial and tribal kinship systems weighed? Does the obligation extend throughout the world? How are the mechanisms for supply and demand of the market provided?

These questions are just some of the issues that will be touched upon in the ensuing chapters. The intent is to provoke thought and discussion as the basis for further reasoned analysis. The process of weighing alternatives and reaching decisions on ethical issues in health care is only another manifestation of the human right to freedom of choice.

## THE NATURE AND LIMITS OF RIGHTS IN A BIOMEDICAL CONTEXT

An assumption of the assertion of the right to life, the right to live as persons, and the right to health care is that the meaning of rights must be made clear. Rights, however, have variously been defined as needs, interests, powers, claims, and entitlements [24, 31]. The concept of rights has many facets. Each of these definitions has strengths and difficulties [6, 7]. These defining terms are not, however, necessarily identical with rights. Human needs and the interests of powerful groups may directly conflict. It may be a person's right but not in his or her interest to eat and drink excessively [9]. Some physicians, nurses, and other health professionals occasionally make claims to omnipotence and infallibility as if it were their right to do so, and some patients make claims about their physicians that deify them. Nevertheless, the emphasis upon rights, even though variously defined, provides a focus for biomedical issues.

Human rights are singled out as "fundamentally important" [13, 14]. They are seen as essential to "a decent and fulfilling human life" [30], more important than any other rights [7], and "shared equally by all human beings" [14, pp. 84–85]. Human rights are considered "an independent standard of political criticism and justification" [26, pp. 3–4].

To compound matters, it is not as if one could take a position in favor of rights or be opposed to them. For there are positions about the meaning and limits of rights that vary and that reflect positions about moral values other than rights. A further issue concerns the status of rights. To some writers [12] rights are regarded as prima facie, which means that in interest-balancing situations rights can be overturned or set aside. If, for example, one has a right to one's house and the state needs the property for a four-lane highway, then one's right to one's house may be overridden. A patient's right to a bed in a particular hospital at a specific time or to a particular day in the operating room for elective surgery may be contingent on the availability of that hospital's space and time and its "right" to set its priorities of care. Or if a person has a right to live but in a given case it would be too expensive to apply exotic lifesaving therapy to save his or her life, for example, the life of an alcoholic derelict, that person's right to life is said to be overridden without being violated. The right to life of the derelict person has in effect been *lost*. Or if a Jehovah's Witness has a right to decide what happens to his or her body and in a biomedical emergency refuses a blood transfusion, his or her right may be set aside by members of the community, including judges, physicians, nurses, and others who decide that the "right to live" of a Jehovah's Witness overrides his or her right to decide what happens to his or her body. According to this view, rights may be annulled, canceled, lost, or set aside. Rights are not absolute.

Opposed to the prima facie view of rights is the view that rights are unconditional, absolute, inalienable, indefeasible, and exceptionless. According to the absolute view of rights, the First Amendment guarantee of freedom of expression, for example, cannot be overturned by an appeal to the national interest [28]. According to this view, slaves in ancient Greece had a right to be free regardless of the legal norms and institutions existing at that time [9]. Similarly, Jews who were gassed by the Nazis had a right to live, a right that could never be overturned. Their right to live, however, was violated, thus constituting "a grave affront to justice" [13, p. 51]. In this view, the right of Jehovah's Witnesses to decide what happens to their bodies is absolute and exceptionless. There may be occasions in which a lifeboat ethic, that is, the belief that a few must die to save the many, takes precedence over other values, but in the absolute view of rights, violating a person's rights is never condonable or excusable by appeal to the ethical doctrine of "necessity." If a right to life or the right to be free exists, violating these rights is unjust no matter what the necessities, circumstances, or consequences. At the very least such injustice, according to this view, should be acknowledged with serious regret [14, p. 75]. As with attempts to define rights, both these positions on the strength of rights have advantages and drawbacks.

Despite the different positions taken on rights, some people consider rights as unimportant. To health professionals who see their relationship with the patient as primarily characterized by trust, kindness, and compassion, the emphasis on rights is considered a deterrent. These critics of rights contend that talk about rights only promotes "adversarial relations," which create animosity between the client and health professional, and that the widespread discussion about rights only raises the cost of malpractice insurance. At best, talk about patients' rights is regarded as useless; at worst, talk about rights is considered malicious and is thought to undermine the practice of the health professions.

The views held by the editors coincide with the views eloquently expressed by the writers who regard rights as an indispensably valuable moral possession [9, 14, 20, 31]. In this view, to have rights of importance is to fulfill some necessary conditions for being a person and distinguishes one who is at least to some extent free from one who is not. Rights distinguish one who is free and autonomous from one who is anomic or heteronomous. To have rights is to have dignity and respect and is the basis of self-respect as well.

To have a right is to have not just a claim, need, or interest but to have also a justification for claiming one's due. As Joel Feinberg [14] points out, a person with rights does not need to beg, plead, or supplicate for his or her own due or be thankful or show gratitude afterward. One need not be thankful for one's paycheck, for example. And if that to which one has a right is not forthcoming, one is put in a position to claim one's due effectively and with rightful indignation. According to another writer, Maurice Cranston, a right of importance, a human right, "is something of which no one may be deprived without a grave affront to justice" [13].

An advantage of rights is that to have a right is to be immune to the charge of wrongdoing. It may be unwise, foolish, or imprudent to exercise one's right, but it is never immoral or illegal. Nor does one need to exercise the right one has. To have a right is to be free to choose not to exercise it, and there is no wrongdoing attached to either exercising it or not.

A further advantage of rights is that they are not mere liberties. Rights imply corresponding duties on others to enable right-holders to exercise their rights. "Rights are necessarily the grounds of other people's duties" [14, p. 58]. Rights provide a justification for other people's obligations. Rights are a justification for one's claims and provide a justification for imposing corresponding obligations on other people.

In this view of rights, it is better to live in a world with assertive, aggressive, or even hostile persons than to live in a world of quiet, docile, and obedient patients who can be drugged or persuaded into submission. To have rights in this view is to be free of a master-slave relation. Values associated with rights, such as dignity, self-respect, decency, and autonomy, are important to a society that respects rights. We do not, however, wish to convey the view that rights are the only value in health care.

## ALTERNATIVE MORAL PRINCIPLES

Rights in health care comprise only a part of the domain of moral values. They are sometimes regarded under the heading of juridical values. There are, however, other important moral values such as love, trust, kindness, mercy, pleasure, joy, happiness, sympathy, understanding, duty, responsibility, intelligence, wit, sound judgment, and knowledge, to name a few. To appreciate the scope and limits of rights is to see their role in the context of other values.

Although the Greeks did not have an articulated conception of rights [18], they had well-developed views of the closely related conception of justice and of each person receiving his or her *due*. The early Greeks, especially Plato and Aristotle, saw the role of justice in relation to the larger context of moral values, notably the relation of justice to the good life and to the healthy life.

To *Plato* there could not be a good or healthy life without justice. To Plato justice consists in never harming anyone [27, p. 6]. To be just and to be treated justly without harm being done presupposes the knowledge of what is just and what will not harm a person. Plato (427—347 B.C.), perhaps the first Western philosopher to articulate the concept of functionalism, maintained that insofar as one functions in any capacity "strictly speaking" one does so without error. By this definition, Plato meant that one who functions knows what he or she is doing. Moreover, to know what one is doing consists in doing no harm. To Plato [27] "knowledge is virtue," and ignorance is a vice. Knowledge underlies all virtuous action, including just acts. Plato identifies justice with "the excellence of the soul, ... [and] the just soul [as with] the just man then will live well" [27, p. 27].

A second major alternative moral value to the orientation of rights is Aristotle's doctrine of the mean [4], which stresses happiness as an "activity in accordance with virtue" and identifies happiness, in part, with "external goods," including reasonable wealth, health, friendship, and general "well-being." To Aristotle each species has its function. The function or purpose of human beings is happiness. For Aristotle "every art ... science ... action and choice, seem to aim at some good" [4, p. 3]. [And] "the good ... [is] that at which all things aim" [4, p. 3]. To Aristotle being happy consists in "living well" and "doing well" [4, p. 6]. To Aristotle one cannot be healthy without external goods, including friends, adequate wealth, and good health. Nor can one be happy if one is "ugly ... or ill born" or if one "lives all by himself and has no children" [4, pp. 21—22]. "Happiness ... requires completeness in virtue as well as a complete lifetime" [4, p. 23].

Aristotle's appeal to students of the natural sciences is enduring. His concern with induction and his concern with not only one fixed form of the good but rather with many goods, including human happiness and a life not devoid of pleasure, make Aristotle's conception of a good life, a life in which one aims for the reasonable and apt mean between excesses and defects, an appealing view.

A third alternative to an ethics oriented by rights is the ethics of love, first enunciated by Jesus of Nazareth, St. Paul, and later by St. Francis, and known as the "agapeistic" way of life [11, 21]. According to the ethics of love, there is one commandment and that is "to love." First, "Thou shalt love the Lord thy God with all thy heart, and with all thy Soul and with all thy mind. This is the first and great commandment." And the second is "Thou shalt love thy neighbor as thyself. On these two commandments hang all the law and all the prophets" (Matt. 22:37–40, 15:56–57). St. Francis (1182–1226) of Assisi extends this love to the love of – and the kissing of – lepers, love of the poor, love of birds, and love of all living things [19, pp. 149–152].

The virtue of love is sometimes portrayed in contrast to the juridical virtues. Note, for example, Emil Brunner's statement quoted by Frankena:

The sphere [of] just claims, rights, debits and credits, and in which justice is . . . the supreme principle, and the sphere in which the gift of love is supreme, where are no deserts, where love, without acknowledging any claims, gives all – these two spheres lie as far apart as heaven from hell . . . If ever we are to get a clear conception of the nature of justice, we must get . . . a clear idea of it as differentiated and contrasted with love [16, p. 2].

Other writers do not regard love and justice as antithetical, but they place love and compassion in a sphere of its own. The relation between the juridical virtues and love and other values may be pictured as a collection of separate circles within a larger circle called the Realm of Moral Values (see figure). The circles are not meant to be mutually exclusive. As with Wittgenstein's "family resemblances," or family games, there may be other values, and there are, of course, overlappings and crisscrossings.

The ethics of love applied to health care means that nurses and physicians show love, trust, and kindness to their patients. The word "care" is implied by the term "love."

A fourth moral value, a value that emphasizes neither knowledge of the good nor happiness but one's duty and responsibility instead, is found in the ethics of Kant [22]. To Kant duty and responsibility presuppose freedom. One cannot act well or badly or heed a moral obligation nor be responsible without being free to decide. There may or may not be freedom, but it has to be assumed in order to impute responsibility to human behavior.

To Kant [22] no value is good without exception except a good will. Intelligence, wit, judgment, happiness, courage, pride, or charity – none of these attributes is good without qualification, for one could use all of these goods for bad ends. Only a good will is good without exception.

Kant distinguishes two kinds of imperatives or commands, hypothetical and categorical. A hypothetical imperative consists of an if-then relation, in which a

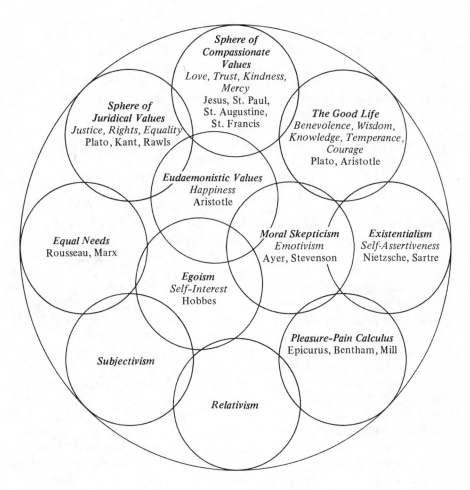

Sphere of
Compassionate
Values
Love, Trust, Kindness,
Mercy
Jesus, St. Paul,
St. Augustine,
St. Francis

Sphere of
Juridical Values
Justice, Rights, Equality
Plato, Kant, Rawls

The Good Life
Benevolence, Wisdom,
Knowledge, Temperance,
Courage
Plato, Aristotle

Eudaemonistic Values
Happiness
Aristotle

Equal Needs
Rousseau, Marx

Moral Skepticism
Emotivism
Ayer, Stevenson

Existentialism
Self-Assertiveness
Nietzsche, Sartre

Egoism
Self-Interest
Hobbes

Pleasure-Pain Calculus
Epicurus, Bentham, Mill

Subjectivism

Relativism

Realm of Moral Values

person decides which goal he or she wishes to achieve, such as losing weight or getting a college degree or going to the movies. The "if" part of the command provides the condition(s) for achieving his or her goal. But this kind of command is amoral, since it appeals to a person's *inclination* or desire as an end state, not his or her sense of *obligation*. For a hypothetical imperative, a person decides his or her own goal, and the facts determine the appropriate means.

Distinct from a hypothetical imperative is a categorical imperative, which holds that one ought always to do that which everyone else also ought to do in the same or similar situation. According to Kant, "Act only on that maxim [or rule] whereby thou canst at the same time will that it should become a universal law" [22, p. 38].

The categorical imperative, sometimes referred to as the "universalizability principle," states that the right thing to do regardless of anyone's inclination, impulse, convenience, or expedience, or even of the general welfare, is to do what is rational, universal, and desirable for the whole human race, independent of anyone's pleasure and no matter what happens, and without exception. If, for example, a tank with an innocent child tied to it fires in our direction, it is wrong to throw a hand grenade at the tank if it would kill the innocent child.

Kant cites four examples of the categorical imperative. The first example rules against suicide on the grounds that, if everyone in a state of dissatisfaction did so, there would be no human race and the will "to impel to the improvement of life would contradict itself" [22, p. 39].

A second example is the obligation to keep one's promises and never break a promise out of convenience or advantage to oneself. If making promises did not entail their being kept, the idea of promise keeping "would necessarily contradict itself" [22, pp. 39–40].

The universalizability principle in effect says: Act on principles. The making of a promise, for example, entails the obligation to keep it. Principled morality, which underlies all undeviating morality, is illustrated in the application of equality to social life. The promise to treat people equally implies the obligation to institute social and legal practices consistent with that promise, and without compromise.

A third example of the categorical imperative is that one ought not will to squander one's talents. "For as a rational being he necessarily wills that his faculties be developed" [22, p. 40]. This rule can also be applied to the principle that one should not live an unhealthy life.

A fourth example of the categorical imperative is the requirement not to let avoidable misery go unhelped. This may be called the Good Samaritan imperative, which holds: If one has enough, help another in need. This imperative provides a basis for the right to assistance and is a source of altruistic motivation for health professionals.

A practical application of the categorical imperative for Kant says: Treat humanity "whether in thine own person or in that of another as an end withal, never as means only" [22, p. 46]. According to this principle, one ought to respect every person. Health care ought to be based on voluntary and informed consent by a patient or subject of experimentation, not coercion, deception, or brainwashing. In practice, this principle also says not to exploit clients or other health professionals, such as persuading people to have surgery who do not need it, or in any way use a person as a means only of someone else's aims.

The categorical imperative is sometimes confused with the Golden Rule, which it superficially resembles. The Golden Rule says to treat others as one would like to be treated. The categorical imperative says to treat others as one and everyone else in a similar situation ought to be treated. If, for example, one enjoys smoking, on grounds of the Golden Rule, one may justifiably impose cigarettes on others; not so on the rational grounds of the categorical imperative.

Opposed to the orientation of duty (called "deontological") is John Stuart Mill's utilitarian [25] view of a right action, sometimes called "the greatest happiness principle." According to Mill, "actions are right in proportion as they tend to promote happiness; wrong as they tend to produce the reverse of happiness" [25, p. 10]. To Mill, when one judges one's happiness against others' "utilitarianism requires him to be strictly impartial as a disinterested and benevolent spectator. In the golden rule of Jesus of Nazareth, we read the complete spirit of the ethics of utility 'To do as you would be done by' and 'to love your neighbor as yourself' constitute the ideal perfection of utilitarian morality" [25, p. 22].

Although Mill [25] invokes the ethics of Jesus, utilitarianism is concerned with doing the good for the greatest number. To revert to our tank example, if more lives are to be saved by throwing a hand grenade at a tank with an innocent hostage tied to it, on utilitarian grounds that act is not wrong.

Similarly, in triage problems in which only some lives can be saved, utilitarian ethics emphasizes help to the greatest number. The subordination of the care of some, such as the terminally ill, to the care of the many who will recover in a situation of limited resources of time, staff, and money is an application of Mill's principle of the happiness of the greatest number.

A sixth and different view from any of the other positions is forwarded by Jean Paul Sartre [29], which emphasizes the need to make decisions in the context of the "here and now" without slavish dependence on any principles. The emphasis is on making one's own *authentic, inner-directed* choice, not a choice that is outer-directed or dictated by pressures from others. One has to "take charge," be responsible for one's decision, and take the consequences. Lying to oneself and lying to others as well as ignorance, unconsciousness, or failure to take responsibility for one's decisions are instances of bad faith. The decisions one makes must be conscious and felt. This ethic, which is an ethic of personal honesty, holds that one must be true and responsible to oneself and others.

The Existentialist view emphasizes the role of decision making in health care and is dramatically illustrated in emergency and crisis situations, where nurses and physicians must reach instant decisions. Principles count, but they have to be internalized and authentic.

A seventh ethical position, which emphasizes values other than rights, is identified with the major principle of distributive justice identified with Marx [23]. This position may be referred to as the equal needs principle, according to which one should give "to each according to his needs and from each according to his ability" [23].

Marx showed impatience with rights and contended that "rights" are a middle-class bourgeois phenomenon created by the ruling class. In place of rights, Marx [23] emphasized the aspects of capitalism that block fulfillment of real economic needs, especially of downtrodden, exploited people. The Marxist principle applied to health care means the equal distribution of health resources to those in need, so that all have equal access to the human resources currently enjoyed by the most advantaged members of society.

An eighth moral point of view emphasizes each person's or group's deciding what is right or wrong. In one form this position is known as Subjectivism, and in another form it is known as Relativism. A third and more recently developed variation is known as Emotivism [5]. This view holds that no one really knows what is right or wrong. According to this view, a statement only has meaning if it is true or false and if it can be appropriately verified as either true or false, either deductively or from experience. Since ethical statements cannot be shown to be either true or false, they simply have no meaning.

On the basis of this perspective, it is a short route from the view that science helps us decide what the facts are to individuals and groups deciding what is right or wrong. There are no standards of right and wrong, only standards of truth and falsity. Emotivism does not hold that any moral decision is right or wrong. But the absence of a standard of right and wrong has the effect, quite possibly unintended, of reinforcing the relativist position that each person or group decides what is right or wrong.

The relativist moral view has its advocates in health care. According to this view, what one does morally to another person, whether participating in abortion, euthanasia, sterilization, or any procedure or experiment, such as psychosurgery, has no moral consequences, since it is up to each person or group to decide what is right or wrong.

In one respect, Emotivism [5] harks back to the first of the views presented, namely, the necessity of knowledge underlying ethics. For both Plato [26] and contemporary emotivists [5], there can be no right action without knowledge. The difference is that Plato [26] believes there is knowledge of the good and of right actions, whereas emotivists (or logical positivists) [5] contend that there is no such knowledge, since there are no true or false propositions in ethics.

An unfortunate by-product of Emotivism and also of Subjectivism and Relativism, which are otherwise different from Emotivism, has been that health care decisions that were not scientific or technological were interpreted as "value-free" and personal, thus suggesting that no criteria of criticism and justification applied to moral issues in health care. We believe, however, that knowledge is important to ethics but that the criteria of ethical knowledge need not be as rigorous as Plato requires.

The conditions of knowledge required in ethics based on science have given way to autonomous but not arbitrary conditions of ethical justification. In recent years the most formidable effort to show that science and morality are not synonymous is found in John Rawls's *A Theory of Justice* [28]. To Rawls, as truth is the first

virtue of thought, "justice is the first virtue of social institutions" [28, p. 3].
Rawls's work aims to do much else besides, notably to integrate the virtues of
classical Platonic and Aristotelian thought on justice with modern social contract
theory. He includes the Kantian emphasis on uncompromisable moral principles
with the utilitarian concern for providing maximum benefits to everyone in a society.

Rawls believes that justice is a fair relation between rational persons. The social
contract is between rational persons, and it is hypothetically assumed that people
want to live a rational way of life. On that assumption, justice has priority in the
scale of values; it regulates the relations between the rights and duties of individuals.

To Rawls the principles of justice take priority. "To respect persons is to recognize that they possess an inviolability founded on justice that even the welfare of
society as a whole cannot override" [28, pp. 3–4]. "The loss of freedom for some
is never made right by a greater good shared by others . . . Rights secured by justice
are not subject to political bargaining or to the calculus of social interests" [28,
pp. 3–4]. If 85 percent of a population is better off by removing the rights of
15 percent, to Rawls, as with Kant, that is not made right. Similarly, slavery is
never made right or just by the greater economic good it brings others.

The principles of justice "provide a way of assigning rights and duties in the basic
institutions of society and they define the appropriate distribution of the benefits
and burdens of social cooperation" [28, p. 4]. To Rawls there are two principles
of justice to which rational persons would agree. These principles are that (1) each
person has "an equal right to the most extensive system of liberties compatible with
a similar system of liberty for all" [28, p. 302]; and (2) social and economic inequalities are arranged "so that they are both (a) to the greatest benefit of the least
advantaged" [28, p. 302], consistent with the advantage of all, and (b) attached to
"positions open to all under . . . fair equality of opportunity" [28, p. 302].

Inequalities are justified if they are of advantage to all and particularly to the
least advantaged, according to (a). But, according to (b), if merit benefits a whole
society, rewarding it by an unequal wage system is made just. The principle in (b)
is that inequality is justified if it accrues to everyone's benefit and if it helps the
least advantaged, consistent with "the just savings principle." (According to this
principle, one is to save what he would have expected his predecessors to save for
him.)

The Rawlsian principle of providing for the least advantaged calls for the just
distribution of "primary goods." This principle seems to imply for Rawls a floor of
support, which includes food, shelter, clothing, health care, and education. Implicit
in Rawls's system seems to be that a just society minimizes the difference between
the least advantaged and the most advantaged consistent with the advantage of all.
If it is to the advantage of all to minimize this difference, rights to the primary goods
of life, including health care, apply equally to all persons.

The first principle has priority for Rawls. Everyone's "equal right to the most
extensive total system of equal basic liberties compatible with a similar system for
all" is never compromised even for the sake of the second principle, namely, justly
arranging and lessening social and economic inequalities.

CONCLUSION

We think these positions show values in addition to the values of rights in health care that have a cutting edge on the search for the good life. Knowledge, happiness, love, a sense of duty and responsibility, commitment to one's values, needs, and skepticism, in addition to the juridical values of rights based on justice, have a place in the good life. These values impinge on the relation between health consumers and health providers in significant ways.

There are conflicts between these values. One cannot embrace Kant's [22] ethics oriented to deontological duty simultaneously with Mill's ethics [25] oriented to utilitarian happiness in pinch cases. Deciding who lives and dies, who is punished and harmed, and who gets health care are differently answered by these positions. But, hopefully, there will be no health consumer or health provider who, after carefully taking stock, will become a total advocate of either position. Nor need one choose eclecticism out of mere convenience. Rather, there are times and places where one view will seem appropriate, and often these views will serve as checkpoints for consideration instead of ready-made conclusions.

Take abortion, for example. Difficulties are encountered by a person who considers only her body or her convenience in deciding whether or not to abort a fetus. The right to life of a fetus has defendants. Yet the antiabortionist has also to consider the right to live, to eat, and to work of the people currently alive in an over-populated country as well as the likelihood of repeated famines. Likewise, for people who advocate informed consent, difficult choices exist involving unconscious or comatose patients and consent by proxy. Similarly, many difficulties are encountered by people who advocate euthanasia as an act of kindness without adequate rational safeguards against the progressive murder of unwanted beings, such as the senile, the mentally retarded, and the chronically mentally ill.

The subject of behavior control and imprisonment raises problems about the use of punishment. Avoiding the use of punishment may encourage crime. On the other hand, some forms of punishment are "crimes" and appear to lead to habitual criminal behavior, while rehabilitation measures seem ineffective.

Difficulties in the concept of informed consent are numerous. Give a child with a tumor in need of amputation a choice and he or she may make the wrong choice. Another difficulty for one who believes that either the right to freedom or the right to life is uppermost is to decide what to do about a Jehovah's Witness who refuses a lifesaving blood transfusion. Does the duty to preserve life override a person's freedom to decide? One can see clashing rights as well as existentialist values in conflict with the right to life or to self-preservation.

A major problem also exists in the issue of the right to health care. Some people maintain that the present system of financing and delivering health care is grossly inadequate and *unfair* to large segments of the population of low or moderate income. Other people point out that most health problems include a voluntary behavioral component, such as excessive smoking, eating, drinking, and driving too fast. These critics point to the inherent unfairness of a social contract that requires society to pay all the costs of the consequences of self-destructive practices. The overriding question, however, is of financial priorities. In a world of finite resources, how much are we willing to spend on health at the expense of education, recreation, housing, and other factors contributing to the quality of life?

Each of these positions, oriented by knowledge, happiness, duty, authenticity, needs, and skepticism, provides values to consider in addition to the juridical virtues. A life of rights based on justice but without consideration of these other values would be impoverished. Furthermore, rights cannot be isolated from these other values. Rather, as with Wittgenstein's language games, there are overlappings and crisscrossings between rights, needs, love, duties, knowledge, happiness, and authenticity that defy efforts at a final classification.

The issues to be examined are not resolved by resorting to the mere granting of rights. In our view, other values are highly desirable. Love is desirable, but love is not enough. Love can be ignorant. Love can be unfair. We think rights are a guide for choices to be made, not infallible but a guide. Furthermore, rights are a justification for claiming one's due and for justifying the obligations of others to the rightholder. Rights can provide a guarantee when other virtues are not forthcoming. For these reasons, a world of rights without love is empty, and love without rights is blind; and neither love nor rights can thrive without knowledge and justice.

## REFERENCES

 1. American Hospital Association. *Statement on a Patient's Bill of Rights.* Chicago: American Hospital Association, 1975.
 2. A.N.A. Committee on Ethical, Legal and Professional Standards. *Code for Nurses with Interpretive Statements.* Kansas City, Mo.: 1976.
 3. Annas, G. J., and Healy, J. The patient rights advocate. *J. Nurs. Adm.* 4:25, May–June, 1974.
 4. Aristotle. *Nicomachean Ethics* (translated by M. Ostwald). Indianapolis: Bobbs-Merrill, 1962.
 5. Ayer, A. J. *Language, Truth and Logic.* New York: Dover, 1950.
 6. Bandman, B. Human Rights of Patients, Nurses, and Other Health Professionals. This volume, Chap. 47.
 7. Bandman, B. Some legal, moral and intellectual rights of children. *Educational Theory* 27:169, 1977.
 8. Battistella, R. M. The right to adequate health care. *Hosp. Prog.* 55:36, 1974.
 9. Benn, S. Rights. In P. Edwards (ed.), *The Encyclopedia of Philosophy* (vol. 7). New York: Macmillan, 1966.
10. Bentham, J. *The Principles of Morals and Legislation.* New York: Hafner, 1948.
11. Braithwaite, R. An Empiricist's View of the Nature of Religious Belief. In J. Hick (ed.), *The Philosophy of Religion.* Englewood Cliffs, N.J.: Prentice-Hall, 1964. Pp. 429–439.
12. Brandt, R. *Ethical Theory.* Englewood Cliffs, N.J.: Prentice-Hall, 1959.
13. Cranston, M. Human Rights, Real and Supposed. In D. D. Raphael (ed.), *Political Theory and the Rights of Man.* Bloomington: Indiana University Press, 1967.
14. Feinberg, J. *Social Philosophy.* Englewood Cliffs, N.J.: Prentice-Hall, 1973.
15. Frankena, W. *Ethics.* Englewood Cliffs, N.J.: Prentice-Hall, 1974.
16. Frankena, W. The Concept of Social Justice. In R. B. Brandt (ed.), *Social Justice.* Englewood Cliffs, N.J.: Prentice-Hall, 1962.

17. Fried, C.   Rights and health care — Beyond equity and efficiency. *N. Engl. J. Med.* 293:241, 1975.
18. Golding, M. P.   The Concept of Rights: A Historical Sketch. This volume, Chap. 4.
19. Golding, M. P.   Obligations to future generations. *Monist* 56:85, 1972.
20. Golding, M. P.   Towards a theory of human rights. *Monist* 52:521, 1968.
21. Jones, W. T.   *A History of Western Philosophy* (vol. 2). New York: Harcourt Brace, 1969. Pp. 149–152.
22. Kant, I.   *The Foundations of the Metaphysics of Morals* (translated by L. W. Beck). Indianapolis: Bobbs-Merrill, 1959.
23. Marx, K., and Engels, F.   In L. S. Feuer (ed.), *Basic Writings on Politics and Philosophy.* Garden City, N.Y.: Anchor Books, 1959.
24. McCloskey, H. J.   Rights. *Phil. Quar.* 15:59, 1965.
25. Mill, J. S.   *Utilitarianism.* Indianapolis: Bobbs-Merrill, 1957.
26. Nickel, J.   Are Social and Economic Rights Real Human Rights? Paper presented at Society for Philosophy and Public Affairs, City University of New York, Graduate Center, New York City, March 1977.
27. Plato.   *The Republic* (translated by G. M. A. Grube). Indianapolis: Hackett, 1974.
28. Rawls, J.   *A Theory of Justice.* Cambridge, Mass.: Harvard University Press, 1971.
29. Sartre, J. P.   *Existentialism and Human Emotions* (translated by B. Frichtman). New York: Philosophical Library, 1947.
30. Scheffler, S.   Natural rights, equality and the minimal state. *Can. J. Phil.* 6:64, 1976.
31. Wasserstrom, R.   Rights, Human Rights and Racial Discrimination. In A. I. Melden (ed.), *Human Rights.* Belmont, Calif.: Wadsworth, 1970.

# CHAPTER 1

# The Nature and Value of Rights
*Joel Feinberg*

1

*I would like* to begin by conducting a thought experiment. Try to imagine Nowheresville — a world very much like our own except that no one, or hardly any one (the
qualification is not important), has *rights*. If this flaw makes Nowheresville too
ugly to hold very long in contemplation, we can make it as pretty as we wish in
other moral respects. We can, for example, make the human beings in it as attractive
and virtuous as possible without taxing our conceptions of the limits of human
nature. In particular, let the virtues of moral sensibility flourish. Fill this imagined
world with as much benevolence, compassion, sympathy, and pity as it will conveniently hold without strain. Now we can imagine men helping one another from
compassionate motives merely, quite as much or even more than they do in our
actual world from a variety of more complicated motives.

This picture, pleasant as it is in some respects, would hardly have satisfied
Immanuel Kant. Benevolently motivated actions do good, Kant admitted, and
therefore are better, *ceteris paribus,* than malevolently motivated actions; but no
action can have supreme kind of worth — what Kant called "moral worth" — unless
its whole motivating power derives from the thought that it is *required by duty*.
Accordingly, let us try to make Nowheresville more appealing to Kant by introducing the idea of duty into it, and letting the sense of duty be a sufficient motive for
many beneficent and honorable actions. But doesn't this bring our original thought
experiment to an abortive conclusion? If duties are permitted entry into Nowheresville, are not rights necessarily smuggled in along with them?

The question is well-asked, and requires here a brief digression so that we might
consider the so-called "doctrine of the logical correlativity of rights and duties."
This is the doctrine that (i) all duties entail other people's rights and (ii) all rights
entail other people's duties. Only the first part of the doctrine, the alleged entail-
ment from duties to rights, need concern us here. Is this part of the doctrine correct?
It should not be surprising that my answer is: "In a sense yes and in a sense no."
Etymologically, the word "duty" is associated with actions that are *due* someone
else, the payments of debts *to* creditors, the keeping of agreements with promises,
the payment of club dues, or legal fees, or tariff levies to appropriate authorities or
their representatives. In this original sense of "duty," all duties are correlated with
the rights of those *to* whom the duty is owed. On the other hand, there seem to be
numerous classes of duties, both of a legal and non-legal kind, that are *not* logically
correlated with the rights of other persons. This seems to be a consequence of the
fact that the word "duty" has come to be used for *any* action understood to be
*required*, whether by the rights of others, or by law, or by higher authority, or by
conscience, or whatever. When the notion of requirement is in clear focus it is likely
to seem the only element in the idea of duty that is essential, and the other component notion — that a duty is something *due* someone else — drops off. Thus, in this

From *The Journal of Value Inquiry,* Vol. 4 (1970), 243–257, by permission of the author and
publisher.

widespread but derivative usage, "duty" tends to be used for any action we feel we *must* (for whatever reason) do. It comes, in short, to be a term of moral modality merely; and it is no wonder that the first thesis of the logical correlativity doctrine often fails.

Let us then introduce duties into Nowheresville, but only in the sense of actions that are, or believed to be, morally mandatory, but not in the older sense of actions that are due others and can be claimed by others as their right. Nowheresville now can have duties of the sort imposed by positive law. A legal duty is not something we are implored or advised to do merely; it is something the law, or an authority under the law, *requires* us to do whether we want to or not, under pain of penalty. When traffic lights turn red, however, there is no determinate person who can plausibly be said to claim our stopping as his due, so that the motorist owes it to *him* to stop, in the way a debtor owes it to his creditor to pay. In our own actual world, of course, we sometimes owe it to our *fellow motorists* to stop; but that kind of right-correlated duty does not exist in Nowheresville. There, motorists "owe" obedience to the Law, but they owe nothing to one another. When they collide, no matter who is at fault, no one is accountable to anyone else, and no one has any sound grievance or "right to complain."

When we leave legal contexts to consider moral obligations and other extra-legal duties, a greater variety of duties-without-correlative-rights present themselves. Duties of charity, for example, require us to contribute to one or another of a large number of eligible recipients, no one of whom can claim our contribution from us as his due. Charitable contributions are more like gratuitous services, favours, and gifts than like repayments of debts or reparations; and yet we do have duties to be charitable. Many persons, moreover, in our actual world believe that they are required by their own consciences to do more than that "duty" that *can* be demanded of them by their prospective beneficiaries. I have quoted elsewhere the citation from H. B. Acton of a character in a Malraux novel who "gave all his supply of poison to his fellow prisoners to enable them by suicide to escape the burning alive which was to be their fate and his." This man, Acton adds, "probably did not think that [the others] had more of a right to the poison than he had, though he thought it his duty to give it to them."[1] I am sure that there are many actual examples, less dramatically heroic than this fictitious one, of persons who believe, rightly or wrongly, that they *must do* something (hence the word "duty") for another person in excess of what that person can appropriately demand of him (hence the absence of "right").

Now the digression is over and we can return to Nowheresville and summarize what we have put in it thus far. We now find spontaneous benevolence in somewhat larger degree than in our actual world, and also the acknowledged existence of duties of obedience, duties of charity, and duties imposed by exacting private consciences, and also, let us suppose, a degree of conscientiousness in respect to those duties somewhat in excess of what is to be found in our actual world. I doubt that Kant would be fully satisfied with Nowheresville even now that duty and respect for law and authority have been added to it; but I feel certain that he would regard

[1]H. B. Acton, "Symposium of 'Rights'," *Proceedings of the Aristotelian Society,* Supplementary Volume 24 (1950), pp. 107–108.

their addition at least as an improvement. I will now introduce two further moral practices into Nowheresville that will make the world very little more appealing to Kant, but will make it appear more familiar to us. These are the practices connected with the notions of *personal desert* and what I call a *sovereign monopoly of rights*.

When a person is said to deserve something good from us what is meant in parts is that there would be a certain propriety in our giving that good thing to him in virtue of the kind of person he is, perhaps, or more likely, in virtue of some specific thing he has done. The propriety involved here is a much weaker kind than that which derives from our having promised him the good thing or from his having qualified for it by satisfying the well-advertised conditions of some public rule. In the latter case he could be said not merely to deserve the good thing but also to have a *right* to it, that is to be in a position to demand it as his due; and of course we will not have that sort of thing in Nowheresville. That weaker kind of propriety which is mere desert is simply a kind of *fittingness* between one party's character or action and another party's favorable response, much like that between humor and laughter, or good performance and applause.

The following seems to be the origin of the idea of deserving good or bad treatment from others: A master or lord was under no obligation to reward his servant for especially good service; still a master might naturally feel that there would be a special fittingness in giving a gratuitous reward as a grateful response to the good service (or conversely imposing a penalty for bad service). Such an act while surely fitting and proper was entirely supererogatory. The fitting response in turn from the rewarded servant should be gratitude. If the deserved reward had not been given him he should have had no complaint, since he only *deserved* the reward, as opposed to having a *right* to it, or a ground for claiming it as his due.

The idea of desert has evolved a good bit away from its beginnings by now, but nevertheless, it seems clearly to be one of those words J. L. Austin said "never entirely forget their pasts."[2] Today servants qualify for their wages by doing their agreed upon chores, no more and no less. If their wages are not forthcoming, their contractual rights have been violated and they can make legal claim to the money that is their due. If they do less than they agreed to do, however, their employers may "dock" them, by paying them proportionately less than the agreed upon fee. This is all a matter of right. But if the servant does a splendid job, above and beyond his minimal contractual duties, the employer is under no further obligation to reward him, for this was not agreed upon, even tacitly, in advance. The additional service was all the servant's idea and done entirely on his own. Nevertheless, the morally sensitive employer may feel that it would be exceptionally appropriate for him to respond, freely on *his* own, to the servant's meritorious service, with a reward. The employee cannot demand it as his due, but he will happily accept it, with gratitude, as a fitting response to his desert.

In our age of organized labor, even this picture is now archaic; for almost every kind of exchange of service is governed by hard bargained contracts so that even bonuses can sometimes be demanded as a matter of right, and nothing is given for nothing on either side of the bargaining table. And perhaps that is a good thing; for consider an anachronistic instance of the earlier kind of practice that survives, at

[2] J. L. Austin, "A Plea for Excuses," *Proceedings of the Aristotelian Society,* Vol. 57 (1956–57).

least as a matter of form, in the quaint old practice of "tipping." The tip was originally conceived as a reward that has to be earned by "zealous service." It is not something to be taken for granted as a standard response to *any* service. That is to say that its payment is a *"gratuity,"* not a discharge of obligation, but something given apart from, or in addition to, anything the recipient can expect as a matter of right. That is what tipping originally meant at any rate, and tips are still referred to as "gratuities" in the tax forms. But try to explain all that to a New York cab driver! If he has *earned* his gratuity, by God, he has it coming, and there had better be sufficient acknowledgement of his desert or he'll give you a piece of his mind! I'm not generally prone to defend New York cab drivers, but they do have a point here. There is the making of a paradox in the queerly unstable concept of an "earned gratuity." One can understand how "desert" in the weak sense of "propriety" or "mere fittingness" tends to generate a stronger sense in which desert is itself the ground for a claim of right.

In Nowheresville, nevertheless, we will have only the original weak kind of desert. Indeed, it will be impossible to keep this idea out if we allow such practices as teachers grading students, judges awarding prizes, and servants serving benevolent but class-conscious masters. Nowheresville is a reasonably good world in many ways, and its teachers, judges, and masters will generally try to give students, contestants, and servants the grades, prizes, and rewards they deserve. For this the recipients will be grateful; but they will never think to complain, or even feel aggrieved, when expected responses to desert fail. The masters, judges, and teachers don't *have* to do good things, after all, for *anyone.* One should be happy that they *ever* treat us well, and not grumble over their occasional lapses. Their hoped for responses, after all, are *gratuities,* and there is no wrong in the omission of what is merely gratuitous. Such is the response of persons who have no concept of *rights,* even persons who are proud of their own deserts.[3]

Surely, one might ask, rights have to come in somewhere, if we are to have even moderately complex forms of social organization. Without rules that confer rights and impose obligations, how can we have ownership of property, bargains and deals, promises and contracts, appointments and loans, marriages and partnerships? Very well, let us introduce all of these social and economic practices into Nowheresville, but *with one big twist.* With them I should like to introduce the curious notion of a "sovereign right-monopoly." You will recall that the subjects in Hobbes's *Leviathan* had no rights whatever against their sovereign. He could do as he liked with them, even gratuitously harm them, but this gave them no valid grievance against him. The sovereign, to be sure, had a certain duty to treat his subjects well, but this duty was owed not to the subjects directly, but to God, just as we might have a duty to a person to treat his property well, but of course no duty to the property itself but only to its owner. Thus, while the sovereign was quite capable of *harming* his subjects, he could commit no wrong against them that they could complain about, since they had no prior claims against his conduct. The only party *wronged* by the sovereign's mistreatment of his subjects was God, the supreme

[3]For a fuller discussion of the concept of personal desert see my "Justice and Personal Desert," *Nomos VI, Justice,* ed. by C. J. Chapman (New York: Atherton Press, 1963), pp. 69–97.

lawmaker. Thus, in repenting cruelty to his subjects, the sovereign might say to God, as David did after killing Uriah, "to Thee only have I sinned."[4]

Even in the *Leviathan,* however, ordinary people had ordinary rights *against one another.* They played roles, occupied offices, made agreements, and signed contracts. In a genuine "sovereign right-monopoly," as I shall be using that phrase, they will do all those things too, and thus incur genuine obligations toward one another; but the obligations (here is the twist) will not be owed directly *to* promises, creditors, parents, and the like, but rather to God alone, or to the members of some elite, or to a single sovereign under God. Hence, the rights correlative to the obligations that derive from these transactions are all owned by some "outside" authority.

As far as I know, no philosopher has ever suggested that even our role and contract obligations (in this, our actual world) are all owed directly to a divine intermediary, but some theologians have approached such extreme moral occasionalism. I have in mind the familiar phrase in certain widely distributed religious tracts that "it takes three to marry," which suggests that marital vows are not made between bride and groom directly but between each spouse and God, so that if one breaks his vow, the other cannot rightly complain of being wronged, since only God could have claimed performance of the marital duties as his *own* due; and hence God alone had a claim-right violated by nonperformance. If John breaks his vow to God, he might then properly repent in the words of David: "To Thee only have I sinned."

In our actual world, very few spouses conceive of their mutual obligations in this way; but their small children, at a certain stage in their moral upbringing, are likely to feel precisely this way toward *their* mutual obligations. If Billy kicks Bobby and is punished by Daddy, he may come to feel contrition for his naughtiness induced by his painful estrangement from the loved parent. He may then be happy to make amends and sincere apology to *Daddy;* but when Daddy insists that he apologize to his wronged brother, that is another story. A direct apology to Billy would be a tacit recognition of Billy's status as a right-holder against him, someone he can wrong as well as harm, and someone to whom he is directly accountable for his wrongs. This is a status Bobby will happily accord Daddy; but it would imply a respect for Billy that he does not presently feel, so he bitterly resents according it to him. On the "three-to-marry" model, the relations between each spouse and God would be like those between Bobby and Daddy; respect for the other spouse as an independent claimant would not even be necessary; and where present, of course, never sufficient.

The advocates of the "three-to-marry" model who conceive it either as a description of our actual institution of marriage or a recommendation of what marriage ought to be, may wish to escape this embarrassment by granting rights to spouses in capacities other than as promisees. They may wish to say, for example, that when John promises God that he will be faithful to Mary, a right is thus conferred not only on God as promisee but also on Mary herself as third-party beneficiary, just as when John contracts with an insurance company and names Mary as his intended beneficiary, she has a right to the accumulated funds after John's death, even though

[4]II Sam. 11. Cited with approval by Thomas Hobbes in *The Leviathan,* Part II, Chap. 21.

the insurance company made no promise to her. But this seems to be an unnecessarily cumbersome complication contributing nothing to our understanding of the marriage bond. The life insurance transaction is necessarily a three party relation, involving occupants of three distinct offices, no two of whom alone could do the whole job. The transaction, after all, is defined as the purchase by the customer (first office) from the vendor (second office) of protection for a beneficiary (third office) against the customer's untimely death. Marriage, on the other hand, in this our actual world, appears to be a binary relation between a husband and wife, and even though third parties such as children, neighbors, psychiatrists, and priests may sometimes be helpful and even causally necessary for the survival of the relation, they are not logically necessary to our *conception* of the relation, and indeed many married couples do quite well without them. Still I am not now purporting to describe our actual world, but rather trying to contrast it with a counterpart world of the imagination. In *that* world, it takes three to make almost *any* moral relation and all rights are owned by God or some sovereign under God.

There will, of course, be delegated authorities in the imaginary world, empowered to give commands to their underlings and to punish them for their disobedience. But the commands are all given in the name of the right-monopoly who in turn are the only persons to whom obligations are owed. Hence, even intermediate superiors do not have claim-rights against their subordinates but only legal *powers* to create obligations in the subordinates *to* the monopolistic right-holders, and also the legal *privilege* to impose penalties in the name of that monopoly.

2

So much for the imaginary "world without rights." If some of the moral concepts and practices I have allowed into that world do not sit well with one another, no matter. Imagine Nowheresville with all of these practices if you can, or with any harmonious subset of them, if you prefer. The important thing is not what I've let into it, but what I have kept out. The remainder of this paper will be devoted to an analysis of what precisely a world is missing when it does not contain rights and why that absence is morally important.

The most conspicuous difference, I think, between the Nowheresvillians and ourselves has something to do with the activity of *claiming*. Nowheresvillians, even when they are discriminated against invidiously, or left without the things they need, or otherwise badly treated, do not think to leap to their feet and make righteous demands against one another though they may not hesitate to resort to force and trickery to get what they want. They have no notion of rights, so they do not have a notion of what is their due; hence they do not claim before they take. The conceptual linkage between personal rights and claiming has long been noticed by legal writers and is reflected in the standard usage in which "claim-rights" are distinguished from other mere liberties, immunities, and powers, also sometimes called "rights," with which they are easily confused. When a person has a legal claim-right to $X$, it must be the case (i) that he is at liberty in respect to $X$, i.e. that he has no duty to refrain from or relinquish $X$, and also (ii) that his liberty is the ground of other people's *duties* to grant him $X$ or not to interfere with him in respect to $X$. Thus, in the sense of claim-rights, it is true by definition that rights logically entail other people's duties. The paradigmatic examples of such rights are the creditor's right to be paid a debt by his debtor, and the landowner's right not to be interfered with by anyone in the exclusive occupancy of his land. The creditor's right against his debtor, for example, and the debtor's duty to his creditor, are precisely the same relation seen from two different vantage points, as inextricably linked as the two sides of the same coin.

And yet, this is not quite an accurate account of the matter, for it fails to do justice to the way claim-rights are somehow prior to, or more basic than, the duties with which they are necessarily correlated. If Nip has a claim-right against Tuck, it is because of this fact that Tuck has a duty to Nip. It is only because something from Tuck is *due* Nip (directional element) that there is something Tuck *must do* (modal element). This is a relation, moreover, in which Tuck is bound and Nip is free. Nip not only *has* a right, but he can choose whether or not to exercise it, whether to claim it, whether to register complaints upon its infringement, even whether to release Tuck from his duty, and forget the whole thing. If the personal claim-right is also backed up by criminal sanctions, however, Tuck may yet have a duty of obedience to the law from which no one, not even Nip, may release him. He would even have such duties if he lived in Nowheresville; but duties subject to acts of claiming, duties derivative from the contingent upon the personal rights of others, are unknown and undreamed of in Nowheresville.

Many philosophical writers have simply identified rights with claims. The dictionaries tend to define "claims," in turn as "assertions of right," a dizzying piece of circularity that led one philosopher to complain — "We go in search of rights and are directed to claims, and then back again to rights in bureaucratic futility."[5] What then is the relation between a claim and a right?

As we shall see, a right *is* a kind of claim, and a claim is "an assertion of right," so that a formal definition of either notion in terms of the other will not get us very far. Thus if a "formal definition" of the usual philosophical sort is what we are after, the game is over before it has begun, and we can say that the concept of a right is a "simple, undefinable, unanalysable primitive." Here as elsewhere in philosophy this will have the effect of making the commonplace seem unnecessarily mysterious. We would be better advised, I think, not to attempt definition of either "right" or "claim," but rather to use the idea of a claim in informal elucidation of the idea of a right. This is made possible by the fact that *claiming* is an elaborate sort of rule-governed *activity*. A claim is that which is claimed, the object of the act of claiming. . . . If we concentrate on the whole activity of claiming, which is public, familiar, and open to our observation, rather than on its upshot alone, we may learn more about the generic nature of rights than we could ever hope to learn from a formal definition, even if one were possible. Moreover, certain facts about rights more easily, if not solely, expressible in the language of claims and claiming are essential to a full understanding not only of what rights are, but also why they are so vitally important.

Let us begin then by distinguishing between: (i) making claim to . . . , (ii) claiming that . . . , and (iii) having a claim. One sort of thing we may be doing when we claim is to *make claim to something*. This is "to petition or seek by virtue of sup-

[5]H. B. Acton, *op. cit.*

posed right; to demand as due." Sometimes this is done by an acknowledged right-holder when he serves notice that he now wants turned over to him that which has already been acknowledged to be his, something borrowed, say, or improperly taken from him. This is often done by turning in a chit, a receipt, an I.O.U., a check, an insurance policy, or a deed, that is, a *title* to something currently in the possession of someone else. On other occasions, making claim is making application for titles or rights themselves, as when a mining prospector stakes a claim to mineral rights, or a householder to a tract of land in the public domain, or an inventor to his patent rights. In the one kind of case, to make claim is to exercise rights one already has by presenting title; in the other kind of case it is to apply for the title itself, by showing that one has satisfied the conditions specified by a rule for the ownership of title and therefore that one can demand it as one's due.

Generally speaking, only the person who has a title or who has qualified for it, or someone speaking in his name, can make claim to something as a matter of right. It is an important fact about rights (or claims), then, that they can be claimed only by those who have them. Anyone can claim, of course, *that* this umbrella is yours, but only you or your representative can actually claim the umbrella. If Smith owes Jones five dollars, only Jones can claim the five dollars as his own, though any bystander can *claim that* it belongs to Jones. One important difference then between *making legal claim to* and *claiming that* is that the former is a legal performance with direct legal consequences whereas the latter is often a mere piece of descriptive commentary with no legal force. Legally speaking, *making claim to* can itself make things happen. This sense of "claiming," then, might well be called "the performative sense." The legal power to claim (performatively) one's right or the things to which one has a right seems to be essential to the very notion of a right. A right to which one could not make claim (i.e. not even for recognition) would be a very "imperfect" right indeed!

Claiming that one has a right (what we can call "propositional claiming" as opposed to "performative claiming") is another sort of thing one can do with language, but it is not the sort of doing that characteristically has legal consequences. To claim that one has rights is to make an assertion that one has them, and to make it in such a manner as to demand or insist that they be recognized. In this sense of "claim" many things in addition to rights can be claimed, that is, many other kinds of proposition can be asserted in the claiming way. I can claim, for example, that you, he, or she has certain rights, or that Julius Caesar once had certain rights; or I can claim that certain statements are true, or that I have certain skills, or accomplishments, or virtually anything at all. I can claim that the earth is flat. What is essential to *claiming that* is the manner of assertion. One can assert without even caring very much whether anyone is listening, but part of the point of propositional claiming is to *make sure* people listen. When I claim to others that I know something, for example, I am not merely asserting it, but rather "obtruding my putative knowledge upon their attention, demanding that it be recognized, that appropriate notice be taken of it by those concerned. . . ."[6] Not every truth is properly assertable, much

[6]G. J. Warnock, "Claims to Knowledge," *Proceedings of the Aristotelian Society,* Supplementary Volume 36 (1962), p. 21.

less claimable, in every context. To claim that something is the case in circumstances that justify no more than calm assertion is to behave like a boor. (This kind of boorishness, I might add, is probably less common in Nowheresville.) But not to claim in the appropriate circumstances that one has a right is to be spiritless or foolish. A list of "appropriate circumstances" would include occasions when one is challenged, when one's possession is denied, or seems insufficiently acknowledged or appreciated; and of course even in these circumstances, the claiming should be done only with an appropriate degree of vehemence.

Even if there are conceivable circumstances in which one would admit rights diffidently, there is no doubt that their characteristic use and that for which they are distinctively well suited, is to be claimed, demanded, affirmed, insisted upon. They are especially sturdy objects to "stand upon," a most useful sort of moral furniture. Having rights, of course, makes claiming possible; but it is claiming that gives rights their special moral significance. This feature of rights is connected in a way with the customary rhetoric about what it is to be a human being. Having rights enables us to "stand up like men," to look others in the eye, and to feel in some fundamental way the equal of anyone. To think of oneself as the holder of rights is not to be unduly but properly proud, to have that minimal self-respect that is necessary to be worthy of the love and esteem of others. Indeed, respect for persons (this is an intriguing idea) may simply be respect for their rights, so that there cannot be the one without the other; and what is called "human dignity" may simply be the recognizable capacity to assert claims. To respect a person then, or to think of him as possessed of human dignity, simply *is* to think of him as a potential maker of claims. Not all of this can be packed into a definition of "rights"; but these are *facts* about the possession of rights that argue well their supreme moral importance. More than anything else I am going to say, these facts explain what is wrong with Nowheresville.

We come now to the third interesting employment of the claiming vocabulary, that involving not the verb "to claim" but the substantive "a claim." What is to *have a claim* and how is this related to rights? I would like to suggest that *having a claim consists in being in a position to claim, that is, to make claim to or claim that.* If this suggestion is correct it shows the primacy of the verbal over the nominative forms. It links claims to a kind of activity and obviates the temptation to think of claims as *things,* on the model of coins, pencils, and other material possessions which we can carry in our hip pockets. To be sure, we often make or establish our claims by presenting titles, and these typically have the form of receipts, tickets, certificates, and other pieces of paper or parchment. The title, however, is not the same thing as the claim; rather it is the evidence that establishes the claim as valid. On this analysis, one might have a claim without ever claiming that to which one is entitled, or without even knowing that one has the claim; for one might simply be ignorant of the fact that one is in a position to claim; or one might be unwilling to exploit that position for one reason or another, including fear that the legal machinery is broken down or corrupt and will not enforce one's claim despite its validity.

Nearly all writers maintain that there is some intimate connection between having a claim and having a right. Some identify right and claim without qualification; some define "right" as justified or justifiable claim, others as recognized claim, still others as valid claim. My own preference is for the latter definition. Some writers, however, reject the identification of rights with valid claims on the ground that all claims as such are valid, so that the expression "valid claim" is redundant. These writers, therefore, would identify rights with claims *simpliciter*. But this is a very simple confusion. All claims, to be sure, are *put forward* as justified, whether they are justified in fact or not. A claim conceded even by its maker to have no validity is not a claim at all, but a mere demand. The highwayman, for example, *demands* his victim's money; but he hardly makes claim to it as rightfully his own.

But it does not follow from this sound point that it is redundant to qualify claims as justified (or as I prefer, valid) in the definition of a right; for it remains true that not all claims put forward as valid really are valid; and only the valid ones can be acknowledged as rights.

If having a valid claim is not redundant, i.e. if it is not redundant to pronounce *another's* claim valid, there must be such a thing as having a claim that is not valid. What would this be like? One might accumulate just enough evidence to argue with relevance and cogency that one has a right (or ought to be granted a right), although one's case might not be overwhelmingly conclusive. In such a case, one might have strong enough argument to be entitled to a hearing and given fair consideration. When one is in this position, it might be said that one "has a claim" that deserves to be weighed carefully. Nevertheless, the balance of reasons may turn out to militate against recognition of the claim, so that the claim, which one admittedly had, and perhaps still does, is not a valid claim or right. "Having a claim" in this sense is an expression very much like the legal phrase "having a *prima facie* case." A plaintiff establishes a *prima facie* case for the defendant's liability when he establishes grounds that will be sufficient for liability unless outweighed by reasons of a different sort that may be offered by the defendant. Similarly, in the criminal law, a grand jury returns an indictment when it thinks that the prosecution has sufficient evidence to be taken seriously and given a fair hearing, whatever counter-vailing reasons may eventually be offered on the other side. That initial evidence, serious but not conclusive, is also sometimes called a *prima facie* case. In a parallel *"prima facie* sense" of "claim," having a claim to $X$ is not (yet) the same as having a right to $X$, but is rather having a case of at least minimal plausibility that one has a right to $X$, a case that does establish a right, not to $X$, but to a fair hearing and consideration. Claims, so conceived, differ in degree: some are stronger than others. Rights, on the other hand, do not differ in degree; no one right is more of a right than another.[7]

Another reason for not identifying rights with claims *simply* is that there is a well-established usage in international law that makes a theoretically interesting

[7]This is the important difference between rights and mere claims. It is analogous to the difference between *evidence* of guilt (subject to degrees of cogency) and conviction of guilt (which is all or nothing). One can "have evidence" that is not conclusive just as one can "have a claim" that is not valid. "Prima-facieness" is built into the sense of "claim," but the notion of a "prima-facie right" makes little sense. On the latter point see A. I. Melden, *Rights and Right Conduct* (Oxford: Basil Blackwell, 1959), pp. 18–20, and Herbert Morris, "Persons and Punishment," *The Monist*, Vol. 52 (1968), pp. 498–9.

distinction between claims and rights. Statesmen are sometimes led to speak of "claims" when they are concerned with the natural needs of deprived human beings in conditions of scarcity. Young orphans *need* good upbringings, balanced diets, education, and technical training everywhere in the world; but unfortunately there are many places where these goods are in such short supply that it is impossible to provision all who need them. If we persist, nevertheless, in speaking of these needs as constituting rights and not merely claims, we are committed to the conception of a right which is an entitlement *to* some good, but not a valid claim *against* any particular individual; for in conditions of scarcity there may be no determinate individuals who can plausibly be said to have a duty to provide the missing goods to those in need. J. E. S. Fawcett therefore prefers to keep the distinction between claims and rights firmly in mind. "Claims," he writes, "are needs and demands in movement, and there is a continuous transformation, as a society advances [towards greater abundance] of economic and social claims into civil and political rights . . . and not all countries or all claims are by any means at the same stage in the process."[8] The manifesto writers on the other side who seem to identify needs, or at least basic needs, with what they call "human rights," are more properly described, I think, as urging upon the world community the moral principle that *all* basic human needs ought to be recognized as *claims* (in the customary *prima facie* sense) worthy of sympathy and serious consideration right now, even though, in many cases, they cannot yet plausibly be treated as *valid* claims, that is, as grounds of any other people's duties. This way of talking avoids the anomaly of ascribing to all human beings now, even those in pre-industrial societies, such "economic and social rights" as "periodic holidays with pay."[9]

Still for all of that, I have a certain sympathy with the manifesto writers, and I am even willing to speak of a special "manifesto sense" of "right," in which a right need not be correlated with another's duty. Natural needs are real claims if only upon hypothetical future beings not yet in existence. I accept the moral principle that to have an unfulfilled need is to have a kind of claim against the world, even if against no one in particular. A natural need for some good as such, like a natural desert, is always a reason in support of a claim to that good. A person in need, then, is always "in a position" to make a claim, even when there is no one in the corresponding position to do anything about it. Such claims, based on need alone, are "permanent possibilities of rights," the natural seed from which rights grow. When manifesto writers speak of them as if already actual rights, they are easily forgiven, for this is but a powerful way of expressing the conviction that they ought to be recognized by states here and now as potential rights and consequently as determinants of *present* aspirations and guides to *present* policies. That usage, I think, is a valid exercise of rhetorical licence.

I prefer to characterize rights as valid claims rather than justified ones, because I suspect that justification is rather too broad a qualification. "Validity," as I

[8]J. E. S. Fawcett, "The International Protection of Human Rights," in *Political Theory and the Rights of Man,* ed. by D. D. Raphael (Bloomington: Indiana University Press, 1967), pp. 125 and 128.

[9]As declared in Article 24 of *The Universal Declaration of Human Rights* adopted on December 10, 1948, by the General Assembly of the United Nations.

understand it, is justification of a peculiar and narrow kind, namely justification within a system of rules. A man has a legal right when the official recognition of his claim (as valid) is called for by the governing rules. This definition, of course, hardly applies to moral rights, but that is not because the genus of which moral rights are a species is something other than *claims*. A man has a moral right when he has a claim the recognition of which is called for — not (necessarily) by legal rules — but by moral principles, or the principles of an enlightened conscience.

There is one final kind of attack on the generic identification of rights with claims, and it has been launched with great spirit in a recent article by H. J. McCloskey, who holds that rights are not essentially claims at all, but rather entitlements. The springboard of his argument is his insistence that rights in their essential character are always *rights to,* not *rights against:*

My right to life is not a right against anyone. It is my right and by virtue of it, it is normally permissible for me to sustain my life in the face of obstacles. It does give rise to rights against others *in the sense* that others have or may come to have duties to refrain from killing me, but it is essentially a right of mine, not an infinite list of claims, hypothetical and actual, against an infinite number of actual, potential, and as yet nonexistent human beings . . . Similarly, the right of the tennis club member to play on the club courts is a right to play, not a right against some vague group of potential or possible obstructors.[10]

The argument seems to be that since rights are essentially rights *to,* whereas claims are essentially claims *against,* rights cannot be claims, though they can be grounds for claims. The argument is doubly defective though. First of all, contrary to McCloskey, rights (at least legal claim-rights) *are* held *against* others. McCloskey admits this in the case of *in personam* rights (what he calls "special rights") but denies it in the case of *in rem* rights (which he calls "general rights"):

Special rights are sometimes against specific individuals or institutions — e.g. rights created by promises, contracts, etc. . . . but these differ from . . . characteristic . . . general rights where the right is simply a right to . . .[11]

As far as I can tell, the only reason McCloskey gives for denying that *in rem* rights are against others is that those against whom they would have to hold make up an enormously multitudinous and "vague" group, including hypothetical people not yet even in existence. Many others have found this a paradoxical consequence of the notion of *in rem* rights, but I see nothing troublesome in it. If a general rule gives me a right of noninterference in a certain respect against everybody, then there are literally hundreds of millions of people who have a duty toward me in that respect; and if the same general rule gives the same right to everyone else, then it imposes on me literally hundreds of millions of duties — or duties towards hundreds of millions of people. I see nothing paradoxical about this, however. The

[10]H. J. McCloskey, "Rights," *Philosophical Quarterly,* Vol. 15 (1965), p. 118.
[11]*Loc. cit.*

duties, after all, are negative; and I can discharge all of them at a stroke simply by minding my own business. And if all human beings make up one moral community and there are hundreds of millions of human beings, we should expect there to be hundreds of millions of moral relations holding between them.

McCloskey's other premise is even more obviously defective. There is no good reason to think that all *claims* are "essentially" *against,* rather than *to.* Indeed most of the discussion of claims above has been of claims *to,* and we have seen, the law finds it useful to recognize claims *to* (or "mere claims") that are not yet qualified to be claims *against,* or rights (except in a "manifesto sense" of "rights").

Whether we are speaking of claims or rights, however, we must notice that they seem to have two dimensions, as indicated by the prepositions "to" and "against," and it is quite natural to wonder whether either of these dimensions is somehow more fundamental or essential than the other. All rights seem to merge *entitlements to* do, have, omit, or be something with *claims against* others to act or refrain from acting in certain ways. In some statements of rights the entitlement is perfectly determinate (e.g. *to* play tennis) and the claim vague (e.g. *against* "some vague group of potential or possible obstructors"); but in other cases the object of the claim is clear and determinate (e.g. *against* one's parents), and the entitlement general and indeterminate (e.g. to be given a proper upbringing). If we mean by "entitlement" that *to* which one has a right and by "claim" something directed at those against whom the right holds (as McCloskey apparently does), then we can say that all claim-rights necessarily involve both, though in individual cases the one element or the other may be in sharper focus.

In brief conclusion: To have a right is to have a claim against someone whose recognition as valid is called for by some set of governing rules or moral principles. To have a *claim* in turn, is to have a case meriting consideration, that is, to have reasons or grounds that put one in a position to engage in performative and propositional claiming. The activity of claiming, finally, as much as any other thing, makes for self-respect and respect for others, gives a sense to the notion of personal dignity, and distinguishes this otherwise morally flawed world from the even worse world or Nowheresville.

# CHAPTER 2

# A Postscript to the Nature and Value of Rights (1977)
*Joel Feinberg*

*I would like* to take this opportunity to supplement the brief account [in Chap. 1] of the role of rights in human life and to correct some of its emphases. First, it appears in several places as though *having* rights is what is necessary for self-respect, dignity, and other things of value. Actually, it is not enough to have the rights; one must know that one has the rights. In fact, the poor benighted citizens of Nowheresville do have various rights, whether they know it or not. They could not possibly know — or understand — that they have rights, however, because they do not even have the *concept* of a personal right. Such a notion has never even been dreamed of in Nowheresville. The inhabitants are consequently deficient in respect for self and others, even though, as hypothetical human beings, they have dignity in the eye of our imaginations.

Second, even knowing that one has rights and being prepared to act accordingly are not sufficient (but only necessary) for a fully human and morally satisfactory life. A person who never presses his claims or stands on his rights is servile, but the person who never waives a right, never releases others from their correlative obligations, or never does another a favor when he has a right to refuse to do so is a bloodless moral automaton. If such a person fully understands and appreciates what rights are and invokes that understanding in justification of his rigid conduct, he is a self-righteous prig as well. If he can also truly testify that he always conscientiously performs *his* duties to others and respects *their* rights, he has then achieved "the righteousness of the scribes and pharisees."

The point to emphasize here is that (with some rare exceptions mentioned below) right-holders are not always obliged to exercise their rights. To have a right typically is to have the discretion or "liberty" to exercise it or not as one chooses. This freedom is another feature of right-ownership that helps to explain why rights are so valuable. When a person has a discretionary right and fully understands the power that possession gives him, he can if he chooses make sacrifices for the sake of others, voluntarily give up what is rightfully his own, freely make gifts that he is in no way obligated to make, and forgive others for their wrongs to him by declining to demand the compensation or vengeance he may have coming or by warmly welcoming them back into his friendship or love. Imagine what life would be like without these saving graces. Consider Nowheresville II where almost everyone performs his duties to others faithfully and always insists upon his own rights against others; where debtors are never forgiven their debts, wrongdoers pardoned, gratuitous gifts conferred, or sacrifices voluntarily made, so long as it is within one's rights to refuse to do any of these things. The citizens of Nowheresville II have forgotten, if they ever knew, how to exercise rightful discretion. They have but half the concept of a right; they know how to claim but not how to release, waive, or surrender.

The point I wish to emphasize is not that the saving graces show that there is a limit to the moral importance of rights, but rather that rights are even more important — and important in other ways — than my original article suggests. Knowing that one has rights makes not only claiming (and self-respect) but also releasing (and magnanimity) possible. Without the duties that others have toward one (correlated with one's rights against them) there could be no sense in the notion of one's supererogatory conduct toward other people, for to help others when one has a right to decline is precisely what conduct "above and beyond duty" amounts to. Understanding that one has rights, of course, is not *sufficient* for one to have an admirable character, for one might yet be a mean-spirited pharisee, unwilling ever to be generous, forgiving, or sacrificing. But consciousness of one's rights is *necessary* for the supererogatory virtues, for the latter cannot even be given a sense except by contrast with the disposition always to claim one's rights. Waivers and gratuities can exist only against a background of understood rules assigning rights and duties. Forgiving debts obviously would not be possible without the prior practice of loaning and repaying with its rule-structured complexes of rights and correlative duties. Even in Nowheresville I, I suppose, one person can give a useful thing to another, but he cannot make a *gift* or *gratuity*, since giving more than the recipient can rightly claim (a gift) presupposes that others *can* make rightful claims in some circumstances and that there is such a concept and such a practice.

One final point. Some familiar political rights appear to be exceptions to the assertion above that rights confer liberty or discretion upon their possessors who may always choose, if they wish, not to exercise them. The "right to education," for example, seems to be a kind of "mandatory right" in that children who possess it have no choice whether to go to school or not. Similarly, the legal right of schoolchildren to be vaccinated against certain contagious diseases is entirely coincident (except for the exemption on religious grounds) with their legal *duty* to be vaccinated. I suggest that when we use the language of rights in this way to refer to duties, we do so because we think that some of our duties are so beneficial that we can make *claim* against others to provide the opportunity for, and to abstain from interference with, our performance of them.

Textbooks frequently say that to have a claim-right to do $X$ is (1) to be at liberty with respect to $X$ and (2) for others to have a duty to one to provide or (as the case may be) not to interfere with $X$. When a claim-right is analyzed in this fashion, its component liberty is then said to be simply the absence of a duty *not* to $X$. But this characterization of a liberty, I submit, is misleading. To be at liberty to do $X$ in ordinary speech is to have *discretion* in respect to $X$, to be free *both* of a duty not to do $X$ *and* of a duty to do $X$. To be free of a duty not to do $X$ is to have only a "half-liberty" with respect to $X$ if one should at the same time have a duty *to do $X$*. Thus schoolchildren have "no duty" to stay away from school (a half-liberty with respect to school attendance), though they do have a duty to go to school. They are, therefore, deprived of the other "half-liberty" that would add up to full liberty, or the discretion to decide whether to attend school or not. Most rights to do $X$ are full liberties to do $X$ or not to do $X$ as one chooses, conjoined with duties of other people not to interfere with one's choice. But so-called "mandatory rights" to do $X$ confer only the half-liberty to do $X$ without the other half-liberty not to do $X$. Why then are they called "rights" at all?

The answer is that the rights in question are best understood as ordinary duties with associated half-liberties rather than ordinary claim-rights with associated full liberties, but that the performance of the duty is presumed to be so beneficial to the person whose duty it is that he can *claim* the necessary means from the state and noninterference from others as *his* due. Its character as claim is precisely what his half-liberty shares with the more usual (discretionary) rights and what warrants his use of the word "right" in demanding it.

# CHAPTER 3

# Rights and Claims*
*Bertram Bandman*

*What are rights?* Where do they come from? How are rights related to claims? And what functions do rights and claims have in the law, morals, politics and epistemology?

I will consider some recent work on the language of rights and claims that bears on the above questions. I will begin with a critique of Joel Feinberg's work on the nature of rights. According to Feinberg's analysis of rights and claims, rights play a primary and indispensable role in relation to claims. I shall argue that the concept of claims is primary and that the concept of rights is secondary. Without claims there would be no rights. In the second part of this paper I will propose a conceptual revision of the relation between rights and claims, one that freed of the difficulties in Feinberg's analysis, may have further bearing on our opening questions.

First, however, why the recent concern about rights? Rights, it is thought, may give us reasons for acting one way rather than another. So if we could only fasten down what rights are, we could invoke them to justify at least some of our actions. Why are claims important? Claims enable us to exercise our rights; to stand up if necessary and demand justifiably what is our due. Concerning both rights and claims, we could parody Kant's dictum: rights without claims are empty, and claims without rights are blind.

Feinberg, in a most engaging and important paper, tries to show how vital rights are by asking us to perform a thought experiment, to imagine a world, which he calls Nowheresville, very much like our own but with one difference. It is a world without rights. Nowheresville has duties and even benevolently motivated actions, but no rights [19, 20].[1]

## PART I: CRITIQUE OF FEINBERG'S ANALYSIS OF RIGHTS OF CLAIMS

In Nowheresville the subjects act for the most part, to please their master, but without a whimper when unrewarded for performing exceptional service. They are

From *The Journal of Value Inquiry,* Vol. 7 (1973), 204–213, by permission of the publisher. Page references cited are to Chapter 1 in this book, The Nature and Value of Rights by Joel Feinberg, from *The Journal of Value Inquiry*, Vol. 4 (1970), 243–257.

*A slightly revised version of a paper read at the Long Island Philosophical Society, May 15, 1971. I wish to thank Lowell Kleinman, Alex Orenstein, Peter Manicas and Karsten Struhl for their helpful criticisms.

[1]Several reasons of varying strength could be cited against considering Nowheresville even as a viable thought experiment: 1. At least one person, a sovereign, has rights, even in Nowheresville. 2. Subjects have roles in relation to one another and at least some of these involve rights, such as sex rights, family and kinship rights. That is, subjects have rights among themselves. 3. A sovereign has a duty not to kill or harm all or most of his subjects wantonly, which are their rights. 4. If times get bad and extreme scarcity sets in at Nowheresville, some people who may know nothing of rights, may get the idea that it pays to be pushy. And in due course, some Nowheresvillians are liable to think they have rights, like the right to live.

The point here, however, is that Feinberg uses this hypothetical conditional, this world without rights in contrast to our world of rights, to reveal the importance of rights. But when Feinberg asserts the primacy of rights in relation to claims, gnawing questions arise. How, for example, did the sovereign of Nowheresville get his rights? How do subjects get or wrest their rights?

not conscious of their rights or of their due. Nowheresvillians have no reason to make complaints, claims or demands. No matter how "badly treated," they "do not think to leap to their feet and make righteous demands . . ." [24].

According to Feinberg, the main difference between Nowheresvillians and ourselves is that we have rights and they don't. "They have no notion of rights" and so no "notion of what is their due; hence they do not claim before they take" [24]. According to Feinberg, one has to have a notion of rights before one has a notion of what it is to make claims, demands, grievances, complaints. To Feinberg, rights are primary.

Feinberg's conception of the primacy of rights is additionally illustrated, directly or indirectly with these three further assertions.

1. "Having rights . . . makes claiming possible" [27].
2. Claiming is a "rule governed activity" [25].
3. "To have a claim" puts one "in a position to engage in performative and propositional claiming" [27, 31].

As to the first, "having rights . . . makes claiming possible," if we didn't have rights, we wouldn't go around making claims and demands and shouting, marching and the rest. I think he is wrong in thinking that one can only make claims if he has rights. Rights may make some claims possible but rights are not presupposed in all acts of claiming. The making of claims may outrun rights held. People have been known to make all sorts of claims to rights. To assert that there can't be claims without rights is based on the view that claiming cannot precede the existence of rights.

Feinberg says, secondly, that "claiming . . . is [a] rule-governed activity" [25]. To Feinberg there are no claims without rights and rules. Feinberg correctly identifies some claims within a legal system, but he does not account for claims, demands, complaints and grievances that a person might make that are pre-legal or extra-legal, or illegal, or on the borderline. And in ignoring the pre-legal, he ignores a possible explanation of how we got our rights.

I would say, contra Feinberg, that the activity of claiming goes on outside as well as within the range of rule-governed activity. The rubric "rule-governed" does not cover all acts of claiming. People who make claims do not always stay within the rules. Some claimants are said to stretch the rules, and for still others there are no rules.

We come now to Feinberg's third assertion, "having a claim" puts one "in a position to engage in performative and propositional claiming" [31]. Feinberg distinguishes between "(i) making claim to . . . (ii) claiming that . . . and (iii) having a claim" [25].

A claim *to* has the legal power "to make things happen" [26]. A claim *that* or propositional claim is a mere descriptive commentary which obtrudes itself on our attention but without the legal power to make things happen. To claim *that* is clearly a weaker sister.

This brings us to the strongest of the three, "having a claim." To have a claim is to have a *prima facie* case, one with at least minimal plausibility, one that deserves "a fair hearing and consideration" [28]. A claim *to* has the legal power to make things happen but only if one *has* a claim. Claims can only be claimed by those who *have* them [26]. And so, "having a claim," which presupposes rights and rules, puts one in a position to make a claim *to* or to claim *that*, that is, to engage in performative and propositional claiming.

Feinberg correctly identifies the notion of "having a claim" as rule governed and he may be correct that having rights makes claiming of that kind possible. But he is not correct to so refer to the entire class of acts of claiming which is not bound with previously established rights and rules.

My objection then to Feinberg's analysis of rights and claims is to his view that claiming cannot precede the existence of rights and rules.

But even if he didn't mean this, I think it is entirely the other way around. Claiming makes rights possible and claiming need not be a rule governed activity. I would turn Feinberg's point around completely. Without claims there could be no rights. It is not rights that makes claiming possible. It is claims that make rights possible.


## PART II: RIGHTS AND CLAIMS

How do we get rights? What accounts for their generation and expansion? Rights didn't just come about or come *sui generis* into our world as if by a magic wand. We got our rights by claiming them, by fighting, marching, demanding, clamoring and claiming them. Without claiming rights we wouldn't have them.

I will try to explain how we got our rights, their generation and expansion in the following two stages: (i) We make claims, including claims to rights. (ii) A legal system generates rights and rules for sustaining claims.

### 1. We Make Claims, Including Claims to Rights
#### A. Claims

We have in stage one the fact of our having needs, wants, interests and desires which we express with claims which, to hark back to the etymology of "claims," clamor or cry out for attention. To claim is to cry out. Claiming in its primitive sense expresses our conative strivings.[2]

To William James, every claim cries out as demand made by a concrete, live person, and in that account alone, "creates . . . an obligation." To James, every claim which any person may make, "ought . . . to be satisfied" unless some other claim can be shown to conflict with it.[3] James shows us the use of a claim other than the one Feinberg presents.

[2]Whether it is a primitive man or the baby, the cry of want or pain is one of the first and fundamental forms of expression to come from humans. It is there. We hear it. We sympathize or agonize with or about it. It could be us doing the crying. Unsatisfied over too long a period it becomes anguish. A person trapped in a cave with a howling blizzard and no food whines and bellows as with a person trapped in an elevator with no food, water or relief. A person's desire to live and to give an outcry whenever his life seems threatened is for a person the basis for making claims. For that reason, claiming may be regarded as a verbally primitive expression of the co-native.

Spinoza held that conatus is "the effort by which each thing endeavors to persevere in its own being." (See also P. Caws, *Science and the Theory of Values*, Random House, 1967, p. 56.)
[3]W. James, *The Will to Believe*, Dover, 1956, pp. 194–195.

To make a claim, contra Feinberg, is not (necessarily) to have the legal power to make things happen; it is *to express a desire to have the power to make things happen.* Such a desire is not necessarily unaccompanied by the force requisite to cause things to happen *legally.*[4]

Claiming in its primitive sense is not rule-bound or a species of rights talk. Claiming in its primitive sense is conative.

### B. Claims to Rights

Among the claims we make are claims to rights, which do not just cry out. These claims express a desire for something relatively more permanent, not just a desire for a piece of bread, but a desire to eat unimpeded by others.

The origin of rights can be traced to the way the sort of sovereign referred to by Feinberg gets his "rights." He claims the right to have rights, which he is then loath to distribute to others without having to. In a similar connection, Martin Golding points out that "an historically important factor in the generation of rights" is "the claims and demands put forward by individuals and groups of individuals against others."[5]

A person's claim to a right, however, is not a right, for he can claim more than his rights.

The notion of claims to rights accommodates Feinberg's "manifesto writers," who speak of claims as needs of the downtrodden, for example, that cry out; but these are not rights. Nor are these "mere claims" either. They are claims to rights, which may express frustrations, aspirations, hope, despair or protest.

The "manifesto writers' " use of "rights" can be interpreted as persuasive definitions of rights aimed at extending the meaning of rights to include the satisfaction of more and more human needs. One used to hear, for example, of the right to a free education. More recently, one hears of the right to health care and the right to a guaranteed annual income and periodic holidays with pay. But these are not rights. They are claims to rights with varying degrees of support.

A claim to a right marks out what we do to get others to recognize as our right, but it is not necessarily our right.

What then are rights? This brings us to another important, but different stage.

### 2. A Legal System Generates Rights and Rules for Sustaining Claims
### A. Rights

The enormous numbers of claims to rights that rain down on us cannot all be met. Some have to give way.

A legal system is then established as a basis for judging from among the competing claims to rights those that merit being designated as "rights." Thus, some claims to rights get privileged status and become identified as rights. For a function of a legal system is to stabilize a community by pruning down the endless procession of claims and claims to rights that members of a community make on one

---

[4]Claiming is not, I think, illocutionary. It is perlocutionary. A claim effects a future. I hope to have been reasonably faithful to J. L. Austin's rendering of claims, which he calls "exercitives" in *How To Do Things With Words,* Oxford University Press, 1965, pp. 155–159.

[5]M. Golding, "Towards a Theory of Human Rights," *Monist,* 1968, Vol. 52, No. 4, p. 521.

another; and it carries out this function, in part, by judging what claims to rights qualify as rights.

Rights thus generated by a legal system are conferred, granted, given, awarded, bestowed or even "gained," but they are not taken, stolen, captured, made or gotten *only* by being claimed. The rights persons have, hold or possess have been *given* or *granted* by a legal system.

Rights conferred — and they are conferred within a community[6] — are not, however, sacrosanct, fixed or final like three sided triangles. One can lose one's rights. Rights are revokable. Nor, however, are rights as temporary as the clothes we put on and take off every day. Rights are (somewhat) like renewable season passes, licenses, passports, authorizations, titles, deeds, "entitlements," — H. J. McCloskey calls them.[7]

*B. Rules*

Rules are needed to give teeth to rights. Otherwise, one can not exercise rights that have been granted. As Jeremy Bentham put it . . ., "the law cannot grant a benefit" or "create a right in favor of one" without imposing at the same time "some burden" or "corresponding obligation."[8]

A legal system also provides rules for recognizing and adjudicating claims presented. A legal system thus provides three kinds of rules, rules of recognition, adjudication and enforcement.

The first of these are rules of recognition.[9] To give recognition to a case means that a court is willing to hear a case as one that falls within its jurisdiction.

The following anecdote illustrates the role of such rules. *The New York Post* some years ago reported a man who lost several suits against a city in Florida. He slipped and broke his hip and sued the city and lost. After several similar mishaps and court losses in which he was told that these accidents were legally "acts of God," this accident prone person filed a claim against God. A judge thereupon ruled the case out on the grounds that "the claim was beyond the Court's jurisdiction." The point illustrated here is that a claim unrecognized by a legal system cannot be judged either way.

There are secondly rules of adjudication for resolving conflicts between claimants. Adjudication consists of rules for judging the merits of claims made, and provides rules either for upholding or defeating claims that have been presented to a court. These rules involve canons of evidence and other epistemic considerations that bear on "the facts of a case," such as the reliability of eyewitness reports.

There are thirdly rules of enforcement that provide for the application of appropriate sanctions in the event that a legal offense is committed. A claim upheld

---

[6]To some writers like Marx there are no rights outside the economic system and the cultural components conditioned by it. I refer here to Marx's *Critique of the Gotha Program,* in L. Feuer, Marx and Engels, Doubleday, 1959, p. 119.

[7]H. J. McCloskey, "Rights," in *The Philosophical Quarterly,* Vol. 15, No. 59, April, 1965, p. 118.

[8]J. Bentham, "Principles of the Civil Code" in M. R. Cohen and F. Cohen, *Readings in Jurisprudence and Legal Philosophy,* Prentice-Hall, 1950, p. 606. Also E. A. Hoebel in *The Law of Primitive Man* (Harvard University Press, 1954, pp. 47–48), applying Hohfeld's reciprocal legal relations to a study of the origin and nature of rights in primitive tribes, points out that where any person A has duty to another person B, there is a corresponding right.

[9]See H. L. A. Hart, *The Concept of Law,* Oxford University Press, 1961, pp. 92–93, 113.

by a court means nothing if it cannot be enforced. In this connection, Charles
Peirce once remarked that the law is only as the power of a sheriff to put his
hands on a violator and herd him off to jail. For a judge to say, "I recognize and
I grant your right to enter your premises, but I cannot provide any way for you to
exercise that right" is indeed an empty right.

A legal system, consisting of rules of recognition, adjudication and enforcement,
accordingly enables rights and powers to be granted and protected. Without these
(or similar) kinds of rules, one could make claims, but one could not appraise them,
nor could any one make any claim, that was ever upheld, "stick."

## C. Having the Right to Make a Strong Claim

A resulting legal system confers rights bound by rules, which enables a person to
exercise his rights. A legal system of rights and rules, thus generated, gives a person
the right to make — for want of a better expression — *a strong claim*. The right to
make a strong claim means that there is a legal support for such a claim.

We can now interpret "having a claim" to mean having the right to make a strong
claim. "Having a claim" puts one in a position not to make just an ordinary claim.
It puts a person in a position to make a strong claim.

To have a claim is, I think, not a claim at all, but the right to make a strong
claim. It is true that one speaks of "having a claim" to some land or other, but that
seems to be another way of saying that one has a title, deed or right to the land.
"Having a claim" can be shown to involve having a title, deed or right. In that way,
we can see how rights back claims.

To have a claim is to have the *right to make a strong claim,* one that has, at least,
an initial degree of credibility or plausibility in its favor.[10] To have a claim check,
for example, is to have the (presumptive) right to make a strong claim (or demand)
for the goods or services specified by the claim, one that can be sustained, if necessary, in a court of law.

The right to make a strong claim does not mean that such a claim will inevitably
be upheld as one that is valid (or indefeasible). It is a claim that can be sustained
or defeated with a prospect of being sustained. To have the right to make a strong
claim means that one is entitled to make a claim which merits recognition and consideration, even though it is not yet judged valid, and even though it may be judged
invalid. The argument for it is promising, although there is no assurance as to the
outcome by a court. For a strong claim, as distinct from a weak or unrecognized
claim, has at least an initial presumption in its favor. I refer to the sorts of examples
Feinberg cites, like a claim check, a baggage stub, an I.O.U. or a bank check.

Making and having claims are different. One can make a claim, but the degree
of support for a claim, the conditions for its favorable appraisal requires appropriate
evidential backing of the kind indicated by saying of someone that he "has a claim."
To say "he has a claim" is as much to appraise and endorse a claim as to predict for
it the likelihood of its success.

[10]For an analogous epistemic account of the id. . of "systematic import" with statements
having degrees of credibility within a system of statements, see N. Goodman, "Sense and
Certainty," in C. Landesman, *The Foundations of Knowledge,* Prentice-Hall, 1970, pp. 173–
180; also I. Scheffler, "On Justification and Commitment" in *The Journal of Philosophy,* 1954,
pp. 180–190; also his Science and Subjectivity, Bobbs-Merrill, 1967, pp. 116–124.

There are all kinds of claims that are made and, in fact, easily rebutted in daily life. In this connection, A. N. Prior, for example, points out that to say, "Mind that tiger!" (is a claim that) loses its justification if there is no tiger.[11] The point is we make and rebut claims all the time. We don't have to impose rigid qualifications as to what counts as a claim. One doesn't have to have a claim or a strong claim before making a claim. Prior's example shows how a weak claim is rebutted. That is what we do. We make claims and with appropriate rights we can also sustain or rebut them.

### D. Claiming to and Claiming That

We can go on to reinterpret a claim *to* and a claim *that* to accord more with common usage as follows: A claim *to* implies a claim *that*. Without the activity of claiming there would be no (propositional) claims. A claim as the outcome of the act of claiming, specifies a content. As there is no claim without the act of claiming, there is no claim that is about nothing. One can't make a claim about nothing just as one can't have dinner but have nothing for dinner.[12] Every claim is about something. This is its content. To make a claim is at the same time to specify *the content of the claim.*

Claiming to and claiming that are different facets of claiming. The one refers to the act of claiming in a more or less obtrusive manner.[13] The other facet refers to the content of the claim. But neither a claim *to* nor a claim *that* presupposes the rights and rules of a legal system.

To make a claim *to* implies a claim *that*, and having a *strong* claim means we are in a position to give favorable, initial appraisal to a claim thus made. We are in that position not because of a claim we have, but because we have the right to make a strong claim.

### E. Two Senses of a Claim

There are two difficulties I wanted to try to guard against. One difficulty consists in defining claims too narrowly either as rights or as presupposing the prior existence of rights. This is the problem with Feinberg, placing excessive restrictions on the concept of claims. He only countenances rule-governed and rights-dependent claims. A second difficulty is to identify claims *too* widely with virtually any needs, interests, desires or demands, where the mere making of a claim is thought to entail its own justification.

To offset both these difficulties, I suggest a distinction between two senses of a claim. The first or primitive sense of a claim refers to the act of making a claim, a

---

[11] A. N. Prior, *Logic and the Basis of Ethics,* Oxford University Press, 1956, p. 79.

Having a strong claim, I want to say, is not a condition for making a claim. For we make claims which are either strong, valid, weak or to which some other predicate attaches. Some claims are termed, by Karen Horney, for example, as "neurotic" or "exaggerated" or "gigantic" (in her *Neurosis and Human Growth,* Norton, 1950, pp. 40–63). We qualify claims in this and other ways as exaggereated, unfair or false, frivolous (See M. Golding, "Obligations to Future Generations," *Monist,* 1972, forthcoming), improper or mistaken (See S. Toulmin, *The Uses of Argument,* Cambridge University Press, 1958, pp. 57–62), fraudulent or weak.

[12] I. Scheffler uses a similar example in a slightly different connection in *The Language of Education,* Thomas, 1960, p. 38.

[13] A claim can be expressed to, for or against someone; or it can be for or against some state of affairs or other. To say one "has a claim on" is, I think, tantamount to saying that one has a strong claim.

crying out, a demand for one's supposed due or right; and is expressed with the intention of obtruding on the attention of others. There is, however, no way to judge such a claim. A claim in this sense is expressed, but is not *presented for appraisal.* The unmet claims of a poor Mexican orphan are claims in this primitive sense of a claim. There is unfortunately no one to hear them. The activity of claiming in its primitive sense does not presuppose the rules of a system. It is only those claims presented before a court and governed by an elaborate set of rules that are open to appraisal.

A second sense of a claim accordingly refers to the appraisal of claims made. A claim in this sense can be sustained or rebutted by appealing to a right within a rule-governed system. One way to appraise a claim is to say it is a strong claim. Thus, my right to vote gives me a strong claim against anyone who should interfere, threaten, or obstruct that right. A legal system that confers rights enables us to submit claims before a court for appraisal. For claims thus presented meet the admissibility requirements of a legal system.

We may accordingly refer to two senses of a claim. In the first sense, claims are the cries and demands that people make. In a second sense, a claim is presented for appraisal. Only at this stage do questions about rights arise.[14]

[14]There may be some who are left essentially unsatisfied with this account of rights, lamenting that legal rights and claims are not moral rights and claims and who object to ignoring or excluding the question of the morality of rights and claims – and well they might be. For without justice and morality, as St. Augustine says, a state is nothing but a band of robbers. For those thus dissatisfied one could provide a third stage after stage 1 (making claims) and stage 2 (A legal system generates rights and rules for sustaining claims). A possible third stage consists in the generalizability of (some) legal rights and rules into moral rights and obligations.

The generalizability of legal rights and rules into moral rights and obligations also involves a shadowy counterpart of legal rules of recognition, adjudication and enforcement, and with appropriate modifications. A "Court of Morality" – if there even is one, is less visible; enforcement is less by external control than by internal self-control; recognition involves something more like (in T. H. Green's words) "the conscience of a community" than a set of rules of jurisdiction. Thus, for example, readers of *Huckleberry Finn* can recognize his claim to morality in refusing to turn his friend Jim in as a runaway slave.

Another approach to the moral generalization of rights and claims, however, is to consider moral rights and claims more a matter of degree that accompanies the making of claims and the granting of rights and not add stages beyond necessity. For there is no reason to mark this third state off as distinct from the moral generalizability that can and does occur in stages (i) and (ii) – and again for the sort of reason St. Augustine gave.

"Manifesto rights" or claims to rights, like those of abolitionists or pre-legal desegregationists, or the would-be rights proclaimed in the U.N. Declaration of Human Rights (like the equal right of all persons to periodic holidays with pay) as well as the claims of a Mexican urchin can be moral as well as pre-legal. Similarly, a legal system can also provide moral rights and obligations to those upon whom it confers legal rights and rules. Legal or pre-legal claims and rights are not necessarily either amoral or immoral.

One can examine the morality of pre-legal as well as legal rights and claims by considering the degree and extent to which a community maximizes some principal features commonly associated with morality. Thus, a pre-legal or legal right or claim is generalized into a moral right or obligation to the degree and extent to which that right or claim is (1) reversible (Baier), that is, applicable in principle, to all or nearly all persons and where exceptions are appropriately justified; (2) rational (P. Taylor), that is free, enlightened and impartial; (3) beneficial, that is likely, on the whole, to bring about good rather than harm or injury or needless pain or extinction (Classical Utilitarianism): (4) capable of appealing to a wide range of economic needs, interests and wants (Classical Marxist Humanism); and (5) capable of inspiring conscious and deliberate recognition and acceptance by the members of a community (T. H. Green).

But even here a difficulty may arise. A claim may be made in a primitive sense. But on occasion a claim in this sense (brute demand) is converted by seemingly imperceptible degrees into a claim in a secondary sense, one that appears to have suitable backing (to be found, for example, in the language of advertising, including philosophical advertising). This shift from a primitive to a secondary sense may be regarded as "capricious." For claim-making is not *ipso facto* claim-sustaining.

Every claim is or starts out as a primitive claim but some are appropriately distilled as claims in a secondary sense. A reason for distinguishing claims is that the claims we make are not automatically self-sustaining. To be judged either way, a claim cannot remain outside the door of a claims court.

This distinction between these two senses of a claim enables us to be most lenient and tolerant as to the first, letting in any claim whatsoever, but being appropriately and duly stringent with the latter, in the appraisal and scrutiny of claims made. This distinction might not be worth making except for the breaches to it.

*Conclusion*

By way of conclusion, I have tried to show that rights do not come from nowhere, that is, rights are not *sui generis.* They come from claims. Rights do not make claims possible; rather claims make rights possible. For out of claims come claims to rights and from the welter of such claims to rights a legal system is established which, after sifting and refining, accepts some claims to rights and dignifies these as deeds, titles, rights and rejects others; and provides rules enabling persons to exercise their rights. A system of rights and rules thus generated gives one the right to make strong claims. Although having a right is not a condition for making a claim, having a right is necessary to sustain and appraise a claim. Appealing to rights enables us to distinguish weak from strong claims. For rights may sustain or rebut claims though they are not themselves claims.

How can we appraise claims? A claim *to* implies a claim *that,* the latter being an outcome of the former. If the resulting claim is open to appraisal of the sustain/reject or true/false kind, then it is a claim in a sense other than a primitive cry in the wild. If one can go on to say of a claim that is open to appraisal that one has a right to make such a claim or that one has a strong claim, this is to give favorable, initial appraisal to a claim thus made; and is a claim not in a *primitive* but in a *secondary* and ultimately more significant sense.

# CHAPTER 4

# The Concept of Rights: A Historical Sketch

*Martin P. Golding*

*Any discussion of rights* and health care must sooner or later confront general issues in ethical analysis and theory: What are rights and what is the basis of rights? These questions, to be sure, are difficult, and they are the subject of much debate, for important philosophical matters are at stake. In these remarks we shall be concerned with some aspects of the history of the concept of human or moral rights — what these rights are — not with theories of the basis of rights. When do we begin to get a crystallized notion of rights as a term of ethical discourse? Where are the first definitions of the concept found? What changes have taken place in the meaning of the concept in the course of its development?

## OPTION RIGHTS AND WELFARE RIGHTS

It is useful to begin by distinguishing between *option rights* and *welfare rights* [5] . Briefly put, option rights essentially involve the notions of freedom and choice and welfare rights the idea of being entitled to some good or benefit. Instances of welfare rights are: the right to an elementary school education, decent housing, and adequate health care — assuming, of course, we have such rights, for the fact is that many people lay claim to goods or benefits to which they are neither morally nor legally entitled.

The concept of an option right is more complicated. A trivial example of an option right, perhaps, is the right to wear any kind of tie one chooses. This right corresponds to a sphere of freedom, a sphere of action subject to the options — choices — of the individual who possesses such a right. Within this sphere the individual is a kind of "sovereign," and he may act as he chooses. Of course, even with respect to the right to wear any kind of tie one chooses, one's sovereignty may in fact be limited. For instance, it is doubtful that an elementary school teacher has the right, legal or moral, to wear a tie with a racially derogatory remark printed on it. It is sometimes maintained that one has the right to do with one's body as one wishes, that one is sovereign over one's own body. But it is plain that even the right to do with one's body as one chooses is a limited option right and that one is only a limited sovereign, as it were. One does not have the right, legal or moral, to inject oneself with the bacteria of a contagious disease, for example.

This last statement, however, should not be taken as implying a general principle to the effect that one has the right to do as one pleases as long as no one else is harmed. The validity of this principle is a contested point in normative ethical theory and social and political philosophy. Whatever its validity, the principle does not follow from the concept of option rights as such. As far as the concept itself is concerned, the possibility is left open (1) that there may be things a person has no right to do even if others are not harmed and, contrariwise (2), that there may be some things a person has the right to do even if others are likely to be harmed. In a free society the risk is taken that an individual might do things that have harmful consequences to himself or others. Nevertheless, option rights, and the spheres of freedom that correspond to them, must be limited in various respects; it would be self-defeating if everyone in a society had the right to do whatever he pleased.

Another important feature of option rights is that they involve not only the freedom to control one's own actions and affairs, but they also imply a degree of control over others, directly or indirectly. One person's sovereignty over himself, in effect, limits another person's sphere of freedom. If I have the right to behave in certain ways if I so choose, your freedom to interfere with me is thereby restricted; to this extent I have a degree of indirect control over your behavior. But the control over others that is implied in some particular option right can be direct, too. Parents, for example, have a kind of direct sovereignty over their minor child. That is, parents in general have the legal and moral competence to make choices for their child. Parents thus have the right to directly limit their child's freedom and also the right to determine whether their child shall receive certain benefits or goods, for these, too, are often subject to the choices of parents. The extent of these parental rights, however, is much debated today. An analogous problem arises with respect to the mentally ill and criminals. Do we have the right to change their behavior, Dr. Gaylin asks? This question is about choice and the right to make choices for others. Or as Gaylin puts it, it raises the question of the right to limit the rights of others [4].

Welfare rights are rights to certain benefits or goods, while option rights concern the scope of the individual's freedom to choose and to act on the basis of his choices. A number of issues arise as to the relationship between welfare rights and option rights. Two issues can be noted briefly here. First, are welfare rights *mandatory* rights, rights that cannot be renounced, rights of which the individual cannot refuse to take advantage? Or are welfare rights subject to the choice (an option right) of the individual to refuse or to accept the benefits that correspond to them [3]? It may be suggested that there is no general "Yes" or "No" answer to these questions: Some welfare rights may be subject to the option of the individual, while others are not. If, as is maintained in some quarters today, everyone is entitled to a college level education as a matter of right, there seems to be no reason why an individual should not be free to refuse to take advantage of this (welfare) right. On the other hand, some welfare rights may be so important that an individual has no (option) right to refuse the benefits that correspond to them; that is, no right to renounce them.

The second issue is related to the issue just mentioned. Which type of rights is ethically primary? Does the recognition of option rights depend on welfare rights, perhaps? This issue goes beyond the analysis of rights and into the ethical basis of rights. It may be suggested that the resolution of this issue turns on a conception of the good life, a conception of the human good. Such a conception will determine the relative roles of freedom and well-being, or the role of freedom as an ingredient of well-being, in the good life.

Which type of rights has historical priority? Which type achieved the earlier recognition as a term of moral and sociopolitical discourse? Or did they appear simultaneously?

These questions will probably seem strange to a modern audience. The terminology of rights pervades so much of our everyday moral and political discourse that it is virtually inconceivable to us that such discourse could ever have gotten along without this terminology. We meet with the language of rights whenever we open up the daily newspaper. Is it imaginable that civilized human beings could have carried on a moral discussion, could have engaged in ethical or political debate, or could have delved into the problems of moral and social philosophy without using this kind of terminology?

## HISTORICAL ORIGIN OF RIGHTS

The fact of the matter appears to be that the concept of rights did not play a significant role in moral and political discourse until the later Middle Ages. It is plain that the grand ethical systems of Plato and Aristotle do not give the concept of rights any prominence, and the same is true of ancient Greek law. The concept of rights, if present at all, remains below the level of consciousness from the time of the Greek philosophers until late medieval times. For example, in Plato's dialogue, the *Crito*, Socrates is faced with a question that *we* would put in the following manner: Does a person who believes himself to have been unjustly convicted of a crime have the right to escape from jail? An examination of Plato's text, however, shows that this dialogue is *not* formulated in the language of rights. There is no term in the text that literally translates into "a right." Instead, Socrates is concerned with whether it would be right or just for him to escape from jail. Now it might seem, at first blush, that there is only a subtle difference of language between asking whether an act is the right thing to do, on the one hand, and between asking whether an individual has the right to do it, on the other. Yet behind this subtlety lies a momentous difference of substance.

A second example comes from Thomas Aquinas (1224–1274), the most influential natural law thinker of the Middle Ages. While Aquinas defends the doctrine of natural law, nevertheless, he has no doctrine of natural rights, despite the fact that the concept of rights was in the air. Aquinas, as did Socrates in the *Crito*, poses a question that we might well put in the language of rights: Does a starving man have a right to steal a loaf of bread? Aquinas does not phrase the question in this way. He asks, instead, whether it is licit to steal from stress of need [1, p. 137]. Again, some of Aquinas's translators have him discussing the subject of "the right to property" and whether "the right to private property" derives from natural or human right. But what Aquinas actually asks is whether it is natural for men to possess external things; and he answers that private possession is not opposed to natural right, or natural justice [1, p. 130]. Aquinas did not say, indeed he almost could not have said, as John Locke did, that men have natural rights to property. For Aquinas there is natural right but no catalogue of specific natural rights.

It remains for us to inquire what was understood by the concept of rights — what rights are — as that concept increasingly came to be used. For this clarification we must turn to the writings of philosophers and jurists. It seems that no one was minded enough to offer a definition of "a right" until the fourteenth century. First, though, let us turn to the understanding of the concept of rights in the eighteenth century, the heyday of natural rights thinking.

An excellent statement of the meaning of "rights" in the grand tradition can be found in the work of the English jurist William Blackstone, whose *Commentaries on the Laws of England* (1765 et seq.) had a profound influence in the American colonies and in the early days of the republic. Blackstone tells us that the "principal objects of the laws are Rights and Wrongs" and that the subject matter of rights is of two kinds, the rights of persons (*jura personarum*) and rights of things (*jura rerum*). He then goes on:

> The absolute rights of man, considered as a free agent, endowed with discernment to know good from evil, and with the power of choosing those measures which appear to him to be most desirable, are usually summed up in one general appellation, and denominated the natural liberty of mankind [2, p. 125] .... The rights themselves ... will appear from what has been premised, to be no other, than that *residuum* of natural liberty, which is not required by the laws of society to be sacrificed to the public convenience; or else those civil privileges, which society has engaged to provide in lieu of the natural liberties so given up by individuals [2, p. 128].

Blackstone tells us that these are the rights of all humankind. Certain key phrases in Blackstone's text reveal what his conception of rights is: "power of choosing," "power of acting," "free will," and "residuum of natural liberty." Blackstone conceived of rights as *option rights*, as spheres of personal sovereignty, as spheres of freedom in which the individual is free to act as he pleases and is free from the interference of others in his actions.

This identification of rights with option rights is characteristic of other jurists and philosophers of the seventeenth and eighteenth centuries, and it carries through and beyond such important political documents as the revolutionary French *Declaration of the Rights of Man and Citizens* (August 26, 1789). Writers as far apart as Thomas Hobbes (1588–1679) and John Locke (1632–1704) on issues in political philosophy share in the identification of rights with spheres of personal freedom. Thus, in his *Dialogue on the Common Law,* Hobbes states that "my right is a liberty left to me by the law," and in Chapter 14 of the *Leviathan*, he writes that "Right consisteth in liberty to do or forbear." In a context that is not political or ethical, Locke asserts that "the idea of liberty is the idea of a Power in any agent to do or forbear any particular action" [8, p. 316], which is conjoined with his view of man's God-given capacity for rational choice to render Locke's notion of rights in his *Second Treatise on Government.* But though the concept of rights – what rights are – was substantially the same for writers in the seventeenth and eighteenth centuries, it does not follow that they were in agreement on the issue of the basis of rights or on the issue of what the concrete rights are that men actually have. (The English expression "human rights" seems to have appeared for the first time in Thomas Paine's translation of the French *Declaration*; in his own work, *The Rights of Man,* Paine uses the more conventional term "natural rights.")

The concept of rights as option rights represents in a way the theoretical endorsement by philosophers and jurists of a notion of rights that goes back to the early part of the thirteenth century. (The Magna Carta of 1215 conjoins the two terms *iura* and *libertates,* rights and liberties.) The concept of rights was not only in the air, but rights were also conceived of as option rights, as spheres of freedom.

When did the concept of rights first receive "official" endorsement, so to speak, from a philosopher? Who was the first philosopher, or writer of any kind for that matter, to define *ius* (right) in the sense of a personal right, that is, a right of the kind that can be the possession of an individual? This question has been extensively researched by Professor Michel Villey, of the University of Paris, and treated in many of his publications. Villey attributes to the English Franciscan philosopher William of Ockham (1280/90–c. 1349) the credit of having been the first writer to have explicitly defined *ius* in the sense of a right in contrast to *ius* in the sense of law, justice, or what is right [11]. Villey, in fact, seems to go as far as to claim that Ockham was the first writer to employ the notion of *ius* in the former sense. This thesis, however, is debatable, although no one appears to have *defined* the term in this way prior to Ockham.

The most explicit passage occurs in Chapter 65 of Ockham's *Opus Nonaginta Dierum* (Work of Ninety Days): "Natural [divine?] right is nothing other than a power to conform to right reason, without an agreement or pact; civil right is a power, deriving from an agreement, and sometimes conforms to right reason and sometimes discords with it" [9, p. 579]. The crucial term in this definition is "power," which is a problematic notion and would have to be treated in relation to a number of aspects of Ockham's philosophy. Perhaps it is sufficient for our purposes to note that in Ockham's definition we have the ancestor of Blackstone's view of a right as "a power of acting as one thinks fit, without any restraint or control, unless by the law of nature." We might also add that the context of Ockham's definition is the problem of Franciscan poverty: Is not anything that is possessed by someone who is sworn to poverty unlawfully possessed by him? Ockham is therefore concerned with the meaning of "rights," particularly property rights and rights involved in the transfer and use of property.

How did the definition of *ius*, as a right, get from William Ockham in the fourteenth century to the natural rights theorists of the seventeenth and eighteenth centuries? It would be difficult, and perhaps impossible, to trace any direct route. In order to do so, we would have to examine the distinctive nuances in the phraseology of definitions of *ius* in many writers. These nuances take on their own meaning in virtue of the particular philosophical theories in which they are embedded. By the time we get to the end of the sixteenth century, we find that the terminology has shifted to some extent from Ockham's *potestas*, power, to *moralis facultas*, moral faculty or capacity, and *qualitas moralis*, moral quality. But the essential view of the concept of rights as option rights, spheres of personal freedom and sovereignty, remains nonetheless.

The close of the late medieval and Renaissance tradition on rights is probably best represented by the Spanish writer Francisco Suarez (1548–1617). Suarez distinguishes different senses of *ius*, one of which refers to *facultas quaedum moralis*, a sort of moral capacity [10, p. 24], from which he proceeds to develop a doctrine of natural rights. Perhaps more influential than Suarez, however, is the Dutch jurist and humanist Hugo Grotius (1583–1645), who is known as the "father" of international law. In his work *De jure belli ac pacis* (The Law of War and Peace), Grotius says that *ius* has many meanings, "one of which concerns the person, in which sense right is a moral quality of a person competent to have something or do

something" [6, p. 35]. The transfer of rights through a promise is described by
Grotius as the alienation of part of our liberty — in effect, giving up options.

In the classical period the concept of rights was accorded virtually no significance
at all. Defined in an explicit manner first by Ockham, perhaps, the concept of rights
was fully crystallized by the end of the sixteenth century, and "rights" became one
of the vital terms of moral and political discourse. In the seventeenth and eighteenth
centuries it was a watchword of revolution, and its rhetorical force is today an important factor on both the domestic and international scenes. It will have hardly
escaped attention, however, that the historical concept of rights is one of option
rights, rights as spheres of personal freedom or sovereignty.


EMERGENCE OF WELFARE RIGHTS

What accounts for the fact that the classical idea of right (objective right, what is
right) partially gave way in the late Middle Ages and Renaissance to a full-blown
concept of rights, that is, option rights? When did the concept of welfare rights
emerge, and what accounts for its emergence?

These questions are difficult, and answers can be put forward merely as tentative
suggestions. First, it may be suggested that the antique (though not antiquated)
conception of objective right proved inadequate as a buffer against the growing
power of the secular nation-state, whose consolidation began in the Middle Ages.
The concept of option rights seemed to mark out a preserve of personal sovereignty
for the individual against the claims to supremacy of the sovereign, all-powerful
state. The development of the idea of the rights of the individual, with rights in
private property as the paradigm case, is hardly surprising. It was then that capitalism, with the individualism it entails, and the growth of the middle class began.

The shift from the centrality of objective right, what is right, to option rights,
which are rights of choice, is, therefore, no mere matter of verbal form but rather
one of momentous substance. An uneasy tension, however, exists between the
concept of what is right and the concept of (option) rights. But it is because of
the partial displacement of the notion of what is right by the concept of rights
that we can today ask whether a person can have the right to do what is morally
wrong, however this question is answered.

Second, it may be suggested that just as the classical idea of right needed to be
supplemented in the late Middle Ages by the idea of rights, so too did option rights
eventually need to be supplemented by another type of rights: welfare rights. It is
far from clear that Blackstone's conception of rights was adequate to cover everything he regarded as rights. For example, is the right to personal security to be fully
understood as an option right or even a set of such rights? The idea of option rights
was no more adequate to all the rights contained in the French *Declaration of the
Rights of Man and Citizens* than it is to all the rights listed in the *Universal Declaration of Human Rights* of 1948. It was the nineteenth century, with an entrenched
Industrial Revolution and fully established nation-states, that saw the explicit recognition of a new type of rights.

It is not entirely certain as to who was the first to offer explicitly a definition of
"rights" in terms of welfare rights. Perhaps the palm should go to the German

jurist and legal historian, Rudolf von Jhering in his book *Der Zweck im Recht*, published in 1887. There, von Jhering defined "a right" as a legally protected interest [7, p. 50]. Though this definition is criticized as too narrow, the introduction of the notion of an "interest," which puts the focus on well-being, is as important as the older emphases on free will, power, and capacity. It also can be argued that Jeremy Bentham and his utilitarian followers, as well as T. H. Green and his liberal followers, have a greater claim than von Jhering as the first definers of rights in terms of welfare rights.

CONCLUSION

We have come a long way from the questions of rights in health care that gave rise to the treatment of rights presented in this chapter. Our purpose was to set in historical perspective the conceptual aspects of these questions. Normative questions — such as, what is the ethical basis of rights and what are our rights — were left entirely aside. Undoubtedly, these questions are the crucial issues. It may be that just as there has been an uneasy tension between option rights and what is right, so too there is a host of tensions between option rights and welfare rights. Just as one man's freedom is secured at the price of another man's freedom, so too is one man's benefit secured at the price of another man's freedom. More than that, one man's benefit is secured at the price of another man's benefit. Claims to rights are in conflict, and hard decisions need to be made. Perhaps we have had an inflation in our moral economy, an inflation in demands put in terms of rights, to the point that the concept of rights is beginning to lose its value as moral tender. We seem to need a new concept in order to be able to think clearly about these hard decisions. But what could this new concept be?

REFERENCES
1. Aquinas, T. *The Political Ideas of St. Thomas Aquinas* (translated by the Fathers of the English Dominican Province). New York: Hafner, 1953.
2. Blackstone, W. *Commentaries on the Laws of England*. London: T. Cadell and J. Butterworth, 1825.
3. Feinberg, J. Voluntary Euthanasia and "the Inalienable Right to life." Paper given at The Brooklyn Center, Long Island University, April 8, 1976.
4. Gaylin, W. The Functions of Prisons and the Rights of Prisoners. Paper given at The Brooklyn Center, Long Island University, April 8, 1976.
5. Golding, M. P. Towards a theory of human rights. *Monist* 52:521, 1968.
6. Grotius, H. *The Law of War and Peace* (translated by Francis W. Kelsey). Indianapolis: Bobbs-Merrill, 1925.
7. Jhering, R. von. *Der Zweck im Recht* (translated by I. Husik as *Law as a Means to an End*). New York: Macmillan, 1924.
8. Locke, J. *Essay on Human Understanding* (vol. 1). New York: Dover, 1959.
9. Ockham, W. *Opera Politica* (vol. 2). Manchester: University of Manchester Press, 1963.
10. Suarez, F. *De Legibus* (vol. 1). Madrid: Consejo Superior de Investigaciones Cientificas, 1971.
11. Villey, M. *Seize Essais de Philosophie du Droit*. Paris: Dalloz, 1969. Pp. 140–178.

# CHAPTER 5

# Option Rights and Subsistence Rights

*Bertram Bandman*

*Some people argue* that the concept of rights presupposes freedom and that without freedom there are no rights [16]. Other people argue that freedom is desirable for an adequate account of rights, which are identified as option rights, but that freedom based rights or option rights are not essential. In addition, they argue that there are also other kinds of rights that those who are not free may nevertheless have, namely, "rights of recipience" [24] or "welfare rights" [13, 14]. A third group argues that freedom is not only a necessary condition of rights but also is a sufficient condition and that there are no rights other than those rights based on freedom [23, 26]. I shall call these positions (1), (2), and (3) and argue for a fourth position (4), namely, that freedom is a necessary condition of any rights but that, contra positions (1) and (3), freedom does not provide a sufficiently enriched basis for rights [7, pp. 267–278]. Contra positions (2) and (3), however, freedom is not only desirable but also essential to any rights, but so are subsistence rights of well-being or welfare rights. According to the fourth position, welfare rights, sometimes called subsistence rights* [29], or rights of social and economic justice limit the scope of options and are presupposed in option rights; that is, there is no freedom without well-being. The fulfillment of subsistence rights is essential to the exercise of option rights.

## A PRESUPPOSITION OF ALL RIGHTS

A condition for having any rights at all has been well stated by H. L. A. Hart; he believes that if there are any rights at all, there is at least the equal right of all persons "to be free" [16]. According to Hart, having a right to be free means one may not be unjustifiably coerced. The right to be free provides a justification against anyone else's unjustified interference. The right to be free is the right to one's *domain,* whether it is one's body, one's life, one's property, or one's privacy. It is the area of one's life over which, as Joel Feinberg aptly puts it, one is the "boss," so to speak [10].

The right to be free also means one does not have to do whatever one has a right to do. Furthermore, to exercise one's right is to be immune to the charge of wrongdoing. It may be foolish, unwise, or imprudent to exercise one's right, but it is never immoral or illegal to do so.

For health recipients, the right to be free importantly includes the right not to be brainwashed, lied to, kept ignorant, deceived, tricked, involuntarily put to sleep, or otherwise unjustifiably coerced. This right is absolute in the sense that one cannot have these things done and still be free to live as a person [11, p. 97]. This definition means that one's regard for a patient implies the injunction to respect a person's regard for rationality by telling him or her the truth, as Elsie

Part of this paper is adapted from my paper, Some Legal, Moral and Intellectual Rights of Children, *Educational Theory* 27:169, Summer 1977.
*The title of this chapter is owed, in part, to the title and theme of Shue's paper. For convenience in this chapter, I will use the expressions "rights of subsistence" and "rights of welfare" interchangeably and leave for another occasion the need to disentangle glossed-over features.

Bandman points out [3]. The Tuskegee syphilis case, the United States Army LSD study, or Nurse Ratched telling Murphy in *One Flew over the Cuckoo's Nest* to take his pills are all violations of a person's right to be free. This domain in which one has authority to determine one's own acts without outside interference is a necessary condition for having any rights at all.

In an important recent article, Martin Golding calls rights based on freedom "option rights" [15] and argues, contra Hart, that some human beings, including infants, the mentally incompetent, the comatose, and the aging who are no longer capable of choice, nevertheless have rights of another kind. These rights are sometimes called rights to receive assistance or rights of recipience or welfare rights. The right to live and the right to acquire and receive whatever means are necessary to live may be regarded not as option rights but as welfare rights [14, 15].

Golding acknowledges that the right to be free is part of an adequate and desirable theory of rights, but that freedom is not a necessary condition for having a right. To Golding there are two kinds of rights, option rights and welfare rights. People who may not have option rights may nevertheless have rights of another kind, namely, rights of recipience or welfare rights.

In defense of Golding's contention that option rights are not indispensable, there are rights that people who are not free are nevertheless said to have, such as the right to live. Prisoners in captivity have been willing to be enslaved in order to live. Slaves have been known to have rights, including the right not to be killed and even the right to revolt. If slaves, prisoners, the unconscious, and the aged who are incapable of choice have rights, it would seem that one does not have to be free to have rights.

However, in defense of the view that rights presuppose freedom, as Hart contends [16], there is a distinction between freedom applied to a class of beings and freedom applied to the members of that class. This difference is sometimes known as the collective-distributive distinction. To attribute freedom to the class of beings having rights does not entail that each and every member of the class of right-holders has to be free in order to have rights. For one thing, rights can be exercised on behalf of others in the form of advocacy. One can extend some sorts of rights to individuals who are not free as long as those who exercise rights on behalf of others are free to do so. The requirement that only those individuals "capable of choice" [16] have rights imposes an unnecessary restriction on having rights. However, as both Mill and Frederick Douglass have pointed out, those best suited to protect one's interests and rights are those who are capable of choice [21]. The strength of rights weakens with the growth of advocacy and representation. A political representative's capacity to represent the interests of all his constituents weakens in proportion to the number of people represented. Similarly, the ratio of right-protectors to incapacitated right-holders determines the effectiveness of representing the interests of those right-holders.

Nevertheless, it is possible to have rights without exercising them directly. Other people have been known to defend a person's rights. The point to consider, however, on the basis of the fact that rights weaken in proportion to their reliance on being represented by others, is that without someone being free to exercise the rights of others there could be no rights.

Thus to apply the type-token or collective-distributive distinction to rights, the type, class, or collectivity of rights presupposes freedom, even though each and every right-holder need not be free. On this loosening of the restriction, the attribution of rights to all beings without freedom would, however, be a contradiction of what it means to have rights. To have a right, including even a right to live, means that someone has to be free to exercise or effectively claim the right. A right that no one can ever exercise or effectively claim, is, on principle, no right at all. The freedom to exercise and effectively claim rights on someone's part is a necessary condition of all rights.

## THE CORRELATIVITY BETWEEN RIGHTS AND DUTIES

A freedom worth having, such as a right, has a price. Someone has to provide and protect that freedom. In this connection, an often overlooked point about any right is that there are no rights without correlative duties. Although some "special rights" also imply duties on the right-holder, all rights imply duties on the part of others as well. As Joel Feinberg has put it, "rights are necessarily the grounds of other people's duties" [11, p. 58].

H. L. A. Hart cites four kinds of special rights: promises, authorizations involving consent, professional and civic associations involving "mutuality of restraints," and family and child-parent relations [16]. The special rights ascribed to individuals in these relations imply duties on right-holders as well as reciprocal duties on others. In any event, there are no rights without duties.

We may consider three criteria Maurice Cranston proposes for testing whether or not certain rights, which he calls "human rights," imply correlative duties: Is the right practicable or feasible, a right that can be acted on? Is it universal or impartial in application? Is it singled out and regarded as being of "paramount importance" [6, pp. 49–51]? The first criterion rules out rights that are impossible to achieve; the second rules out unequal rights and privileges; and the third rules out "frivolities and luxuries," such as "fun fairs" and "expensive gifts," and emphasizes rights to the most urgent needs, such as "ambulances" and "fire engines." These criteria are intended to restrict the duties that rights imply on others to provide. Helping a drowning person is important, but if the only one to rescue such a person is a person who cannot swim, the right a drowning person has to be saved, important as it is, is impracticable. To Cranston a human right must pass all three tests in order to imply corresponding duties on others.

## A THIRD CONDITION: RIGHTS PRESUPPOSE JUSTICE

Although freedom and correlative duties are both necessary conditions for having rights of any kind, these duties are not sufficient for an adequate theory of rights.

Considerations of justice are unavoidably necessary in rationally assigning appropriate liberties and duties to persons. One does not need to be undervalued or study those who are or were to know that the older option rights associated with freedom lead without rational restrictions to unfair discriminations and inequities. Even from Cranston's criteria, there is no basis for deciding which interests are of the most "paramount importance," nor does he provide a clue as to the extent to which legislatures should be prodded to make needed goods available to everyone. He, too, restricts rights to option rights [6].

About such rights Woodrow Wilson made this reference to one's home, which was turned into a coal mine with hundreds of laborers toiling and daily risking their lives:

It was not the business of the law in the time of Jefferson to come into my house and see how I kept house. But when my house, when my so called private property, became a great mine, and men went along dark corridors amidst every kind of danger in order to dig out of the bowels of the earth things necessary for the industries of the whole nation, and when it came about that no individual owned these mines, that they were owned by great stock companies, then all the old analogies absolutely collapsed and it became the right of government to go down into these mines to see whether human beings were properly treated in them or not . . . . We are in a new world struggling under old laws. As we go, inspecting our lives together today, surveying this new scene of centralized complex society we shall find many more things out of joint [31].

The unequal distribution of the older, narrowly selected privacy rights, previously vested in a small number of people, does not imply corresponding duties people can justifiably be obligated to accept.

### An Enriched Basis for Rights
It is no longer possible to think or speak only of the older political rights without reference to the newer social and economic rights of recipience or sometimes identified as subsistence welfare rights [9, 15, 24]. Golding's view [14, 15], which attempts to combine both option and welfare rights, seems a more adequate and promising account of rights than a view based solely on the minimal right to be free, a view, for example, favored by Robert Sade [26]. Recognition of the newer social and economic rights, including the right to health care, is also necessary to provide an adequate and just account of rights.

### What Are Subsistence or Welfare Rights?
Welfare rights or rights to assistance, according to the *Universal Declaration of Human Rights,* Articles 22–27, call for the universal right to a job with equal pay for equal work and the right to an adequate standard of living, including nutritious food, clothing, housing, medical care, and education [11, pp. 94–95].

### A Difficulty with Subsistence or Welfare Rights
Some of the rights in the *Universal Declaration of Human Rights* have been caricatured, since they provide not only for everyone's right to a job but also for everyone's "right to security in the event of unemployment, sickness, old age," "rest and leisure," and even to "periodic holidays with pay" [11, pp. 94–95].

A serious difficulty, therefore, with welfare rights is the *thinning out* of rights. There are claims made for the rights of students, teachers, nurses, physicians, policemen, the poor, the aged, the comatose, the mentally ill, the retarded, prisoners, fetuses, unborn generations, animals, and trees [1, 2].

The declaration of and demand for more and more rights, social, economic, educational, ecological, and medical, can only mean a declining possibility of imposing correlative obligations. People somewhere, sometime, have to accept comparable duties implied by the rights of others, as Cranston and other writers have noted, including Fried [12]. The expansion of rights can run the danger of trivializing and making each right less valuable, since other people are increasingly apt to be unable as well as unwilling to accept the duties such rights imply [6, 12, 13].

The expansion of rights beyond even the most remote chance of their being fulfilled violates a well-argued principle stated by Cranston, namely, that a right implies duties that are practicable and achievable [6]. It is not possible to impose duties on every welfare right put forward. According to Cranston, "India, for example, simply cannot command the resources" to house, feed, clothe, educate, employ, and provide health care for every citizen, let alone provide "periodic holidays with pay for everyone" [6, p. 51; 7]. Feinberg points out that if there are not enough jobs to go around, not everyone can have a job. How is one to cope with the growing numbers of rights, including the right to publicly financed health care for everyone?

The expansion of these newer social and economic rights in and to health care, such as the rights enunciated in the *Universal Declaration of Human Rights* and in the World Health Organization Preamble [32], are defensible only if those who press for them mark out the corresponding duties such rights imply in answer to the question: "Who shall justifiably provide?" This question can be asked even where appeal is made to policymakers with enlightened principles of justice.

*A Further Objection: Welfare Rights as "Mandatory Rights"*

A further objection to recipient rights or welfare rights is made by antipaternalists. Joel Feinberg, for example, has recently argued that these recipient rights or welfare rights or rights to assistance are really "mandatory rights," such as "the right to an education" or "the right to be inoculated or vaccinated" [9, 10].

One interpretation of the right to live, for example, rules out the right of a person to refuse a lifesaving blood transfusion, which a Jehovah's Witness may invoke. This right turns out not to be much of a right if the right to life means that one has no choice but to live. About such mandatory rights, Feinberg contends, "You have no choice in the matter. Whether you like it or not you have to exercise these rights" [9, 10, 14], such as "the right to be vaccinated," the right to drink uncontaminated water, or the right to live. Feinberg does not think that this kind of a right, about which one has no choice, is very much of a right. Indeed, on grounds marked out by various writers, such a right lacks the freedom for the right-holder to exercise it or not, which seems conceptually essential to any right. Without this freedom of choice there really is no right at all, if I have understood Feinberg's argument correctly.

Since welfare rights are mandatory and mandatory rights lack the characteristic of freedom with which all rights are conceptually identified (mandatory rights are justly considered paternalistic and otiose), there can be no mandatory rights and, hence, no welfare rights. The argument goes:

If there are welfare rights, then they are all mandatory rights.
*But there are no mandatory rights.*
Therefore, there are no welfare rights.

*A Response to the Thinning Out of Rights and to Antipaternalism*
Although rights imply freedom and corresponding duties, some rights, like the right
to clean water or air, impose additional limits, increasing constraints, and rules and
duties on everyone, including right-holders. The right to drink uncontaminated
water, for example, imposes an additional duty on everyone not to contaminate
the water supply. Similarly, the right to health care, education, or a job imposes
duties that were not required of the older option rights. The right to be vaccinated,
for example, implies that every eligible person has to be inoculated. In such welfare
rights, the coercive or mandatory aspect, however, leaves one without a choice. To
become free from certain diseases, one has to do something, like going to a clinic,
waiting, and submitting forms in triplicate before receiving the benefit. These steps
are the preconditions for the right to be vaccinated that one is then said to have.

Some welfare or recipient rights may appear as two-faced *janus* rights, which
impose chains on the right-holder as a precondition for freeing him or her. But
perhaps these rights may be regarded as "double-edged rights," in which one has to
do something to receive something.

Some rights may involve duties on the part of right-holders, but that condition is
not, by itself, an anomaly for some kinds of rights. Requirements are not peculiar
to welfare rights. Even the special rights to make wills, contracts, and get married
imply rules. In terms of degree, more rules may be imposed on welfare rights than
on option rights.

The moral correlativity thesis holds that a right-holder also has responsibilities
[4, p. 102; 11, pp. 61–62]. Although the moral correlativity thesis need not
characterize all rights — indeed it ought not to — it may be that all or nearly all
welfare rights imply correlative obligations. These obligations would not only be
incumbent on others and the state to protect such rights, but also on right-holders
themselves as a precondition for the exercise of the very rights they are then said
to have, such as the right to an education.

These triple-directed duties are not only peculiar to "special rights" in groups
calling for "mutuality of restraint," for example, duties found in associations, clubs,
institutions, and nation-states. They are also evident in relations between different
generations or between members of a family as well as people in a community.
These triple-directed or triple-barreled duties involve (1) duties directed to oneself,
(2) duties directed to others, and (3) duties directed to the state [8]. For example,
the right to enjoy a debris-free highway implies the duty of *everyone* to keep the
highway free of debris. One has a right to a debris-free highway only if one also contributes to keeping the highway free of debris. A patient who receives an intravenous injection, blood test, or x-ray, or has his temperature taken with a thermome-
ter, for example, has to accept the intrusion, remain relatively still, cooperate, and
abide by the rules, such as holding his or her breath during the x-ray, which makes
the exercise of his or her welfare right possible. Thus Feinberg writes that for some
rights "a prior condition for the acquisition or possession of [such] rights is the

ability and willingness to shoulder duties and responsibilities," and for some rights "acceptance of duties is the price any person must pay in order to have rights" [11, p. 61].

### Every Right Is an Option Right to Some Form of Well-Being

The view that every right is an option right seems entirely defensible. A "mandatory right" is a contradiction, but a mandatory or coercive aspect of a right need not be. A more defensible view of rights is that rights imply "a sphere of autonomy" and corresponding duties to oneself, to others, and to the state. However, this sphere or domain is *limited* by the equal rights of others to have an equal sphere of autonomy [18].

Moreover, rights are not mere options; they are options *to* something or against someone. They are always about something needed or wanted as a specifiable interest, like a loaf of bread, a cooking stove, a vaccination, or a scholarly book; something that one needs to live "a decent and fulfilling human life" [28]. The rights cited as welfare rights are rights whose fulfillment is needed to achieve this sort of life.

All rights, therefore, may be regarded as option rights and subsistence rights. Both kinds of rights are not only desirable and important, as Golding contends, but also essential for having any rights at all.

The problem of redistributing rights into smaller and smaller shares may necessitate new or added restrictions, so that everyone can have effective rights. Everyone cannot have health care rights unless some older option rights (defended, for example by Dr. Robert Sade [26]) are limited.

### An Enriched Minimum: A Cluster of Option and Subsistence Rights

A cluster of rights, both option and subsistence rights, is needed to live "a decent and fulfilling human life." The means for living this kind of life, includes health care; a fair share of resources; the right to impartial consideration; the right to one's life and to one's body; the right to be free; the right to speak, to be heard, to vote often and on vital policies; the right to know the truth and the right not to be lied to, deceived, drugged, tortured, or involuntarily restrained except to prevent harm to oneself or others; and the right not to starve. These rights form an enriched basis on which all other rights can be generated.

### But Which Predominates? Option Rights or Subsistence Rights?

A remaining issue concerns the distribution of these rights. Should a small number have all the option and subsistence rights in the enriched minimum or should these rights be distributed equally to all persons?

If we interpret option rights and subsistence rights (for the moment) as rights presently held by a (relatively) few versus rights needed by the many, we have the problem of deciding what is a just distribution of these rights. And even if we limit option rights, we still have the problem of further limiting welfare rights. At any rate, Golding does not quite say whether option or welfare rights predominate [14, 15].

Option rights have long been believed to precede and to be logically prior to welfare rights. These social and economic rights have been thought to be a parasitic outgrowth of the primary option rights. Indeed, I suggest that a search through the Greek, Roman, and medieval historical origins of rights will show that option rights are only a modern seventeenth-century and eighteenth-century middle-class phenomenon, a tribute to Locke, but a conception that for over three centuries only temporarily eclipsed a rival and even more fundamental conception of rights. (This account may be compared with Golding's view [14].)

A consideration of the origins of natural rights theory in Greek, Roman, and Christian thought and practice shows, I believe, that having a right (or one's *due*) depends on natural rights, and that natural rights depend on natural law, which holds importantly that all human beings are equal. (*Ius* = Law according to Nature. According to the New Testament, "Ye are all one in Christ, Jesus." [Gal. 3:26–29].) This conception of natural law, in turn, derives philosophically from Roman and Greek conceptions of justice and shows that rights are secured in the tradition of welfare rights, rights founded on well-being and egalitarian justice [20, pp. 143–153; 25; 30].

This interpretation is not to deny a tradition that emphasizes private option rights (e.g., the Sophists); but that is not the only tradition; it is not even the prevalent tradition of justice in Greek, Roman, and Christian thought. St. Augustine, for example, extended Platonic principles of justice to rights in maintaining that rights flow from justice; without justice there can be no rights [26, 27].*

If there is a case for the thesis that a necessary condition for any right is the right to be free, there is an equally compelling thesis that a further condition of any right is the right to be treated justly.

*Subsistence Rights Limit Option Rights*
An enriched conception of rights includes option rights together with subsistence rights. However, welfare rights or subsistence rights, rights distributed equally to all persons, have priority due to their identification with social and economic justice.

In this connection, Alison Jaggar suggests a broader minimum for basing rights than the stark minimum proposed by Hart. She also uses a social ideal that presents a moral alternative to Hart's position. According to Jaggar, to have any right is to have the right to live "a full human life" and "to whatever means are necessary" to such a life [19].

Subsistence rights, rights oriented by justice, thus limit the sphere of action of option rights. The right of access for anyone in need of expensive medical equipment, such as a resuscitator, dialysis machine, or artificial heart (contra the position presented in [12] and [13]), limits any other person's right to any "luxury" or "frivolity" [6], such as a fourth yellow Jaguar or private golf course.

A difficulty with Jaggar's view is that the right to "a full human life," desirable

---

*Even if I am wrong or lacking in evidence for this presumptuous historical thesis, my argument for the logical and moral priority of welfare rights over option rights does not hinge on the argument by origin on the familiar logical grounds that the genetic argument is inconclusive in any event. It hinges rather on the logical point that there are no rights without reference to equal justice.

as it is, is unachievable for everyone. Samuel Scheffler's more modest notion of a right to "a decent and fulfilling human life" will do nearly as well [28].

In this somewhat more modest vein and also freed of some quite understandable objections noted previously [6, 12], Virginia Held has recently added that "a right to life includes not merely being left alone; it also includes being able to acquire what one needs to live. Anyone who recognizes the right to life of a human child acknowledges this; one . . . cannot . . . leave a baby alone unattended to fend for himself or herself. One must see to it that babies have the necessary food, shelter, and so on. Access to what they need to live must also be included in the rights to life of adults, although this requirement is often overlooked especially by so-called libertarians who suppose that we respect persons' rights to life by merely [leaving them alone and] not attacking them" [17].

To Virginia Held property rights qualify as "rights," but she says, "I do not have a right to hang on to whatever I hold, if others who need and deserve it more have rights to have some of it shared with them." Held agrees "to the taxation of those with more than they need to provide the means to live of those who lack such means; this action may often be required to respect rights to life" [17].

If the equal right to live "a decent and fulfilling human life," and with it an implied cluster of welfare rights, has priority, this concept means such rights limit the scope of individual option rights. Thus, the need to feed, clothe, house, provide universal health care, and educate the citizens of India, for example, *limits* anyone's option rights to own a fourth yellow Jaguar, sailboat, or golf course. With reference to Feinberg's example of too few jobs to go around and thus not being able to give everyone a job, the solution is to divide the jobs as Jesus did with the loaves and fishes (Matt. 14:16–20).

In the interest of social justice, the older option rights have to give way wherever they become eroded into special powers and privileges that depart from a vital part of the heritage of rights to make room for the newer social and economic welfare rights. More people than previously will have to eat at the same table, and a just system makes it possible for the food on the plates to be not merely a matter of chance, love, or happenstance. The politically important right not to be tortured, as James Nickel suggests, may be extended to include the right not to starve [22]. As Samuel Scheffler points out, everyone's right to live does not justify noninterference with some people's liberties [28, pp. 65–66].

*Health Care Rights as Option Rights and Subsistence Rights*
To have any rights at all is not only to have the equal right to be free, but also the equal right "to a decent and fulfilling human life" [28] and to whatever means are necessary to achieve such a life, including health care [5]. Such rights may be called human rights.

A health care program in which human rights are recognized is a program in which a Tuskegee syphilis experiment, Willowbrook hepatitis program, or United States Army LSD experiment, all of which were undertaken without the subject's informed consent, would be recognized as a gross violation of human rights. In addition, neglecting to give health care to those who need assistance would be analogously regarded as a violation. In this kind of health care program, health resources would be distributed equally to all members of society. In a health care system oriented by the triple conditions of human rights, freedom, duties, and social justice, there is recognition that human rights are not confined to option rights, but include rights to assistance as well, and that "a human right is something of which no one may be deprived without a grave affront to justice" [6, p. 52] .

REFERENCES
1. Bandman, B.   Human Rights of Patients, Nurses, and Other Health Professionals. This volume, Chap. 47.
2. Bandman, B.   Some Legal, Moral and Intellectual Rights of Children. *Educational Theory* 27:169, 1977.
3. Bandman, E.   The Right of the Mentally Ill to Refuse Treatment. This volume, Chap. 32.
4. Benn, S., and Peters, R. S.   *The Principles of Political Thought.* New York: Collier Books, 1959. P. 102.
5. Blackstone, W.   On health care as a legal right: An exploration of legal and moral grounds. *Georgia Law Rev.* 10:391, 1976.
6. Cranston, M.   Human Rights, Real and Supposed. In D. D. Raphael (ed.), *Political Theory and the Rights of Man.* Bloomington: Indiana University Press, 1967.
7. Dworkin, R.   *Taking Rights Seriously.* Cambridge, Mass.: Harvard University Press, 1977. Pp. 266–278. (See also MacCormick, D. N., Rights in Legislation. In P. Hacker and J. Raz, *Law, Morality and Society.* Oxford: Clarendon Press, 1977. Pp. 187–210.)
8. Feinberg, J.   Rights. In L. Walters (ed.), *The Encyclopedia of Bioethics.* Detroit: Gale Research, 1976.
9. Feinberg, J.   A Postscript to the Nature and Value of Rights (1977). This volume, Chap. 2.
10. Feinberg, J.   Voluntary Euthanasia and the "Inalienable Right to Life." Paper presented at Conference on Bioethics and Human Rights, Long Island University, April 9, 1976, forthcoming in *Philosophy and Public Affairs,* winter 1978.
11. Feinberg, J.   *Social Philosophy.* Englewood Cliffs, N. J.: Prentice-Hall, 1973.
12. Fried, C.   Equality and rights in medical care. *Hastings Report* 6:28, 1976.
13. Fried, C.   Rights in health care — beyond equity and efficiency. *N. Engl. J. Med.* 293:241, 1975.
14. Golding, M. P.   The Concept of Rights: A Historical Sketch. This volume, Chap. 4.
15. Golding, M. P.   Towards a Theory of Human Rights. *Monist* 52:521, 1968.
16. Hart, H. L. A.   Are There Any Natural Rights? In A. I. Melden (ed.), *Human Rights.* Belmont, Calif.: Wadsworth, 1970.
17. Held, V.   Abortion and Rights to Life. This volume, Chap. 11.
18. Held, V.   Equal Liberty in the Welfare State. In W. Feinberg (ed.), *Equality and Social Policy.* Urbana: University of Illinois Press (forthcoming).
19. Jaggar, A.   Abortion and a woman's right to decide. *Phil. Forum* 5:351, 1973–1974.
20. Joachim, H. H.   *Aristotle: The Nicomachean Ethics.* London: Oxford University Press, 1953.

21. Melden, A. I.   Olafson on the Right to Education. In J. Doyle (ed.), *Educational Judgments*. London: Routledge Kegan Paul, 1973.
22. Nickel, J.   Are Social and Economic Rights Real Human Rights? Paper presented at Society for Philosophy and Public Affairs, City University of New York, Graduate Center, New York City, March 1977.
23. Nozick, R.   *Anarchy, State and Utopia*. New York: Basic Books, 1974.
24. Raphael, D. D.   Human Rights, Old and New. In D. D. Raphael (ed.), *Political Theory and the Rights of Man*. Bloomington: Indiana University Press, 1967. Pp. 54–67.
25. Sabine, A.   *A History of Political Theory*. New York: Holt, 1950.
26. Sade, R.   The right to health care: A refutation. *Image* 7:11, 1974.
27. St. Augustine.   *The City of God* (vol. 19). Garden City, N.Y.: Image Books, 1958. Pp. 468–472.
28. Scheffler, S.   Natural rights, equality and the minimal state. *Can. J. Phil.* 6:64, 1976.
29. Shue, H.   Subsistence Rights and Overpopulation. Paper presented at Society for Philosophy and Public Affairs, City University of New York, Graduate Center, New York City, December 13, 1977.
30. Strauss, L.   *Natural Right and History*. Chicago: University of Chicago Press, 1953.
31. Wilson, W.   The New Freedom. In E. Kent (ed.), *Law and Philosophy*. New York: Appleton Century, 1970. P. v.
32. World Health Organization.   Preamble to *The Constitution of the World Health Organization: Basic Documents* (26th ed.). Geneva: World Health Organization, 1976. P. 1.

# CHAPTER 6

# The Enlightenment and Bioethics
*Gary Marotta*

*More human beings* than ever before have a chance of living to maturity, of bearing normal children, of surviving pain, disease, and suffering. The claims of the eighteenth-century Enlightenment, including the claim that individuals possess a right to health care — as a natural right, among the rights of man — are taken today as serious philosophical and historical propositions.

The philosophical movement of the eighteenth century, which underlay the American and French revolutions, decisively transformed the intellectual and social orientation of Western civilization. The Enlightenment subscribed to a rational and scientific approach to problems. It believed in progress. It focused on the improvement of secular conditions, and it propounded premises for individual "rights." Its ideals of secular humaneness, rationality, and science are the ideals of modernity [5].

## EVOLUTION OF MEDICAL CARE

The Enlightenment made indifference toward health conditions unacceptable. Enlightened regimes, health professionals, philosophers, and scientists pushed for public health: They brought about sanitary reforms in prisons, hospitals, armies, and navies, and they waged effective fights against scurvy, smallpox, and typhus. Jean Jacques Rousseau worked to improve the health conditions of infants and children. Others worked in behalf of mothers, old people, deaf-mutes, and the blind [2, 13, 14].

The Enlightenment confronted issues of social responsibility. It recognized, for example, that mental disorder was a disease, not a manifestation of sin, vice, or crime. Philippe Pinel, a pioneer in the treatment of the mentally ill, in 1794 removed the chains from his patients in the Bicêtre hospital in Paris. The entire field of public health was first surveyed by Johann Peter Frank in his monumental work, consisting of six volumes, *Complete System of Medical Policy.* Social conditions, he thought, were the "mother of disease." Thomas Percival brought about a revival of the Hippocratic interest in the duties and responsibilities of health professionals; in 1803 his *Code of Ethics* introduced the term "medical ethics." That treatise became the model for later codes, including the *Ethical Principles of the American Medical Association,* as adopted and revised since 1847 [1, 14].

The Enlightenment also laid the foundation for modern hospital medicine. Following the French revolution, hospitals became a focal area for the treatment of the sick and the conduct of research. The revolution abolished the old universities, and in 1794 the medical school, l'Ecole de Santé, was opened; its establishment moved France to the forefront of medical achievement, education, and service. French medical education was rebuilt on clinical grounds; and the Paris Clinical School boasted of handling 25,000 cases within five years. This new approach gained prominence in both old and new universities and hospitals — at, among others, Leyden, Edinburgh, Dublin, Guy's Hospital in London, the New Vienna School, and the distinguished College of Physicians in Philadelphia [1, 8, 14].

During the first half of the nineteenth century, hospital medicine was developed on this clinical foundation. But future advances rested upon the ability of health professionals to apply the great discoveries of nineteenth-century basic science. These discoveries were made in the reformed and research-oriented universities by a new type of professional, the "pure" scientist. By the close of the century, health research had moved into the laboratories, where the collaborative efforts of scientists and medical personnel, notably in Germany, led to remarkable progress in histology, pathology, physiology, pharmacology, and diagnostic technique and measurement [2, 14].

Despite these achievements, a stay in a hospital, particularly for surgery, was viewed with utter dread. That attitude was changed by the introduction of asepsis and anesthesia. In the face of resistance from medical conservatives, the tragic Ignas P. Semmelweis of the New Vienna School struggled for cleanliness. Decisive work in pain control was done by Horace Wells, a Connecticut dentist, and the Massachusetts General Hospital. Asepsis and anesthesia led, in turn, to unprecedented work in surgery, obstetrics, and gynecology [15].

Yet the health of the greater proportion of people deteriorated, with disease and death rates increasing rapidly. The new industrial capitalism, with its unhygienic factory conditions, child labor, and exhausting and exploitative work regimen, was matched by massive poverty and squalor in the impacted cities and rural slums. Infectious diseases and epidemics spread from the poor to the upper classes. Inspired by French hygienists and the utilitarian philosophy of Jeremy Bentham, British, German, and American reformers initiated the great public health movement. That movement was accelerated by the professionalization and activism of nurses. The first school opened in Germany in 1836, followed by a school in England organized in 1860 by Florence Nightingale. At Bellevue and the Women's Infirmary in the United States, Elizabeth Blackwell, the first American woman medical doctor, trained nurses for war work. The public health movement also produced, for the first time, personnel to treat the health, not of individuals, but of communities. The statistics showing improvements in modern health tend to reflect a measure of public health which demonstrates the effectiveness of preventive rather than curative medicine [6, 13, 16].

But the most dramatic and eventful achievements — inspired and encouraged by the chemist Louis Pasteur and the physician Robert Koch — came from the new field of bacteriology: They derived from the discovery that epidemic diseases are caused by microorganisms. In rapid succession the principal pathogens of infectious disease were identified and classified; viral vaccines were developed and applied to the prevention of the major contagions. The virtual elimination of an array of diseases became possible: among them, amebic dysentery, gonorrhea, typhoid fever, leprosy, malaria, tuberculosis, cholera, diphtheria, meningitis, tetanus, botulism, bacillary dysentery, and whooping cough. Here, between 1875 and 1906, the most momentous development for medicine and for the health of humanity took place: The status of the health professions and the effectiveness of medicine were firmly established [2, 17].

Impressive work in the field of infectious disease continued into the twentieth century. The arrival of the sulfonamides, then penicillin, and after that a host of specific antibiotics, amounted to a revolution in health. Perhaps the most exciting achievement was the virtual elimination of poliomyelitis. In this century monumental advances were also produced in endocrinology, radiological techniques, vitamin research, chemotherapy, organ transplantation, mechanical apparatus, and in surgery of the chest, brain, and sympathetic nervous system.

The extension of health care to the majority of the population, the bridging of the social gap, has been unprecedented. In the United States, which became preeminent in medicine, health care by 1975 was the nation's second largest "industry." Health manpower training programs, hospital and clinical construction, and research and experimentation were supported on a gigantic scale; publicly funded health care programs were introduced. The development of this system, however, has generated major problems in administration and finance, procedure and litigation, the ordering of priorities and the allocation of resources, the application of advanced technologies, and the articulation of guidelines for making complex ethical decisions [3, 12, 18].

## DAMAGING EFFECTS OF PROGRESS

We know now, as the Enlightenment did not, that progress — material or technological — is implicated in a dialectic of damage. Technological advances, glorious in themselves, can produce damage to primary living systems and ecologies. Furthermore, although the means required to heal the sick are often available, inertia, greed, or the politics of ordering priorities may stand in the way. Often the new technocracy is impotent. Our sense of historical progress, even in the field of health where progress seems acutely in evidence, is no longer linear [9, 10, 18].

Debates on bioethics and human rights are conducted in the language of the Enlightenment. That language echoes, as well, in the Geneva Convention Code of Medical Ethics, the International Code of Nursing Ethics, and the Declaration of Helsinki. But the history of the twentieth century — catastrophic global wars, the gas oven, the Nuremberg disclosures, guilt and self-doubt, moral lassitude, and intellectual skepticism — has dislocated the premises of the Enlightenment [11].

The "miracles" of modern health science invite as much question and fear as delight and wonder. There is an array of biomedical "engineering": spare-part surgery, the use of chemical agents against the degeneration of tissues, preselection of the sex of the embryo, the manipulation of genetic factors toward "social" or "ethical" ends, memory transfer through biomedical transplant, behavior control through electrodes and drugs, and cloning. Are the new miracles of technology moving into direct conflict with the ideals of the Enlightenment [4, 7]?

## CONCLUSION

Some students are beginning to ask publicly what many have pondered in private: Should certain lines of inquiry be pursued at all? Can the humane mind survive the next truths? Ought genetic research to continue if it will lead to truths about a species whose moral, political, and psychological consequences we are unable to accept? Should we pursue neurochemical or psychophysiological investigations into the human cortex, if such study brings the knowledge that ethnic hatreds, war, or impulses toward self-destruction are inherited? It may be that the truths that lie ahead wait in final ambush for humankind. Should we, like some primitive society, opt for statis or mythology?

We cannot turn back. The conviction that inquiry must move forward, that truth is meritorious and in harmony with humane advancement, is intrinsic to the Western temper. Contemporary society's ability to come to terms with its "cultural lag" – the lag in Enlightenment concepts of human rights and biomedical technologies – will be a major test of the resilience, conviction, and intelligence of this generation.

REFERENCES

1. Ackerknecht, E. H. *Medicine in the Paris Hospital, 1794–1848.* Baltimore: Johns Hopkins University Press, 1968.
2. Ackerknecht, E. H. *A Short History of Medicine.* New York: Ronald Press, 1968.
3. American Health Foundation. *American Health Foundation Annual Report.* New York: American Health Foundation, 1976.
4. Carlisle, N., and Carlisle, J. *Marvels of Medical Engineering.* New York: Sterling Publishing, 1966.
5. Cassirer, E. *The Philosophy of the Enlightenment.* Princeton: Princeton University Press, 1951.
6. Donlan, J. A. *Goodnow's History of Nursing.* Philadelphia: Saunders, 1963.
7. Fishlock, D. *Man Modified.* New York: Funk and Wagnalls, 1969.
8. Foucault, M. *The Birth of the Clinic.* New York: Pantheon Books, 1973.
9. Institute of Medicine. *Ethics of Health Care.* Washington, D.C.: U.S. Government Printing Office, 1974.
10. McKinlay, J. B. (ed.). *Politics and Law in Health Care Policy.* New York: Prodist, 1973.
11. National Academy of Sciences. *Experiments and Research with Humans: Values in Conflict.* Washington, D.C.: U.S. Government Printing Office, 1975.
12. New York Academy of Medicine. *Social Policy for Health Care.* New York: New York Academy of Medicine, 1969.
13. Sand, R. *The Advance to Social Medicine.* London: Pelican, 1952.
14. Shryock, R. *The Development of Modern Medicine.* New York: Knopf, 1947.
15. Slaughter, F. *Immortal Magyar: Semmelweis.* New York: Doubleday, 1950.
16. Walker, M. *Pioneers of Public Health.* Freeport, N.Y.: Books for Libraries Press, 1968.
17. Winslow, C. *The Conquest of Epidemic Disease.* Princeton: Princeton University Press, 1943.
18. Yost, E. *The U.S. Health Industry.* New York: Praeger, 1969.

# TOPIC II

# The Right to Life and the Boundaries of Life

*The Declaration of Independence* contains the ringing words " . . . We hold these truths to be self-evident, that all men are created equal, that they are endowed by their Creator with certain unalienable rights, that among these are . . . the right to life, liberty and the pursuit of happiness." All of these rights are at stake in deciding who lives or dies. It was probably no accident that the founding fathers mentioned "the right to life" first. This right has often figured first, not only to John Stuart Mill [16] but also to Locke [15] and before Locke to Hobbes [10] and before Hobbes to St. Thomas Aquinas [1]. All four writers identified this right with self-preservation, self-defense, or self-protection [1, 10, 15, 16].

To be alive is thought to be a gift of nature or of God; a treasure no greater. Religions praise the gift of life and pray for its continuation. Science aims to promote and extend it [3, pp. 82–85]. The supreme importance of life is even regarded by some as the most fundamental, natural, or human right. The right to life is considered so important that prisoners in captivity have not minded enslavement if their lives were spared. Similarly, most persons prefer the loss of a limb or breast or other body part or function, such as vision, to death.

The life preferred is, of course, a full life, but people have been known to give up almost all other values, or parts of their bodies, or even their freedom, pleasure, happiness, and wealth, for an extended bit of life. People are said to be increasingly health conscious, but at the same time they do things that are harmful to their health and, in the long run, life threatening. When life is directly and immediately threatened, however, most people will give up almost any value for life.

Kubler-Ross points out that the stages at the end of a person's life are first denial, then anger, depression, bargaining, and finally acceptance [14]. Death or impending death is so terrible to most people that the first stage of dying is the refusal to acknowledge that it is occurring. Some, however, deny that death is ever accepted.

In The Right to Life and the Boundaries of Life, Topic II, the authors consider not only the meaning of life but also the parameters of a meaningful human life. The right to life and the boundaries of life are discussed from the perspective of human rights and bioethics within the topics of genetics, abortion, and euthanasia.

The first issues concern the rights of future generations, such as the case of persons with retinoblastoma who give birth to children who are at risk for transmitting genes carrying the same propensity for blindness. Questions of the responsibilities of this generation to plan for future generations are significant to society as well as to the afflicted individuals and their families. It is a matter of intense debate as to whether a democratic society committed to human rights can also support a eugenics policy. On the other hand, the consequences to society may be too important to permit individuals to make this decision. To complicate matters further, there is considerable doubt in some quarters that the present state of knowledge in genetics warrants a public policy of controls. Meanwhile, children with retinoblastoma continue to be born and live to procreate.

The process of planning for future generations may involve doing amniocentesis on pregnant women at risk for genetic defects in the fetus, such as Down's syndrome, followed by parental or maternal decisions to seek abortions. Opposing opinion believes abortion to be immoral, the murder of a fetus, and a violation of

69

the right to life as an inalienable right. This position holds that to place no value on the interests of the unborn is immoral and carries consequences for all of society.

At the other extreme is the belief that a woman has exclusive rights to her own body [20]. A recent variation on this position holds that a woman, not a male parent, a physician or a nurse, or members of the community, has a right to decide whether or not she should seek an abortion. This decision is the woman's, not because of her right to her body, but rather, according to this view, recently put forward by Alison Jaggar [12], because a woman has the right to decide based on two principles. The first principle is that the right to live means "the right to a full human life." Jaggar contends that a newborn has the right to a full human life, and if the woman is unable to provide it, she has a right to terminate her pregnancy. Jaggar's second principle is "that decisions should be made . . . only by those importantly affected by them" [12]. Since in our culture the main onus of parenting rests on a woman, she, being most importantly affected, should be the one to decide.

Jaggar's first principle is novel because until recent times the right to life has not implied "the right to a full human life," including necessary means such as a nurturing caretaker and adequate food, shelter, clothing, education, and health care. Her principle seems more of a right than some people are willing to pay for. It is opposed by Maurice Cranston [4] and Charles Fried [7], both of whom object on the practical ground that others are unwilling to provide for a person's "right to a full human life." It is too costly to society. The argument could be advanced that mothers are not expected or required to provide a "full human life." This view appears contrary to the positions put forward for women's liberation and the assumption of a nondependent role by women within the economic and "sexist" society. In Jaggar's argument [12], there is a theme taken up by such writers as Golding [8, 9] and Feinberg [5, 6], who identify rights to a full human life with "welfare rights" [10].

Feinberg and Golding discuss option rights as the rights of persons to be free to make decisions. If one adopts the view that a woman has a right to her body [20], this right would be an example of an option right. But if one sees a woman's right as providing "a full human life" for her offspring, this right would illustrate a welfare right. For a right to a full human life does not only imply forbearance from others but also assistance that involves welfare rights, such as the right to an education, to a pure water supply, and to adequate housing and nutrition, all of which are the means to a "full human life." The contention can be made that without welfare rights, option rights are the privilege of a few, since the rights to adequate food, clothing, shelter, and health care are necessary for the exercise of option rights. One cannot be free if one is sick or too poor to pay for needed health care. What value is still another child to a poverty-stricken, cold, and hungry family with no hope of improving its plight?

In a quite different vein, Newton [17] contends that no analysis of abortion in terms of rights will clarify the abortion issue, since the conflict between maternal rights and the rights of the fetus are irreconcilable. She concludes that compassion must be the determining value and that no general solution of the issue can be found.

At the closing boundary of life, the arguments surrounding euthanasia are conflicting and diverse. One line of reasoning sees a relationship between the claim to life and the duty of others to refrain from killing [21] as analogous to the right to death and the duty of others to avoid prolonging someone's life. Neither the right to life nor the right to death entails the duty of others to either rescue or to kill someone. Therefore, this view concludes, the person has a right to passive euthanasia but not to active euthanasia [21].

James Rachels, on the other hand, sees no distinction between killing and letting someone die in the context of active and passive euthanasia [18]. He cites the act of killing as similar in the case of the greedy uncle killing the rich nephew for his inheritance or of the uncle standing idly by while watching his rich nephew drown in a bathtub.

The existentialists, notably Sartre, could cite Rachels' example of the uncle standing by the drowning nephew without helping as an instance of inauthenticity [19]. This position holds that one is as responsible for what one does not do as what one does; that the omission of an act is as important as the commission. Therefore, for the existentialist, letting the nephew drown when he could have been saved is as evil as actively killing the nephew.

A further dimension of this topic lies in the issue of the lack of a meaningful life as a justification for euthanasia. Kohl argues that Quinlan no longer has a meaningful life [13]. In the Quinlan case it may be easy to cite the presence of particular brain waves as not meeting an important criterion for a meaningful life. It may be much more difficult or impossible to determine (what John Hospers terms) the "quorum features" [11] for distinguishing a meaningful from a meaningless life as a general principle for "allowing" euthanasia or "letting" someone live [13, 18].

The expressions "to allow to die" or "letting die" assume that the person "allowing" and the person requesting "to be allowed" are (1) conscious and (2) voluntary agents. The semantics of "allow" by some writers on euthanasia seems to identify this term with conscious, voluntary persons, such as when one allows one's son to go to the movies, whereas what appears to be true is not that one "allows" but rather "enables" an individual to die. Dr. Smith does not "allow" or "let" Mr. Jones, who is unconscious, die. Instead, Dr. Smith "enables" Mr. Jones to die by discontinuing his intravenus feedings or life-support systems, for example.

This issue has serious implications for health professionals. If a health professional "allows" or "lets" a person die, the onus of responsibility for the decision is on the health professional as the one in authority and in a position to grant permission to die. If, however, a health professional "enables" a person desiring to do so to die, somewhat as the servant assisted Brutus to die, the decision maker is the person or patient and not the health professional. This issue of whether one "allows" or "enables," in turn, affects the issue of establishing criteria for distinguishing a meaningful from a meaningless life. Does the person involved decide, do health professionals decide, does the family decide, or does the whole of society decide? At any rate, if the criteria for a meaningful-meaningless life become of central significance and are so important as to involve the whole society, the final decision would seem to be beyond the scope of the individual concerned or health professionals.

On the other hand, if the decision to die is (1) consciously and (2) voluntarily reached, then and only then, this position holds, may the person be enabled to die if criteria, such as painful and terminal illness, are agreed on. When life is no longer considered tolerable and a person decides to die, that individual does not supplicate or appeal to others to "allow" him or her to die, but rather orders others to kill him or her as Brutus did with his servant, somewhat the way one uses an instrument [21].

The reader is offered a variety of positions from which further questions and additional issues may emerge. The boundaries of life, from its inception, to the question of its continuation, and finally, to its termination, will continue to occupy the forefront of debate and controversy. Human rights are considered in matters of individual decision and public policy. Both option rights of choice and welfare rights of assistance are involved in providing conditions for a full human life [2, 8]. Human rights would seem to be a factor to be weighed in reaching any decision affecting the goals of life and the quality of living between its inception and its termination.

## REFERENCES

1. Aquinas, St. Thomas.  *The Political Ideas of St. Thomas Aquinas* (edited by D. Bigongiari). New York: Hafner, 1953. Pp. 137–138.
2. Bandman, B.  Option Rights and Subsistence Rights. This volume, Chap. 5.
3. Bandman, B., and Bandman, E.  Rights, Justice and Euthanasia. In M. Kohl (ed.), *Beneficent Euthanasia.* Buffalo, N.Y.: Prometheus, 1975.
4. Cranston, M.  Human Rights: Real and Supposed. In D. D. Raphael (ed.), *Political Theory and the Rights of Man.* Bloomington: Indiana University Press, 1967. Pp. 48–53.
5. Feinberg, J.  A Postscript to the Nature and Value of Rights (1977). This volume, Chap. 2.
6. Feinberg, J.  Voluntary Euthanasia and the "Inalienable Right to Life." Paper presented at Bioethics and Human Rights Conference at Long Island University, April 9, 1976, forthcoming in *Philosophy and Public Affairs,* 1978.
7. Fried, C.  An analysis of "equality" and "rights" in health care. *Hastings Report* 6:28, 1976.
8. Golding, M. P.  The Concept of Rights: A Historical Sketch. This volume, Chap. 4.
9. Golding, M. P.  Towards a theory of rights. *Monist* 52:530, 1968.
10. Hobbes, T.  *Leviathan* (edited by M. Oakshot). Oxford: Blackwell, 1928. Pp. 84–85.
11. Hospers, J.  *An Introduction to Philosophical Analysis.* Englewood Cliffs, N.J.: Prentice-Hall, 1967. Pp. 69–74.
12. Jaggar, A.  Abortion and a woman's right to decide. *Phil. For.* 5:351, 1973–1974.
13. Kohl, M.  Karen Quinlan: Human Rights and Wrongful Killing. This volume, Chap. 14.
14. Kubler-Ross, E.  *On Death and Dying.* New York: Macmillan, 1970.
15. Locke, J.  *Two Treatises on Civil Government.* New York: Hafner, 1947. P. 163.

16. Mill, J. S.   On Liberty. In *Utilitarianism, Liberty and Representative Government*. London: Dutton, Everyman's, 1948. Pp. 72–73.
17. Newton, L. H.   No Right at All: An Interpretation of the Abortion Issue. This volume, Chap. 13.
18. Rachels, J.   Active and passive euthanasia. *N. Engl. J. Med.* 292:78, 1975.
19. Sartre, J. P.   *Existentialism and Human Emotions* (translated by B. Frichtman). New York: Philosophical Library, 1947.
20. Thompson, J.   In Defense of Abortion. In J. Feinberg (ed.), *Abortion*. Belmont, Calif.: Wadsworth, 1973. P. 128.
21. Walzer, M.   Consenting to One's Own Death: The Case of Brutus. In M. Kohl (ed.), *Beneficent Euthanasia*. Buffalo, N.Y.: Prometheus, 1975. Pp. 100–105.

# Genetic Responsibilities and the Rights of Future Generations

# CHAPTER 7

# Genetics and the Right to Plan Future Generations
*Lee Ehrman*

*"Do we have* the right to plan future generations?" As is the tendency of biologists, let us first dissect the question: We means whom? We, the putative parents? We, the state? We, the collective federal government of states?

What does planning mean? Quantitative planning, that is, the numbers that constitute our future generations? The numbers of sorts or proportions that constitute future generations? Qualitative planning, that is, the innate health of our young? The environmentally influenced health? Both kinds of health?

The answer to all of these questions should be, though it sounds overtly flippant, "Sure, why not?" And such a quick reply to quantitative planning is certainly at least superficially more reasonable than an equivalent reply to qualitative planning. So let us now abandon unprofitable superficiality and set standards. To do so, two questions are pertinent: What do we know about the genetics of our species? What do we know about the evolution of our species?

## GENETIC LAWS AND TRANSMISSION

The way an organism appears, its *phenotype,* may be conveniently distinguished from the organisms's underlying *genotype,* contributed by parents. The phenotype is far less stable than the genotype, since genes control the manner and the range of reaction to environmental pressures — the *norm of reaction.* Certain laws serve as the foundation, both historically and functionally, of the science of genetics. One concerns *segregation* (particulate units of inheritance, genes, do not contaminate one another in individuals within whom they are borne simultaneously, but they segregate intact when sex cells are formed). The second law deals with *independent assortment* (alleles, variant forms of genes, which recombine and are transmitted independently of one another from one generation to the next). These laws apply to *dominant* genes (one dose produces the trait controlled by a dominant gene), to *recessive* genes (two doses are necessary for phenotypic manifestation), and also to genes controlling neither dominant nor recessive characteristics, that is, intermediate genes (sometimes producing a phenotype somewhere between the phenotypes of the parents) [7, 13].

Because sexually reproducing parents are known to be heterozygous for large numbers of genes, the conceivably possible genotypes outnumber the actually realized genotypes in their limited number of offspring by many orders of magnitude. The probability of two siblings acquiring the same genetic endowment from both or even one of their parents is infinitely small. This probability decreases as more and more pairs of genes are involved [3]. (Identical twins, however, may have the same genotype because they developed from a single fertilized egg.) And the probability of two unrelated persons having the same genotype is utterly negligible! It is a truism that each of us is genetically wholly unique and shall always be so.

"Diabetes mellitus is in many respects a geneticist's nightmare. As a disease, it presents almost every impediment to a proper genetic study which can be recognized" [10]. However, at this point the bulky and complex data from many sources are often consistent with what is known about multifactorial inheritance. Screening via glucose tolerance (and cortisone-glucose tolerance) tests, Neel and his collaborators have shown that in prediabetics or in persons predisposed to diabetes, there are significant deviations from normal glucose tolerance curves in the 10 to 29 age interval. Using appropriate tests and medical as well as familial histories, one can score for "definite" or "potential" diabetics and, of course, for "normal" persons. But diabetes is not a disease of sudden onset, and the dividing line between "latent" and "overt" is apparently obscure.

Congenital cleft lip, and its frequent accompaniment, cleft palate, is a *threshold* character caused by the joint action of multiple genes acting *additively*. For example, if the proband has severe bilateral clefts, the frequency of any sort of cleft in siblings is 6.7 percent. If the index case has just a unilateral cleft, this frequency is only 3.8 percent [14, 15]. Furthermore, sex apparently plays a role. Females with both cleft lip and palate have the highest incidence of affected sibs.

Perhaps the most common condition we can consider here is predisposition to heart disease, that is, to coronary atherosclerosis. The prevalence of this disease among middle-aged men in the United States is at least five percent. Two major factors with definite familial clustering predispose people to coronary atherosclerosis: elevated serum cholesterol and blood pressure levels. We need only consider elevated serum cholesterol here, that is, hyperlipidemia that occurs in four or more versions. Patterson and Slack [12], stimulated by the Framingham study that provided evidence for a linear relationship between the increasing risks of heart disease associated with low to high serum cholesterol levels, tested for lipid abnormalities in survivors of myocardial infarctions. They also surveyed the first-degree relatives of these patients.

"The distribution of cholesterol levels among the relatives suggested that the type-II [Frederickson's familial hyperlipoproteinanemia, high serum cholesterol levels] lipoprotein disturbance in the index patients was determined by the already known single gene of large effect in a minority and that the abnormality in the majority of index patients was multifactorially determined" [12].

Prophylactic measures ranging from diet through chemotherapy to hygienic regimens involving altered work loads and exercise are possible not only after the initial heart attack but should also be preferentially applied to patients appropriately screened, classified, and therefore identifiable as warranting prophylaxis.

In summary, the manipulation of conditions of polygenic origin, though surely the most difficult of all tasks genetically, may be the most rewarding medically and philosophically. The triggering genetic insult, often a family tragedy, need not occur before the resources of an appropriate adviser are recruited. This, for an ultimate example, is indeed the prime type of genetic research and planning applicable to phenomena of human intelligence as well as to phenomena of infrahuman or subhuman intelligence — the former warrants more thorough investigation while the latter warrants more thorough experimentation. However, whether the mental capacities of humans or rats, for example, are under discussion, we cannot even theoretically cross "smart x stupid" and, with a Punnett square, predict or observe offspring in the ratio of 1 smart : 2 average : 1 stupid, or some such absurdity.

## MAN'S EVOLUTIONARY HISTORY

Contemporary man is the end point of a long evolutionary history, as was realized by Darwin [4].* From the fossil record it appears that the Hominidae (human) and Pongidae (ape) families separated during the Eocene geological period, and the differentiation was well established by the Miocene and Pliocene epochs about 10 million to 15 million years ago. In the later part of the Pliocene epoch, about 2 million years ago, the first definite hominid appeared, *Australopithecus.* This remarkable form was characterized by: (1) enlarged cranium and hence neural tissue, although his brain was only slightly larger than that of the modern chimpanzee and hardly more than one-third the size of modern man's; (2) bipedal locomotion, freeing the hands for manipulation; (3) the use of tools, which follows from the development of bipedal locomotion; and (4) the ability to communicate and to hunt in bands and the development of a carnivorous diet. This form existed until well into the Ice Age, perhaps up to 700,000 years ago.

The next major step was the appearance of *Homo erectus* about 600,000 years ago. This species was characterized by a brain size of about 1000 cc, approximately double that of *Australopithecus* and 75 percent of that of modern man. Fossils of *Homo erectus* are often accompanied by stone tools, including axes, which he must have made and used. Since the fossil sites contain bones of large animals, which the hunters apparently killed, the existence of well-organized bands is implied. The existence of both stone tools and organized bands implies a form of speech between members of this species, a much more developed level of communication than that found between apes and other animals. Communication is quite general in the non-human primates, although not verbal communication, like man's. The genetically determined neural substrate is presumably not sufficient to support speech behavior [5]. Evidence from fossil sites indicates that the behavioral traits of *Homo erectus* are closer to modern man's than are those of *Australopithecus.*

The first men, indistinguishable from ourselves, evolved about 35,000 to 40,000 years ago during the last advance of the glaciers. Their appearance was accompanied by a rapid expansion, diversification, and improvement of culture. They buried their dead together with flowers and implements carefully laid around the body, so that it is not unreasonable to assume that they believed in an afterlife and had some form of religion. This was *Homo sapiens* — modern man.

The evolutionary trend, therefore, is toward the development of intellectual capacity — the feature that makes man's position unique. Morphological trends, such as an increase in brain size from about 500 cc in *Australopithecus* to about 1400 cc in *Homo sapiens,* and the development of bipedalism, plus behavioral trends, such as the development of the ability to communicate and make tools, all are in agreement. In modern man, in addition to the advances pioneered by *Homo*

*Professor P. A. Parsons graciously assisted the author by criticizing this section.

*erectus,* we have: (1) advanced toolmaking; (2) elaborate cultural organization; (3) additional increases in brain size; (4) a childhood and adolescence extended in time and providing a longer period in which cultural achievements can be assimilated; and (5) a degree of control of the environment by means of advances in medicine and technology.

The increase in brain size is an example of directional selection that was very rapid in terms of the paleontological time scale. Since it was associated with the development of progressively more advanced intellectual capacity, there must have been a selective advantage in more efficient communication, perhaps related to tool development, the use of fire, and hunting in bands. Quite likely the period when man's brain was increasing most rapidly coincided with the evolution of his ability to invent and use language for communication. Speech is not only essential for these behavioral developments, but it is also basic to the development of ideas and plans for the future. When we look at the recent evolution of man, the importance of behavioral changes in initiating new evolutionary phenomena is undeniable. It can reasonably be assumed that initially there were only minor modifications at the structural level but that the evolution of a morphological change followed a permanent behavioral change. For example, the perfection of bipedalism may well have been speeded up by the preoccupation of the anterior extremities with manipulation, a behavioral characteristic. Similarly, the significant increases in brain size were associated with the development of an efficient system of communication, speech. It is remarkable that the development of language was accompanied by tremendous linguistic diversity and isolation. This was associated with genetic diversity as observed in those regions where tribes of *Homo sapiens* can be still studied.

Quite likely the breeding structure of primitive man affected his evolutionary rate. If the leader of a group had several wives (*polygyny*), he contributed a greater than average share to the genetic composition of the next generation of his group. The tremendous reproductive advantage of a leader in a group or tribe would favor the characteristics of man, since for leadership certain physical and mental traits would inevitably be favored. These traits in turn would depend to a considerable extent on the genotype of the individual, so that the reproductive advantage would ultimately make a maximal contribution to the fitness of the entire group. The actual evidence for polygyny is difficult to obtain, but it may be an original condition in a few living tribes, and it is more or less developed in nearly all anthropoid apes [1].

Modern man, appearing 35,000 to 40,000 years ago, was anatomically indistinguishable from ourselves: There has been little selective pressure since then for altered anatomical features. Has natural selection ceased? The answer must be negative. Man has changed from a species living in small hunting communities to a species of which many members live in large, highly organized communities. Up to this century the rate of population increase in man was quite slow because of an extrinsic factor — disease. There must have been a high premium placed upon genes for resistance to specific diseases. Some of these diseases must have become important because of man and his way of life. For example, Livingstone [9] traced malaria to the slash-and-burn agriculture that opened the forest floor to stagnant pools and brought man into contact with insect vectors and hence malaria. One consequence of malaria was gene pool change. Because of their greater resistance to malaria, carriers of genes for sickle cell anemia, thalassemia, and glucose 6-phosphate dehydrogenase deficiency were favored, and this factor led to polymorphisms in regions where malaria was present.

## ENVIRONMENTAL INFLUENCES

Technological advances have brought man into contact with other diseases [11]. Rodents, attracted to settled populations, brought epidemic diseases. The practice of single-crop agriculture brought nutritional risks (since each cereal has its own limiting amino acids) and a propensity to protein undernutrition and endemic dysentery. Most bizarre is the culturally based occurrence of kuru, a disease contracted by the cannibalistic practice of eating the brains of dead enemies. Today, contagious diseases have receded as a major problem, but a concomitant effect of the control of disease is a rate of population increase reminiscent of the voles during their increase phase. In the voles, emigration of certain genotypes reduced the rate of increase. Until recently, emigration has been a factor in human populations, but for us this phase too is nearly over.

The extreme territoriality of small rodents has been noted. Beginning with population sizes of 32 and 56 Norway rats in different experiments in the presence of adequate food and water, Calhoun [2] showed that rats confined in a 0.25-acre enclosure reached a population size of about 150 at 27 months. One would expect, from the very low adult mortality in uncrowded conditions, a population of 5000 at 27 months. However, at this stage infant mortality was high, and pathological behavior was evident in both sexes, associated with social dominance of certain males. It seems that much of the abnormal behavior observed was a consequence of the disruption of territories and peck orders by the presence of excess numbers. As a consequence, normal behavior patterns broke down, and the predicted population growth curve did not occur because of insufficient space to develop normal social behavioral patterns. If we can argue from the rodents, it seems that space may well become a progressively more limiting factor for man [8]; and as it becomes more limiting, various behavioral changes in populations may occur that may be to some extent under genetic control.

Is there evidence for stress syndromes in man leading to a reduction in the growth rate of human populations? The human pituitary gland responds under stress in a way similar to the response of other mammals, and there is indirect evidence that inmates of concentration camps experienced acute stress syndromes that accounted for their death. Concentration camps may be more analogous to highly congested animal populations than city slums, since even in very crowded cities the poor have some mobility, although the occurrence of street gangs and juvenile delinquency characteristic of crowded cities is a form of social pathology. The increased incidence of atherosclerosis and other cardiovascular pathology associated with urban living and its competitive stresses may be enhanced by crowding. In underdeveloped countries where high birth rates and recently lowered death rates produce a population growth of 2 percent or more per year, the use of health measures increasing life expectancies masks any growth-retarding effect of a stress syndrome. Therefore, direct comparisons with rodents are hardly possible. But although we are far from the rat colonies, space limitations are likely to become progressively more important; indeed, they are already so in the poorer sections of some of the world's largest cities. The tendency is likely to increase. More highly developed technology reduces the number of people needed to produce a given amount of food and encourages the drift of people from rural to urban areas, a tendency that was accelerated by the Industrial Revolution and that is continuing today.

Demographical, medical, and technological changes associated with social change are now occurring so rapidly that populations have difficulty adapting. Therefore, improved understanding of human behavior and its genetic basis is essential, and this is the task of the biologist and psychologist.

A final problem of a general kind is the effect of medicine on the gene pool of man. In any society where either preventive or curative medicine plays a significant role, a greater proportion of zygotes survive for reproduction than survived in previous generations. Genotypes normally eliminated by natural selection continue to survive and reproduce. For example, the fitness of diabetics was dramatically increased by the discovery of insulin. Thus medicine has enabled diabetics to survive, reproduce, and spread their deleterious genes into subsequent generations. The same applies to some behavioral disorders. As a simple example, the IQ of phenylketonurics may be improved by a special diet, which presumably results in an increase of their reproductive rates. Hence the average biological fitness of the population may be expected to fall because of the increased incidence of deleterious genes. In considering the possible future evolution of man, it is important to take account of the effect of medicine on the gene pool itself, as well as the more obvious demographical consequences. But we must be careful, for as we have already seen fitness depends on the environment. Under an optimal environment fitness differences between genotypes may be minimal. If the environment deteriorates in the future on our earth, the effects of natural selection will become more apparent, and fitness differences between genotypes may increase. By environment we mean both the physical and social environment and the rate of change of each. A major consideration is the effect of the various environmental pollutants to which we are exposed; although they may have few short-term effects, their long-term effects may be important [6].

In conclusion and in summation, we do not now possess the genetic armamentarium with which to plan future generations, and so we should not do so.

## REFERENCES

1. Bartholomew, G. A., Jr., and Birdsell, J. B.   Ecology and protohominids. *Am. Anthropol.* 55:481, 1953.
2. Calhoun, J.   Population density and social pathology. *Sci. Am.* 206:139, 1962.
3. Cavalli-Sforza, L. L., and Bodmer, W. F.   *The Genetics of Human Populations.* San Francisco: Freeman, 1971.

4. Darwin, C. *The Descent of Man and Selection in Relation to Sex.* London: Murray, 1871.
5. DeVore, I. (ed.). *Primate Behavior: Field Studies of Monkeys and Apes.* New York: Holt, 1965.
6. Ehrlich, P. R., and Ehrlich, A. H. *Population Resources Environment: Issues in Human Ecology.* San Francisco: Freeman, 1970.
7. Herskowitz, I. *Principles of Genetics* (2nd ed.). New York: Macmillan, 1977.
8. Hoagland, H. Cybernetics of Population Control. In J. B. Bresler (ed.), *Human Ecology: Collected Readings.* Reading, Mass.: Addison-Wesley, 1966. Pp. 351–359.
9. Livingstone, F. B. Anthropological implications of sickle cell gene distribution in West Africa. *Am. Anthropol.* 60:553, 1958.
10. Neel, J., Fajans, S., Conn, J., and Davison, R. Diabetes Mellitus. In *Genetics and Epidemiology of Chronic Diseases* (publication 1163). Washington, D.C.: U.S. Dept. of Health, Education, and Welfare, 1965. Pp. 105–132.
11. Omenn, G. S., and Motulsky, A. G. Biochemical Genetics and the Evolution of Human Behavior. In L. Ehrman, G. S. Omenn, and E. Caspari (eds.), *Genetics, Environment, and Behavior: Implications for Educational Policy.* New York: Academic, 1972. Pp. 129–171.
12. Patterson, D., and Slack, J. Lipid abnormalities in male and female survivors of myocardial infarction and their first degree relations. *Lancet* 2:392, 1972.
13. Strickberger, M. W. *Genetics* (2nd ed.). New York: Macmillan, 1976.
14. Woolf, C. M. Congenital cleft lip: A genetic study of 496 propositi. *J. Med. Genet.* 8:65, 1971.
15. Woolf, C. M. Congenital hip disease: Implications for genetic counseling. *Soc. Biol.* 18:10, 1971.

# CHAPTER 8

# Genetics and Our Obligations to the Future
*Marc Lappé*

*Genetics is going* to revolutionize our lives. Social institutions and legal traditions that we have taken for granted will be shaken to their roots. Our presumptions about the duties that we owe to those who are ill or infirm will be challenged. And new data emerging about the genetic underpinnings of disease states and their inheritance will assault our deepest feelings about ourselves as parents and as persons.

I am convinced that any grasp of these radical shifts will require each of us to become a moral investigator as well as a scientific one. That means going beyond the constraints of science into the world of human emotions and beliefs. I have chosen as my model a case that hopefully will force us to consider our choices judiciously when we deal with policy decisions about reproduction and health care for individuals who are afflicted by genetic disease.

## REPRODUCTIVE RESPONSIBILITIES TO FUTURE GENERATIONS

The paradigm for our problem is that of a person who is afflicted with a dominantly inherited condition that would, in the absence of medical intervention, be fatal. That person is treated successfully, largely with public funds, and survives. Should he now be permitted to reproduce freely in our society knowing that half of his children will be similarly affected? Modern statistical and demographical techniques allow us to project this person's genes into the future and to chart the consequence of his survival and reproduction with relative certainty. But before thinking through the avenues of action that are open to us, I would like to explore briefly how the *genetic* elements of this person's humanhood should be weighed in the policy decisions to follow.

First, you might ask, "Why do we think of genetics at all when we speak of an individual's obligations to future generations?" The answer is deceptively straightforward: For those things that count biologically, genes are the *only* mechanisms that we have for projecting ourselves into the future. Of course, genetics is not the only way human groups make an impact on the future. You may argue that genetics is subordinate to human culture. But whether it is subordinate or not, genetics is the only sure way of knowing where we came from — and of projecting where we are going.

For several hundred descendants of Fletcher Christian and his crew on the Island of Tristam de Cunha or the present-day descendants of the first immigrants to Michigan in the eighteenth century who carried the gene for Huntington's chorea, genetics defines cultural origins — and biological destiny. Each group of descendants carries indelible traces of their biological origins. For the de Cunha islanders, these markers are merely racial. But for the families with Huntington's chorea, each generation has had to live with the prospect of genetically determined death and psychic disfiguration in half their members. But how much of the reality of an *individual's* being do we in fact know when we know something about him genetically? To answer that question we must go beyond what science can now tell us.

## ORDERLINESS OF GENETIC DEVELOPMENT

When thinking about the most striking examples that have moved me to consider genetics as something greater than the rational discipline it is, think of any pair of people that you know to be monozygotic, genetically identical twins. Picture these twins in your mind's eye as if you were confronting them for the first time. What is there about these identical twins that is so striking?

Think of the problem biologically, keeping in mind your picture of the twins. See *how* alike in expression two identical twins are. *How* similar their behavioral idiosyncrasies are. How accidental blemishes seem to be either identical or mirror images of each other in each twin. You might even detect predilections for facial expressions or other seemingly learned behavior. Twins often seem so tightly bound, one to the other, as to defy belief.

Then stop for a moment and ask yourself, "What is the biological problem that a monozygotic twin poses to us conceptually?" In truth, it is the most profound biological problem that I can put to you: How can that much orderliness in development be controlled so completely that it generates two *identical* individuals *every* time twinning occurs?

Think about it as a physical problem. Assume that we live in a mechanical universe. An egg has just split and separates into two cells. Each of those two eggs is going to have to divide perhaps 40 times, making copies of itself in series until it assembles about 200 billion cells into an individual. At the end of embryogenesis, a complete, intricately sculpted face must come into existence with the identical configuration of its twin.

## GENETIC MANUFACTURE

That problem is very different from the problem of manufacture. A model experiment that I have done and which you could do to illustrate this point is to take a bill out of your wallet and consider it the first cell in a fertilized egg. Make a copy of it on the xerox machine. Next take the copy that has come out and use it to make another copy, and then make two, three, four, five, six, seven copies, using each new copy as the template. Now examine the detail you have left in the xerox image of the bill. To your incredulous eyes, you will discover that after 30 duplications or so, the major features of the bill have all but disappeared! Where did all that information go? It has been dispersed by the constraints against orderliness that mechanical duplication always entails. Entropy, the tendency for all physical things to reach their lowest energetic state, means that by the time you use a physical process to make 40 copies of any object, its information content is predictably all but gone. So when you have two identical twins whole cells have passed through a similar process of duplication, the most remarkable thing is that they have retained their identity through birth, adolescence, and even adulthood!

One way to understand this miracle is to consider a critical embryological phenomenon called "buffering." This phenomenon was demonstrated by embryologists like C. H. Waddington in fruit fly embryos about 25 years ago. It works in an analogous way to acid-base systems in a chemistry lab. When you allow drops of acid to fall into an alkaline solution that contains a buffer, the system tends to retain its alkaline chemical characteristics. What does buffering do? It binds excess hydrogen atoms and thus allows you to add more and more acid without changing the pH of the solution.

In an embryo, genetics systems have so many overlays of controls that they too can be said to be buffered. You can add radiation damage, new mutations, and even deleterious genes, and they can, within limits, be so dampened in their effects as to be neutralized. Disorder injected into such an ongoing developmental process in the form of a potential teratogen, for instance, can be absorbed, so that the organism may still maintain its orderly flow of development.

Thus some developmental sequences appear to be so tightly controlled and pro- grammed that the effects of intervening forces are suppressed. Transient changes in oxygen level, the availability of nutrients, or even the appearance of a drug that might otherwise be defect producing can be buffered to some extent by a develop- ing organism so that the deleterious effects are nullified.

## EFFECTS OF DELETERIOUS GENES

Now let us look at genes. When we think about a developing organism we are con- ditioned to think that additions of deleterious genes to that organism invariably cause birth defects or at least drive the organism away from an optimal norm. But recent knowledge has forced us to recognize that genetic systems are unlike any- thing we have studied before. Our mechanical models of the universe may not apply to the processes that are involved in the unfolding of the genetic code in a fertilized egg. Perhaps, most importantly, we need to recognize that the policies with which we are concerned are grossly limited to what we know or think we know about the effects of a very few genes. Indeed all of the rational discussions of policy questions that are currently being proposed or are under active consideration are centered on single genes. Invariably even these simple models are subject to variation. In reality, only statistical generalizations are appropriate for dealing with all but the most lethal genes.

The impetus for the New York State law, which has since been repealed, requir- ing individuals of non-Caucasian, non-Indian, and non-Oriental ancestry to take a blood test before getting a marriage license hinged on a test for the presence of such a single gene, the gene that coded for hemoglobin S (sickle cell trait). For all its limited focus, it was still a rudimentary form of a eugenic law. It implied that the state has a right under its police powers to require individuals to know something about their genetic status — in this case, to know whether or not they are carrying a single dose of a deleterious gene.

In fact there is no evidence to show that health disabilities are associated with carrying a sickle cell gene, except under extraordinary environmental conditions. The gene for hemoglobin S, even when present in a *double* dose (homozygosity), is remarkable for its variability of expression, and decisions about eugenic interven- tions to reduce its frequency are fraught with difficulty [1, 2].

Are there any genes that do generate sufficiently grave effects to justify our examining the moral imperatives of using genetic knowledge to plan future generations? At least one genetic circumstance presents us with unavoidable moral imponderables. This case was brought to my attention by a photograph, which was seemingly just one of several clinical slides reviewed by genetic counselor Hymie Gordon at a genetic counseling conference.

After the photograph was flashed upon the screen, Dr. Gordon asked, like any good teacher, "What's wrong with this picture?" We all looked. "Well," I remember musing, "it's a family picture – people are gathered around; it's a birthday obviously, because everyone is dressed up. Everyone is staring straight at the camera. In fact a couple of the people have stared so straight at the flash that you can see the characteristic red dot reflected in the pupil of their eye." Simple optics: The retina picks up the light of the strobe and reflects it back into the camera lens. But then I saw it. "Ah, one of the persons is not reflecting a red light, but a white light. There's a *white* dot reflected from the eye of one of those individuals."

When Hymie Gordon got to this picture, he recalled that his heart stopped. It was the kind of picture you see through an ophthalmoscope of someone who has a tumor that is developing on the back of the retina. Instead of having the rich vascular redness that the retina should have, this child's tumor has reflected white. He called up the family and discovered that the child indeed had an eye tumor. It is called retinoblastoma, and its removal in this case entailed enucleation of the eye.

Retinoblastoma is a tumor that apparently arises during embryological development from one or more cells that would otherwise be made part of the neural network of the retinal screen. These cells somehow escape the normal controls of development and begin to divide uncontrollably in the back of the eye without the normal constraints that hold other cells in check. It is a rare tumor, but it is invariably malignant. It used to occur in no more than 1 in 30,000 people. It now occurs in England in as many as 1 in every 18,000 births. This exponential increase took place roughly between 1930 and 1960 and almost entirely the result of physicians' being able to detect the tumor, treat it, and allow the individuals to survive and to go on to reproduce. Today 81 to 86 percent of affected individuals survive for four years or longer [4].

It happens that a portion of these tumors are caused by a dominant gene. This example, as with most seemingly simplistic genetic models, is fraught with complexities. In this case, only between 60 to 90 percent of the individuals who receive the dominant gene actually manifest the tumor – this phenomenon is called reduced penetrance. But when they get the tumor, they usually get it in both eyes and would be blinded or die except for new developments in surgery and radiation therapy. Now 70 percent are saved from blindness – and at least four out of five survive [4].

LEGISLATING GENETIC POLICY

The next problem that I would like to propose is imaginary, but it may not be so in the future. Here is a hypothetical memo that the state director of health might receive from someone in the state's genetics unit.

TO:     Michael Hobbs
        State Health Commissioner

FROM:  Dr. Ian Ward
          Genetic Diseases Unit

In response to your request for a memo on our need to develop a policy about genetic disease, I would like to direct your attention to the problem of the transmission of retinoblastoma. Here is the present situation together with a set of alternative policies for the state to follow. I will begin drafting legislation as soon as you give me the word.

Retinoblastoma is a rare and malignant eye tumor that occurs characteristically early in childhood. Until recently, it was invariably fatal. In the 1950s about 5 in every 100,000 children were born with this tumor, and about one in every five carried the gene that caused it. Before effective treatment was instituted 35 years ago, one case in five was genetically caused, the result of a new mutation. Today more cases have a genetic base. They are almost always associated with a family history of retinoblastoma. That is, one parent, but very rarely both parents, will have had the tumor in childhood.

To give you an idea of the statistics that are involved, today fully 36 percent of the cases of retinoblastoma that lead to blindness in the state have family histories, and 20 percent of these individuals will lose both of their eyes. The overall mortality for both forms of this tumor with the best treatments available in 1976 is still 15 percent. This means that 85 percent who might otherwise have died are saved.

This new pattern of retinoblastoma is due almost certainly to the survival and reproduction of those individuals who in previous generations would have died and to an indeterminate extent to an increase in background radiation, which increases the spontaneous mutation rate. When the mutation occurs in the germ plasm, it is dominant. Thus, the risk is that half of all the children of a parent with this genetic form of retinoblastoma will develop the condition, even if the carrier or affected individual is married to a normal spouse. If the tumor-affected individuals are treated successfully and reproduce, there will be a rapid increase in the frequency of this gene beyond what we are presently seeing — with accompanying increments in state costs for dealing with the blind.

One new item of recent data that complicates our policy decision is that the gene for retinoblastoma may also be associated with or closely linked to genes that are involved with intelligence. In tests that we have done in the state, the siblings of affected individuals have been found to have IQ scores that range between 116 and 128, with an average of 120. Therefore, a significant portion of the gifted individuals in this population may well also carry the gene for retinoblastoma; however, they may be among the 10 to 40 percent who do not express its cancer-causing properties.

If we begin an active treatment program for retinoblastoma involving early screening and detection so that we reduce the mortality rate further, the following outcome can be predicted. If the frequency in 1920 was 1 in 100,000, we are now at a point three generations later where the incidence is roughly 4 in 100,000 or 1 in 25,000. By the tenth generation (admittedly some 200 years from now), we will have at least 11 cases per 100,000. Each of those cases now requires approximately $60,000 of surgical care plus follow-up treatment at health facilities in the state.

Our present health system operates, at least passively, on the assumption that if health disabilities are subject to remedial treatment at minimal cost, individuals have a right to medical care. Recently this right has come under closer scrutiny. Some policy advisers maintain that medical care cannot be a universally recognized right, especially for those kinds of care that are exceedingly costly. For example, there was a time when we could not manage or afford to give dialysis treatment to every individual who had renal failure. According to this argument, we still must assign priorities in the delivery of medical care, and we must reassess the question of the proportion of health care that should be carried by publicly financed insurers and the percentage by individuals through private carriers.

The microsurgery required for treating eye cancer of the kind represented by retinoblastoma *is* exceedingly expensive. It may involve lasers, it may involve selective chemotherapy, or it may involve extremely careful surgery if eyesight is to be saved in these presumably gifted individuals. The question then becomes: Should there be *preferential* rescue of retinoblastomic individuals predicated on their likely greater contribution to our society over and against individuals who may require about the same degree of medical care but who are unlikely to contribute as much?

For example, what priority should be assigned to funds for surgical interventions for Down's syndrome, a condition commonly associated with heart defects? Or consider that perhaps 9000 women in the state of New York over the next five years are at high risk for having children with Down's syndrome because of age-related effects. Some will be afforded a screening program for amniocentesis and selective abortion. Assume that it costs roughly the same to *detect* preemptively one case of Down's syndrome as to *treat* one person with retinoblastoma. Where would you place the priority for meeting this expenditure in the state of New York? Before you answer, consider this question: "How would you justify *not* spending a like sum on *saving* an individual with Down's syndrome as on saving one with retinoblastoma?"

## GENETIC POLICY OPTIONS

Now suddenly you are confronted with an ostensibly insoluble genetic dilemma: There *are* some genes that do make a difference in the uniqueness of individuals. They may or may not be distributed systematically among groups of individuals in our society, but we will certainly find individuals who have greater or lesser need for health care because of the genes they carry. We will also be able to find prospective *parents* whose children will pose proportionately larger health burdens to society than will the offspring of "normal" parents. What is our obligation to those children who are at risk for genetic disease — and what are *their parents'* duties to society?

Before considering the policy options, keep in mind that whatever model you use for solving the solution will ultimately have something to do with your attitude, indeed, your moral decision about the responsibility of society as a whole to those who are most disadvantaged over and against those who are advantaged or otherwise normal. What, for instance, are the moral obligations of a society toward those who are least well off? How do these obligations change when there are scarce medical resources to be distributed? Does it make any difference if the individuals who are least well off are *future* members of society and that many of them can be "rescued" and be made "useful" members of society? Others, following rescue, will pose further "risks to society" — should they procreate?

It might be well to recall that virtually every one of the models under which we traditionally operate has some connotation of social utility in it. For example, sociologist Talcott Parsons speaks about the social worth of each individual as being measurable only by the contribution that an individual can make to the social good as a whole. But do we want to make life and death decisions by asking what contribution a given individual might make to society? These questions can only be answered in the context of larger, more fundamental questions about the nature of our society and how we decide what to do with scarce resources.

Let's make our first policy decision hinge on the question of whether persons with retinoblastoma who are found to have family histories strongly indicating a hereditary basis should be as aggressively treated as those cases in which the retinoblastoma is due to environmental factors or sporadic mutations – and are hence *not* transmissible to future generations. Let's assume that both have the same tumor and both are at the same medical risk (although there is a higher incidence of having the tumor in both eyes with the genetic form of the disease). Do you seek out and treat those individuals with the hereditary and nonhereditary forms with equally aggressive medical intervention? From a traditional medical viewpoint, the answer is almost certainly yes. Whether or not the disease is genetic is immaterial to a physician's obligation to relieve suffering in the affected person. Public policymakers may legitimately question how public funds are apportioned where one option (e.g., where the tumor is hereditary) entails greater cost than the other. But here the cost is projected to future generations. How might we incorporate this circumstance into our balancing arguments?

Assume that we do afford the same treatment to individuals who have the genetic form of this eye tumor as to those who have the nongenetic form. Might not one *now* owe us some reciprocal obligation because we have both given him a portion of scarce medical resources for his care and treatment and permitted his survival and potential procreation. Specifically, do genetic retinoblastomics owe society any obligations in terms of their procreation and their planning for future generations? Or might saved retinoblastoma victims have special obligations to use their proportionately higher intelligence and giftedness for the social good?

If you believe there are *some* special obligations for retinoblastoma victims, there are several ways to discharge them. Let us review these alternatives in terms of policy options to the state:

1. Individuals with genetic retinoblastoma will be subject to mandatory genetic counseling prior to marriage. This genetic counseling would be a prerequisite to being given a marriage license by the state.
    *Comment:* The state already exercises a considerable amount of control over the rules under which it will afford individuals a marriage license. How does the state justify the exercise of that power? For the protection of the unborn. It justifies premarital testing on the grounds that individuals who have active syphilis may not jeopardize a child. The state justifies *treating* children for gonorrheal infection of the eyes acquired when the child passes through the birth canal *without* parental consent on similar grounds. Parents may not put an innocent life in jeopardy. So in one way, the state has established its vested interest in controlling the quality (in terms of health) of its offspring. But does it have a right to extend that umbrella of control into genetic matters? Can it require genetic tests as a prerequisite for marriage licenses? Legally, the answer is almost certainly yes.

But such a test might subject those we can now detect with genetic disease to discrimination since we cannot find all such persons.

2. Those who have been successfully treated and have this heritable form of retinoblastoma might be urged in genetic counseling to have one-half or an other arbitrary percentage of the normal number of children.

   *Comment:* Genetic counselors might be given explicit instructions to discourage procreation amongst these individuals and to impose that view on individuals rather than saying, "Well, you're faced with a serious situation, but its treatable .... Here are the consequences if you proceed, but it's up to you."

3. After going through the process of marriage counseling, individuals with retinoblastoma will be requested to submit to sterilization to receive their marriage license.

   *Comment:* If that sounds like undue compulsion, recall that there are 16 states that still have compulsory sterilization statutes on their books. Although very few states have exercised that option, it is still within the purview of many existing state laws to extend the requirements for sterilization. Some might argue that individuals with retinoblastoma who want to get the permission that the state bequeaths on its citizenry to have children through a marriage license might reasonably be asked to submit to sterilization on the grounds that the health burden and dependency represented by an affected child are too grave a health risk to sanction.

Now which, if any, of these policies should we adopt? The argument for sterilization or other compulsory policies hinges in part on the dynamics of gene increase where dominant conditions are involved. One generation after the first successful treatment of retinoblastoma, there would be an actual doubling of the incidence of the gene and an incremental change from then on proportional to the mutation rate. Since 1930 the gene frequency for retinoblastoma has gone up; its incidence now is virtually as high as that for phenylketonuria or PKU. Many states have compulsory testing for all newborns for PKU, which has an incidence of about 1 in 16,400. (Recall that about 1 in 18,000 children will be born with retinoblastoma.)

Can the medical profession justifiably extract, not just a promise, but the acquiescence to sterilization as a requirement for treatment for retinoblastoma? Note that the state does *not* ask the same of women successfully treated for PKU whose children are almost invariably at risk for birth defects because of maternal metabolic imbalance. That is seemingly a novel if not an unreasonable proposal, but if we do not confront it now, someone may well present it to us five years from now in less flexible form. Policymakers will want to know why we cannot require individuals who have dominantly inherited conditions to accept procreative restrictions. Especially where the disease carries a certain social burden and medical weight in terms of expenses should not affected individuals offer some reciprocal payment to society for the public expense of treatment of that condition? What about constraining individuals from procreation by withholding marriage licenses entirely from individuals with a heritable form of retinoblastoma?

My view is that *none* of these policies is defensible. All require an intrusion into the privacy of reproductive decision making by the state, which I believe constitutes a greater harm than leaving such decisions to the couples at risk. Medical care for life-threatening emergencies has been increasingly offered without regard to initial ability to pay, and voluntary genetic counseling has generally been seen as superior to compulsory counseling in both outcome and compliance.

So what might a final policy decision look like? First consider a hypothetical postscript. The memo ends with the following statement:

Mike, these are the options. I don't see how we can go on inflicting this kind of suffering on innocent children one generation after another. If we keep on going as we have been, there are going to be more and more cases. I favor some state restraints in the first set of options I gave you. I am concerned that we know we cannot currently distinguish between genetic and nongenetic forms of the condition in most individuals, but I am also worried about the 20 percent of individuals who carry the genes but who may never develop the symptoms. Blindness is a major health problem in the state; the costs of surgical treatment are escalating; what do we do?

This is a fascinating point because of the linkage of the "good" (high IQ) with the bad (blindness). Such a motif is most likely for any gene that is prevalent in the human population. High gene frequencies for recessive disorders are almost always due to some adaptive advantage accruing to the heterozygous carrier. But the association of a *tumor* with the presence of increased intelligence faces us with a seemingly "lady and the tiger" kind of choice. A health commissioner is then faced with the decision of whether to propose to the state legislature that this disease be singled out for medical attention and care — or that it be eradicated through aggressive state intervention. The importance of this decision is highlighted by the fact that once health legislation is proposed, there is often very little legislative debate over the issues. And if it is cost-effective, it may receive automatic approval.

## GENETIC DECISION MAKING

Before final legislative decisions are made, let us see how we would like individuals to handle the problem. Do parents with genetic diseases indeed have an obligation to future generations? We are not talking about distant generations, to which some claim we owe very few obligations, but the very next generation — our children. Is there not a fundamental obligation to leave the next generation better if not at least as well off as ours? If we can ameliorate the short-term deleterious consequences of *voluntary* actions should we not do so? Do not children deserve to be protected from foreseeable harms? [1] I would propose a technique that seems consistent with the moral traditions of our society. First, recognize that virtually every prospective parent who puts a child at risk for retinoblastoma will have had the same tumor in childhood. He will have been subjected to the treatments and will have experienced the pain, suffering, and other burdens of having that condition. I would assign primacy to this unique experiential basis for judging — over and against rules imposed by society from outside the family unit, insofar as society has the financial resources to cope with the surgical costs of treatment.

In assigning primacy to the individual, I would side with Montaigne, who wrote in his essay on the education of children, "I have never seen a father who failed to claim his son, however mangy or hunchbacked he was. Not that [the father] does not perceive his son's defect . . . but the fact remains, the boy is his [5] ." Thus, no matter how defective a parent's child might be, we shall probably discover the fundamental human condition that *every* parent shares will lead that person to regard his prospective procreative decision seriously. Love and parental bond establish the grounding for a procreative decision in the best interests of the child.

To accept this view, we must acknowledge that human experience is unique, something with which we can only remotely empathize. It is my view that no single answer could be expected from individuals, nor should one be expected. The "truth" here cannot be derived by universal application of the laws of moral analysis [1, 3] . Experience here is assigned the highest priority – and by definition, experience is unique. The parent who would not wish his own fate on his children will refrain from procreating – the one who found his condition bearable might consider otherwise.

What should be done then about our state policy questions concerning retinoblastoma? A treatable condition like retinoblastoma *does* impose a medical burden on society, but is it a burden greater than other similar medical interventions? Retinoblastoma is *preventable,* you might argue. "And with retinoblastoma," you might continue, "even where we treat the disease, the gene remains." To be sure. But the presence of a *gene* cannot be taken to change our duties and obligations to the affected individual, or we would *all* be subject to proscriptions on our procreation, since we each carry three to nine deleterious genes.

"But," you might counter, "the individual cannot ignore a *dominant* deleterious gene in making personal decisions about reproduction. Obligations to their own children demand recognition of the genetic basis of their defect." But even here, however, the heterogeneity of expression, the reduced penetrance, and other complicating features make it impossible for the gene alone to be all determining. As we have seen, a gene far from "fixes" all of a person's characteristics. As many as 40 percent of those who have retinoblastoma will in fact appear normal [4] !

In the face of such ambiguous genetic data, I believe the right to decide must be vested exclusively with the parents that are involved. We would be best off investing the moral authority for making this decision not with the state but with these individuals who primarily bear the human burdens for perpetuating their own genes. To do otherwise would be to place the state over the individual – and to deny that deepest of all feeling that parents have for their children.

## REFERENCES

1. Gustafson, J.  Genetic screening and human values. *Birth Defects* 10:201, 1974.
2. Lappé, M.  Can Eugenic Policy Be Just? In A. Milunsky (ed.), *Prevention of Genetic Disease and Mental Retardation.* Philadelphia: Saunders, 1975.
3. Lappé, M.  Moral obligations and the fallacies of genetic control. *Theological Studies* 33:411, 1972.
4. Lennox, E. L., Draper, G. J., and Sanders, B. M.  Retinoblastoma: A study of natural history and prognosis of 268 cases. *Brit. Med. J.* 2:731, 1975.
5. Montaigne, M. de  *Selected Essays* (translated by Donald M. Frame). New York: Van Nostrand, 1943.

# CHAPTER 9

# Comments on the Presentations of
# Drs. Ehrman and Lappé
*Ernest Nagel*

*The inclusive characterization* of this interdisciplinary conference is "Bioethics and Human Rights"; and the designation for this first session of the conference is "Genetics and the Right to Plan Future Generations." It therefore seemed reasonable to suppose, as indeed I did suppose, that the questions to be discussed this morning would include "What does it mean to have a human right?" and "How can a claim to such rights be established?" However, except for Professor Bandman's brief reference to them in his introductory remarks, my predecessors in this symposium have successfully avoided dealing with these questions. In consequence, in respect to human rights, they have not provided me with anything on which to comment.

Drs. Ehrman and Lappé nevertheless present ample material for clarifying a number of central issues raised by proposals involving genetic engineering. Let me say at once that I feel greatly heartened by the caution they exhibit in discussing eugenic programs, and by the salutory skepticism with which they view quick remedies for genetically transmitted human ills. It is undoubtedly tempting to make some use of a tool just because we happen to possess it, even when we have no clear objective in using it, nor an adequate conception of its capacities for producing good and evil. The late Morris R. Cohen, my great teacher of philosophy when I was in college, once commented on the familiar legend that, when George Washington was a boy he received an axe as a gift and then promptly destroyed a prized cherry tree, by observing that it could have been expected that under the circumstances the boy would do some irremediable damage. Since our present knowledge of genetic mechanisms of human beings and of the biological as well as social effects of genetic manipulations is still fragmentary, and unequal to the task of serving as a basis for reliable planning of future generations, the speakers have wisely refused to endorse comprehensive activist policies of genetic engineering.

1

Despite the absence in the two main presentations of any discussion of human rights, I want to make what I hope will be a few clarifying remarks on this matter.

(a) In the first place, whatever may be understood by the word "right," I doubt whether anyone at this conference would question our right to *plan* for future generations. I will therefore assume that our "right" to plan is not in dispute, and that the question at issue is whether we have the "right" not only to plan for future generations by modifying the genetic constitution of men, but to implement such plans by overtly *acting* on them.

(b) However, it is not at all clear how the term "human right" is to be understood. In one of its senses it means *legal right*, so that on this construal the question under discussion is whether men have the legal right to implement plans for future generations by genetic engineering — or more precisely, whether it is legally *permissible* (but not legally mandatory) for men to do so. To this question the obvious answer is that men are so privileged in some legal systems and at a given time, but not in other legal systems or at different times. For example, men are at liberty to smoke marijuana in Mexico, but at present not in the state of New York.

(c) But I do not believe we have been invited to participate in this session to discuss a legal matter; and I think that the question we have been asked to examine is whether attempts at improving the quality of future generations by genetic engineering are *morally* justifiable. There is, however, no straightforward and unequivocal answer to this question, since every answer presupposes some moral theory and social philosophy; and there is no moral system that all reflective men accept as uniquely correct. In fact, there are influential moral systems (frequently based on religious convictions) that deny that men are ever morally justified in meddling with the genetic constitution of human beings; and there are also widely acknowledged moral systems according to which such activity is morally warranted, provided that its likely consequences are on the whole beneficial to those affected by them. It is evident that an adequate discussion of the morality of genetic engineering would require an examination of a broad spectrum of moral theories — a task which this session of the conference cannot possibly undertake.

Nevertheless, I do want to express briefly my dissent from those moral theories according to which men have a "natural right" to engage in genetic engineering. I readily acknowledge that human beings have diverse needs, desires, and aspirations, which they may claim should be satisfied. However, whether any such claim ought to be satisfied is in my judgment not a matter of natural right, but is a question whose answer depends on how the comparative merits of satisfying those claims are assessed, and on how those satisfactions fit into an inclusive moral economy. Unless I have misunderstood him, this is also Professor Bandman's view in his paper, Rights and Claims, that was distributed to the panelists in this session. I am therefore puzzled by some of his introductory remarks, in which he seems to abandon the position for which he argued so ably in his paper.


2

I must now turn to some of the important issues raised by the main speakers in this session. In one way or another, both of them discuss various conditions that need to be met if planning for future generations and its implementation are to be rational. I want to make these conditions explicit.

(a) It is patently desirable that the objectives to be achieved by planned action are clearly specified. This is frequently an easily satisfied requirement, as in stating one's destination when considering alternative modes of travel for reaching it. It is a difficult assignment when the task involves ascertaining the explicit or tacit objectives of a large number of people, especially when one seeks to determine the goals of an entire society. One important source

of this difficulty is that the human traits one would like to see developed
may not be compatible with one another. This point is brought out forcefully in Dr. Lappé's example of the apparently significant correlation between the occurrence of the genetically determined retinoblastoma and the
occurrence of superior intelligence — the cost of eliminating the former may
be a reduction in the frequency of the latter. There is in fact little general
agreement concerning what heritable human traits it would be desirable to
develop by genetic engineering, and concerning the price men would be
willing to pay for them.

A further difficulty that Dr. Lappé also notes is that planning for future generations involves tacit decisions about the likely needs of still unborn men who will be
living in environments about which we often know little, as well as that assumption
that must be made about how scarce resources are to be used. Moreover, the task
is made even more difficult because of the frequently neglected fact that the desira-
bility of an objective is not independent of the means adopted for achieving it,
since the means used have consequences that are part of the ends actually obtained.
In consequence, the goal sought may have to be repeatedly modified in the process
of deliberating over the means to be employed.

(b) A second requirement for rational planning is that men possess both a
theoretical and a technological mastery of genetic mechanisms in human
beings. Dr. Ehrman's paper provides a clear outline of the present state of
relevant genetic knowledge; but she also makes evident how fragmentary
this knowledge is, and how inadequate it is as a basis for effective and responsible action directed to the realization of hypothetical goals for future
generations. There is every reason for accepting the main conclusion of her
paper that because of the considerations just mentioned, talk of implementing such goals by genetic manipulation is premature.

However, neither of the principal speakers explicitly mentions the familiar and
useful distinction between positive and negative eugenics — the former having for
its aim the development of desirable traits in human beings by genetic means, while
the latter seeks to eliminate heritable traits that are regarded as undesirable. Assuming that the distinction is sound, Dr. Ehrman's paper shows that the present prospects for positive eugenics are negligible. But on the additional assumption that a
reasonable agreement can be reached on the heritable traits that are undesirable
and on the price society would be willing to pay for their elimination, it is not
entirely clear whether her skepticism includes negative eugenics as well. Dr. Lappé's
paper is in fact concerned with negative eugenics; and although he raises an important question concerning the way a program for eliminating retinoblastoma should
be undertaken, he does not seem to doubt that such an elimination (at least in part)
could in fact be made.

Familiarity with evolutionary theory is doubtless desirable for understanding
man's place in the scheme of things. On the other hand, it is not evident that the
theory is relevant to planning the human future. It is therefore not clear why in the
present context of discussion Dr. Ehrman thinks it is important to devote the space
she does to an account of the theory. Unlike her exposition of genetics, I do not

see how her précis of the evolutionary development of the human species contributes anything either to the articulation of plans for coming generations of men or to the assessment of such plans.

(c) A further requirement for rational planning is that the decisions concerning what plans are to be adopted be made in a responsible manner — not only in respect to the substantive content of a plan, but also in regard to the question who is to make the controlling decisions.

Dr. Ehrman quite properly raises this question, but she does not indicate what is her own answer to it. Dr. Lappé, on the other hand, presents alternative policies for eliminating undesirable heritable traits; and he favors the one that vests the final authority for engaging in a eugenic program not in any agency of the state, but in the various individuals who are carriers (or are believed to be carriers) of undesirable genes. On this crucial issue he adopts a position similar to the main conclusion of John Stuart Mill's still influential essay on liberty; and I confess that I wish it were possible to accept it without serious qualifications. In any case, Dr. Lappé offers no reasons for his choice of policy. For my part, I do not believe the view can be maintained that an absolute moral limit can be placed on what the state may do to regulate the conduct of its members — even in such apparently private matters as procreation or providing information concerning latent genetic defects. It is easy to imagine circumstances in which carriers of a severely crippling and socially catastrophic gene, refusing in significant numbers to heed voluntarily the advice of genetic consultants, are forbidden to bear any offspring. The ethical and social issues such circumstances raise are complex and difficult. But they cannot be ignored, and they cannot be resolved by appeals to self-evident principles or natural rights.

# CHAPTER 10

# Eugenics and Human Rights
*Robert Baker*

*Almost all eugenics programs* involve constraints on the reproductive liberty of the parents of potentially defective children [1] ; hence it is incumbent upon proponents of such programs to provide us with some compelling reason to so interfere with the right to liberty. Perhaps the most classic argument of this sort is offered by Justice Oliver Wendell Holmes in the majority decision in *Buck* v. *Bell* (1927:247 U.S.200), which upheld the constitutionality of first wave eugenics statutes.

Mr. Holmes offered two types of argument in favor of sterilization: arguments of expediency and of the rights of the many. The argument from expediency is simply that "it would be better for all the world if . . . society can prevent those who are manifestly unfit from continuing their kind." Now Justice Holmes could consistently offer this argument because he was a realist, a legal pragmatist who did not place stock in human rights. But to anyone who believes in rights, especially the right to liberty, the argument is unacceptable. For what it means to have a right to something is that others are forbidden to interfere with it.* If I have a right to life, others may not take my life; if I have a right to liberty, others may not interfere with my liberty. If I have a right to sing the blues, others may not stop my singing – except, of course, where my right is overridden by another more fundamental right. As Abraham Lincoln noted, my right to swing my fist ends where your nose begins; or to put the point more abstractly, rights to privacy are more fundamental than, and hence override, rights to liberty. So rights are limited by other rights, but nothing else – certainly not by expediency. It may have been, and perhaps still is, expedient for a society to rid itself of its "useless eaters," of Down's syndrome children (mongoloids), of hydrocephalic neonates, of Tay-Sachs's sufferers, and of Werdnig-Hoffmann's infants [5]. But as long as a society is committed to accepting the right to life as fundamental, it cannot follow the precedent set by the Nazis and kill these children. It might be "better for all the world" if these children were killed, but insofar as they have a right to life, even the betterment of the world cannot count as an argument for their extermination. Similarly, insofar as human beings have a right to liberty or a right to pursue happiness, mere considerations of expediency do not provide good grounds for overriding these rights. Only an argument premised on the existence of other rights could hope to be valid.

Whether intentionally or not, Justice Holmes provides two arguments of this kind. Both turn on the point that a eugenics policy is required if we are to protect the rights of the many from being impinged upon by the liberties of the few. One argument is that if a society committed to human rights can (consistently) require innocents to sacrifice their liberty to protect society from hostile populations who menace our rights from *without* its borders, then it can, with equal consistency,

---

*Technically speaking these rights are *negative rights,* rights prohibiting interference; there is another kind of right, *positive rights,* rights requiring aid. Thus to have a negative right to life is to have a right that others not interfere with your living (a right not to be killed), whereas to have a positive right to life is to have a right requiring others to help you to live – by providing you with a kidney machine, for example.

sacrifice the liberty of innocents to reproduce in order to protect society from hostile populations (i.e., the crime-prone progeny of the feebleminded) who will menace our rights from *within* its borders.

Although this argument is valid, it is not necessarily compelling. Much depends upon the extent of the threat that the hostile population poses to rights. Conquest by an alien society poses a profound danger; street crime is seldom comparable. Suppose, for example, that one had fairly definite proof that the relatively rare XYY karyotype had a 50 percent greater chance of becoming criminal than genetically normal males (as people believed just a few years back). Since the karyotype is rare, the crime caused by such karyotypic individuals will be rarer still. Thus sterilizing potential parents of karyotypical individuals will reduce street crime only marginally. When this rather slight benefit is weighed against the extensive violations of maternal rights that a systematic eugenic sterilization policy must entail, the balance must fall against the policy. In general, the genetically defective are not likely to pose a threat comparable to an invading army, and so it is unlikely that the rights of the majority should ever require their eugenic elimination.

Significantly, when the second wave of eugenics legislation was launched (after World War II), proponents tended to adopt a rather different argument of Justice Holmes's, the case for compulsory vaccination. One clear strength of this case is that it is built on a sound factual basis. Just as some people become sick from infectious diseases, others become sick from genetic diseases. Thus, if a society can consistently constrain the liberty and privacy of some possible carriers of infectious diseases to protect the right to health of the majority (by compelling universal vaccination), should it not also, with equal consistency, be able to constrain the liberty and privacy of the actual carriers of genetic disease in order to protect the right others have to health (by compelling eugenic sterilization)? To quote Justice Holmes, is not "the principle that sustains compulsory vaccination . . . strong enough to cover cutting the Fallopian tubes"? In fact, is not the case for cutting the tubes stronger than the case for vaccination? For the person we compel to undergo vaccination may never contract the disease, but the person we require to undergo sterilization will always be a *known* carrier. Actually, the case of a sexually active, fertile, genetic defective seems more like the case of the carrier of an infectious disease, a sort of Typhoid Mary of the gene pool — Tay-Sachs Tilly?

Typhoid Mary was known to have caused 32 cases of typhoid in seven years, three of them fatal. After she was detected she was compelled to live her life in isolation on North Brother Island [2]. If a society that believes in human rights can compel Mary to live in isolation, should it not also be able to isolate her genetic cousin, the Tay-Sachs carrier, by asking her to live in genetic (but not sexual) isolation — that is, should it not require a tubal ligation (or if the carrier is male, a vasectomy) [3, 4]?

The difference between these cases is found in the nature of their victims. Whereas the victims of infectious diseases exist independently of the disease, the victims of genetic diseases are the very progeny of the carrier and so cannot exist unless they inherit the disease — they cannot exist independently of their own genes. The distinction is crucial; for to prevent the transmission of a genetic disease by sterilizing a carrier is to prevent diseased progeny *by eliminating all progeny*. Thus one can only "protect" the health of the 1 in 800 children that might be born with Tay-Sachs disease by violating the right to life of all of the progeny, including the 799 healthy children. One can, of course, "protect" the right to health of any population in a similar way; one can, for example, "protect" potential typhoid victims against typhoid by killing them — but it is doubtful whether any potential typhoid victim would opt for such protection. Why, then, should anyone suppose that a potential sickle cell or Tay-Sachs victim would? Why be protected against a disease at the price of one's existence?

Some might object that, unlike potential typhoid victims, the *potential* victims of genetic diseases have no rights because they do not exist. But this argument cuts two ways. Either the unconceived have rights or they do not have rights. If they do not, their "right" to health can provide no reason to override parental rights to privacy, and hence there will be no justification for a eugenic sterilization program. If the unconceived do have rights, presumably, as is the case for the living, the right to life will be stronger than the right to health. Hence it will be impermissible to prevent their existence as a means of insuring their health, and so there will be no justification for a eugenic sterilization program. So whether or not the unconceived have rights, they provide no justification for a eugenic sterilization program.

REFERENCES

1. Bayles, M.  Harm to the unconceived. *Philosophy and Public Affairs* 5:292, 1976.
2. Ehrenreich, B., and English, D.  *Complaint and Disorders: The Sexual Politics of Sickness.* Old Westbury, N.Y.: Feminist Press, 1974.
3. Haller, M. H.  *Eugenics: Hereditarian Attitudes in American Thought.* New Brunswick, N.J.: Rutgers University Press, 1963.
4. Ludmerer, K.  *Genetics and American Society.* Baltimore: Johns Hopkins University Press, 1972.
5. Roberts, J. A., and Fraser, R.  *An Introduction to Medical Genetics* (6th ed.). London: Oxford University Press, 1973.

# Abortion: The Right to One's Body Versus the Right to Be Born

# CHAPTER 11

# Abortion and Rights to Life
*Virginia Held*

*A discussion of abortion* can help us understand what a right to life *is*. This is true even if we hold, as I do, that human beings have rights to life, but that fertilized ova do not have such rights.

A right to life includes more than merely being left alone; it also includes being able to acquire what one needs to live. Anyone who recognizes the right to life of a human child acknowledges this. One cannot simply leave a baby alone, unattended, to fend for himself or herself. One must see to it that babies have the necessary food, shelter, and so on. Access to what they need to live must also be included in the rights to life of adults, although this requirement is often overlooked, especially by so-called libertarians, who suppose that we respect persons' rights to life by merely not attacking them.

Abortion should be seen, not as an *attack* on the fetus, but as a refusal to provide it with what it needs to live. This refusal ought not to be made toward those who *have* rights to life. But who, then, has rights to life, and what do such rights require? Unfortunately, we do not now recognize the rights to life of vast numbers of human beings, both children and adults. They die of starvation, lack of medical care, and lack of basic necessities because those who could assure them what they need fail to recognize the rights to life of these people. And rights to life are not unlimited: Even where the basic minimum requirements of food, shelter, and medical care are provided through governmental arrangements that tax those lucky enough to have more than they need to provide for those who would otherwise die, rights to life do not give persons valid claims to unlimited amounts and kinds of resources. Thus no matter how much a person may need a new lung or kidney, he may have no right to one. And there are certainly limits to what society should do in the way of mobilizing resources to deal with the right to life of someone with a rare disease, just as there are limits to what society should do to assure a citizen's right not to be attacked.

So rights to life sometimes are not recognized at all, and they are always recognized to have limits. Assuring rights to life often requires the efforts as well as the forbearance of others. But what about abortion?

It seems obvious that any discussion of abortion that fails to include a recognition that being pregnant and giving birth is an enormously exhausting and painful kind of labor does not even need to be given further attention. And yet we still hear discussion after discussion that fails to pay the slightest attention to the role of the woman, beyond referring to her "convenience." In all the vast literature on the subject of abortion, the most consistently overlooked consideration is: What does a woman do and feel when she makes a baby? Babies do *not* make themselves; women make babies. Yet this fact is ignored over and over, as the fertilized ovum is shown turning, as if by itself, into an embryo and then a fetus and then a baby and finally a grown human being.

The first thing we have to do in discussing abortion, then, is to bring clearly into the picture the woman who is making the embryo grow and, by expending a great deal of energy and pain, giving birth and producing a baby. Then the question is: What can justify requiring that she do so, against her will, either legally or morally? Of course the legal and moral questions are separate and different, but some arguments apply to both. And coercion can be of various kinds [5] : To deny Medicaid funds to a woman who cannot afford an abortion may force her to have a child as effectively as would a legal prohibition against abortion.

So far I have spoken only of the work of *producing* a baby rather than that of caring for it. The latter is of course enormously much more exhausting, and it extends over many many years. I strongly believe that the obligations of mothers and fathers are equal in bringing up children [3] and that society ought to provide adequate day care and support, medical care, and education for all children. I also think that, in normal circumstances, no one should be forced to become a parent against her or his will. On this view, it might be morally wrong for a woman to force a man to become a father against his will by refusing to have an abortion if he is strongly opposed to becoming a father. This action would be considerably less wrong than forcing a woman to become a mother, because the man's contribution in making the baby is so trivial in comparison to hers, and his likely contribution to bringing up the child, if she is forced to have one, is slight in comparison to hers. But if he recognizes his obligations as a father, he should not be forced into incurring these obligations against his will.

For the sake of this brief discussion, however, let us limit the issue to *producing* the baby, assuming that the woman can without extraordinary pain give up the child for adoption, or that if she chooses to bring up the child with or without the help of the child's father or the society, she does so voluntarily. So the question remains: Are there ever reasons strong enough to require a woman to suffer the pain and exhaustion of making a baby and giving birth against her will?

Many of those who strongly oppose abortion would not consent to the view that people should be forced to use their bodies for the sake of those in need of them, as in compulsory blood donation or skin grafts. They would not agree that persons should be compelled to do forced labor for others in need, such as for the poor. Except with respect to a woman's ownership of her own body, they have highly developed notions of the sanctity of private property. Although taxation is much less of a curb on liberty than is being forced to do a specific kind of labor, many opponents of abortion even oppose the level of taxation that would be necessary to allow those who are already human beings and fellow citizens to continue to live. The inconsistencies in such positions are startling.

I have less attachment to possessive individualism and a less constricted conception of property than many opponents of abortion. I do not at all share the view of a philosopher like Robert Nozick [6], who believes that whatever money I earn is mine, all mine, and if the government taxes away some of it, I am being forced to surrender my property in a way that is "on a par with forced labor." I think property rights are *rights,* but . . . I do not have a right to hang on to whatever I hold, if others who need and deserve it more have rights to have some of it shared with them [4].

Our rights to own our own bodies are clearly not absolute. Our bodies are not possessions with which we can do whatever we like. Of course we do not have a right to use our bodies to attack and coerce other people as we please. And if a small child is hungry and needs to be fed, and you are her parent, the child has a legitimate claim to have you use your body to feed her. If a small child wants to climb on you, and you are his parent, I doubt that you would think, "get off my property." If the argument that a woman has a right to own her own body has been successful in persuading some judges and others that women should be legally allowed to decide whether to bear children or not, I cannot greatly regret that the argument has been used. But I do regret that the argument about the sanctity of private property is the one that touches so many, since it seems to me one of the weakest of the abortion debate.

## FORCED LABOR

I could easily agree to the taxation of those with more than they need to provide the means to live of those who lack such means; this action may often be required to respect rights to life. And I could even agree to compulsory or forced labor, that is, labor of a specific kind, for certain periods if this really were needed for the same purpose. Hence, if the pain and exhaustion of giving birth were necessary to respect the rights to life of those who *have* such rights, it would, I think, be justifiable to require women to undergo them, as long as others were required to perform comparable types of specific forced labor at specific times and pregnant women were not singled out for special burdens.

However, when we see the problem of forced labor in terms of respecting the rights to life of *all* those who have such rights, the question becomes at least partly one of priorities. We have to ask: Whose rights to life are being most seriously denied and who most deserves access to the means to live? Ideally, we should not weigh assuring the rights to life of some against assuring the rights to life of others, but to some extent such weighing is unavoidable. Research that will help assure the rights to life of those with rare diseases may be given a lower priority than research that will help assure the rights to life of those with common diseases, although if the probability of finding a cure is much higher in the former case than in the latter, the priorities may be reversed.

If we put the problem in these terms, it seems obvious that a person in desperate need, conscious that the cause of his impending death is the callous unwillingness of his fellow human beings to share their surplus with him, is among the *most* deserving of our compulsory labor. To his pain of dying is added the psychological pain of knowing that the cause is quite avoidable and is not being avoided because the fortunate are cruelly indifferent. No fetus can suffer this pain. Hence, if forced labor is to be demanded of us, it should be demanded first for those with the strongest claim to it.

We are not justified in considering the persons within a given nation-state as a closed group whose rights to life we ought to respect but beyond which we need have no concern. However, the ability of persons within a given society to effect respect for the rights to life of persons in other societies is somewhat limited. Although this limitation is often a lame excuse for unjustified inaction, let us assume for the sake of the discussion from here on that we can consider a situation in which the rights to life, including the rights to a basic minimum of what they need to live, *are* assured to *all* persons to whom we *can* assure such rights or have any obligations toward, such as all those within a given society. Then, we can ask, should the concern of the society for respecting rights to life extend to potential persons such as fetuses, embryos, zygotes, or even fertilized ova? Potential persons will surely not suffer from a failure to provide what they need to live in ways comparable to the suffering of conscious human beings, but they might be proper objects of our concern, as are, most surely, infants.

## PERSONHOOD

The view that the early products of conception can be considered persons with any sort of entitlement to life cannot be supported by any arguments that do not dissolve into myths when examined impartially. Religious myths have long influenced the debate by *conferring* personhood on such entities "from the moment of conception." Although believers should be free to consider the early products of conception "persons" on religious grounds if they wish, there are the strongest possible moral arguments against allowing such religious views to be imposed on those who do not believe them. More recently, for some people, genetic codes have taken the place of souls as the essential element of personhood. It has been argued that since the fertilized ovum contains the genetic code for a human being, the entity containing it has a right to life. But any number of cells contain genetic codes from which human beings could, with appropriate technology, be formed, as in cloning. The special status of the fertilized ovum thus disappears.

If we think personhood depends on aspects of human entities other than their possession of a genetic code, as any plausible view demands [1, 7], we must admit that the early products of conception resemble sperm and ova more nearly in all relevant respects than they resemble babies. Then it seems arbitrary to claim that our moral concern, and our legal protection, should extend to zygotes but not to the billions and billions of spermatozoa wasted every day through masturbation.

A more serious argument than some is offered by R. M. Hare, who argues that because most of us who think about it can say, "I am glad that I was not aborted when I was an embryo," we must be able to generalize this into a prescription that others should not be aborted when they are embryos, except for certain overriding reasons [2]. But this presumes what it is intended to prove: That "I" was once an embryo, whereas others might hold that a person does not become "himself" or "herself" until a later stage. And if I was once an embryo, was I not an ovum, and should I then argue that because I am glad that the ovum that was me was fertilized, every ovum should be fertilized? Hare recognizes that his argument implies we have a duty to procreate; he also acknowledges that we ought to limit the population. He goes on to offer no reason for favoring an existing, unwanted, fertilized ovum to a future, wanted, fertilized ovum, *if* the duty to procreate in right amounts is fulfilled either way. And so there seems to be nothing left of his presumption against abortion.

I conclude, then, that if women *choose* to make fertilized ova into human children, that is their right, within the bounds of an acceptable population level. But for women to be *forced* to do so is unjustified forced labor, and a morality that *requires* women to do so is mistaken. Even if technology should advance to a stage where it would be possible for society to take responsibility for the development outside the womb of any unwanted embryo, no matter how young, and thus avoid coercing women into giving birth, it would, I think, be a violation of the rights of the potential mother and father to insist on doing so for an embryo in early stages. It would deny them the choice to become or not become parents, and it would do so for the sake of an entity not yet capable of having rights. It might at some point be a justifiable use of human resources to provide facilities for doing so, for those potential parents and adoptive parents who wished it, but for the foreseeable future, concern for the assurance of other rights should probably have priority.

However, the nearer a fetus approaches the normal stage of birth, the more nearly it is entitled to consideration comparable to that given a human baby. The Supreme Court has recognized this argument in allowing the prohibition of abortion in the third trimester, except when necessary to preserve the life or health of the mother. It is also possible to imagine circumstances in which a society could justifiably require women to bear children: for instance, if birthrates fell to such a low level that the society was threatened with extinction. However, forbidding abortion would still be one of the worst ways to bring about an increased population. There are many ways to encourage women to have more children, such as making the burdens of child care and child support less severe for parents, particularly mothers. These arguments are especially relevant for certain groups who oppose abortion because they wish the numbers of members of their group to grow, or because they want enough babies of a certain kind to be available for adoption.

It is sometimes argued that if we allow abortion, we contribute to a disregard for life, and this disregard will spread to the weak and helpless of any age. It may be argued, on the contrary, that the all too common fixation on the fetus breeds excessive callousness toward the suffering of actual human beings. Making abortion illegal does not significantly reduce its occurrence; instead, it causes large numbers of women to die of complications from illegal abortions. In the Scandinavian countries where abortion has been permitted for some time, the care and concern for children and adults in need is among the best; in Nazi Germany, on the other hand, there were extremely strict laws against and harsh penalties for abortion.

We should, most certainly, have more concern for rights to life than we now have. But we should try to assure, as well, that such concern is directed toward those who need and deserve such concern and is not deflected into a campaign to impose forced labor on women for the sake of entities that have no such rights.

REFERENCES
1. English, J. Abortion and the concept of a person. *Can. J. Phil.* 2:233, 1975.
2. Hare, R. M. Abortion and the Golden Rule. *Philosophy and Public Affairs* 4:201, 1975.
3. Held, V. On the Equal Obligations of Mothers and Fathers. In O. O'Neill and W. Ruddick (eds.), *Having Children: Philosophical and Legal Reflections on Parenthood*. New York: Oxford University Press, 1978.
4. Held, V. John Locke on Robert Nozick. *Social Research* 43:169, 1976.

5. Held, V.   Coercion and Coercive Offers. In J. R. Pennock and J. Chapman (eds.), *Coercion.* Chicago: Aldine-Atherton, 1972.
6. Nozick, R.   *Anarchy, State and Utopia.* New York: Basic Books, 1974. P. 169.
7. Warren, M. A.   On the moral and legal status of abortion. *The Monist* 57:43, 1973.

# CHAPTER 12

# Toward Unraveling the Abortion Problem
*Bernard H. Baumrin*

*Most discussions of* the abortion issue confuse or conflate two entirely separate kinds of consideration — first, the morality of abortions and, second, the legality of abortions.* It seems to me that once these two matters are properly separated, most of the theoretical issues fade away, and we are thrust back to where we were for decades on the fundamental questions.

## ABORTION AND LEGALITY

A. There are many who believe laws should prohibit only acts that are immoral; these thinkers hold that if some act is to be legally proscribed, it must first be determined that the act is immoral. On this view the immorality of an act is a necessary condition for its illegality.

B. There are many who believe that all immoral acts should be legally proscribed. For these thinkers an act's immorality is a sufficient condition for its illegality. Most natural law theorists hold a view of this sort.

C. There are many who believe that acts may be legally proscribed whether they are moral or immoral and that the immorality of an act is neither a necessary nor a sufficient condition for its being legally proscribed. Most (perhaps all) legal positivists hold a view of this sort. On views of this sort, of course, the fact that an act is moral does not exempt it from being legally proscribed either, whereas on either views of sort A or sort B above the fact that an act is moral normally prevents it from being legally proscribed.

On views of sort A the fact that some act is immoral is not enough to legally forbid it — something more is wanted. In this regard views of sorts A and C share something, namely, for these views, there is always some additional consideration that prompts declaring some acts illegal. Usually, though not always, the additional consideration most invoked is state policy couched in terms of the furtherance of the common good as perceived from time to time by monarchs, legislatures, and courts. Clearly for those who believe that illegality is a consequence of policy considerations, when policy considerations change, very likely what is deemed illegal will change also.

Thus if state policy considerations seem to require or involve an increasing population for military, agricultural, or other reasons, infanticide, abandonment, abortion, or even celibacy may be prohibited. Conversely, if a decreasing population is desirable, some or all of the above may be encouraged at least by neglecting to prohibit them, by lifting legal sanctions against them, or even by encouraging new and possibly more inventive population checks. From the point of view of legal positivism, the moral significance of these acts is legally irrelevant, except that on views of sort A legally proscribing childbearing is impossible since the practice of child-

*One signal exception is B. A. Brody's paper, Abortion and the Law, *The Journal of Philosophy* 68 (No. 12):357, 1971.

bearing is not immoral. From the positivist's point of view (C), this fact is quite as irrelevant as any other moral fact, and positivistic governments need have no qualms about the moral limits of their population policy.

All of the foregoing is preliminary to pointing out that the illegality of abortion, or indeed its legality, need not be connected in any way to its morality. Nothing really follows legally about abortion from a determination that it is either moral or immoral, unless of course in a given state we know that the government is unalterably wedded to a given sort of legal philosophy, and then the consequences will vary as the legal philosophy varies. But, as a rule, especially nowadays, the legal philosophies of established governments are pretty wishy-washy and certainly not unalterable even when reasonably clear.

## ABORTION AND MORALITY

So what has the debate been all about? Is it that despite the fact that attitudes toward population expansion have changed, some governments still adhere to the old population growth policy of the nineteenth century and must have their legal policies changed by public pressure? Or is it that some groups, notably Catholic and Jewish, have resisted attempts to legalize abortion because of their adherence to a natural law jurisprudence? While I think these are general factors to be considered in understanding the remarkable jurisprudential shift that has occurred in Western legal systems, I do not think they are the most important facts. The most important fact in my view is a psychological one. Much of the debate on the personhood of the fetus has to do with the fact that nearly everyone, whether proabortion or antiabortion, recognizes that fetal life is (1) life, (2) a necessary precursor to what everyone agrees is a morally relevant state of being, and (3) once ended some possible person will not exist. Put another way, nearly everyone recognizes that killing a fetus at any stage is killing, ending or preventing someone just like us from being, and that only a moral idiot would suppose that the fetus is better off for not having been born.

Although the fetal host, the relatives, the father, and the state may be better off for the shortened growth cycle of the fetus, the fetus itself is not. Hence no moral analysis can be given for making an act of abortion an act in the interest of the aborted. Indeed every analysis of the matter comes to the same thing; that is, from the point of view of the fetus, abortion is immoral, even though it might benefit others. Thus it appears that what proabortion theorists wish to show is that irrespective of jurisprudential stances, abortion should not be illegal because it is not immoral. And, of course, could it be shown not to be immoral the debate would end. It was never enough to prove that abortion should not be illegal, for we all know that many moral acts are legally proscribed and many immoral acts are not illegal.

What is the social difference before and after *Roe* v. *Wade* in the United States? Before this decision, abortionists were committing criminal acts, except in severely circumscribed situations, and persons seeking abortions were doing the same. The sanctions of the state made them aware of the seriousness of their joint venture, and the advantages to both had to outweigh the disadvantages of an unwanted birth as well as the possible threat of heavy penalties. After *Roe* v. *Wade* it would seem that nearly everything had changed, but not quite. There are no legal sanctions to make the parties conscious of the seriousness of their endeavor; only their own moral sensibilities remain.

Yet that is quite enough to bring continually to the fore the moral problems that the possibility of abortion creates; namely, is my convenience, or career, or time, or future prospects, or health, or psychological balance, or whatever, more valuable to me than either this child is valuable to me or this child is valuable to itself? The point here is put rhetorically because the proabortion theorist must present to us some theory of value along with a valuational scale that does not beg the question; that is, a measure that does not place a zero value on the interests of the unborn. If a zero value is placed on their interests, then an argument would have to be presented to show that under no circumstances should the interests of the unborn be calculated into individual and social decision making. On the other hand, if a nonzero value is placed on fetal interests, clearly two consequences follow. First, the unborn have rights, the protection of their interests, and others have duties toward them. Second, from the point of view of the fetus, using the same valuational scale as those who have the power to decide its fate, it is unlikely that the interests of others will be weighed so heavily that the fetus' obligation would be to commit suicide, if it could, in order to facilitate the interests of some other person, even its host.

Divide this question into doctor and patient, forgetting relatives, fathers, the state, and perhaps the childless. Every doctor who performs an abortion has acted as if he has decided that the furtherance of his interests are more valuable to him than the child is valuable to him or than the child is valuable to itself, and while such a decision is perfectly understandable, it is not rendered moral by being easily understood. Every woman who has had an abortion has acted as if she has made the same choice, and again while easily understood, it is not made moral because it is intelligible. I do not intend to admonish doctors about their morals, but it seems clear to me that not every woman has been made fully appreciative of the hard moral choice she is making — a choice that may take its toll in the course of her life at least as heavily as bearing and even rearing a child might.

## ACTING IMMORALLY

I do not mean to suggest, as is often done, that because some act is immoral it must not be done, or even because some act is immoral it should not be done. I do not think either of those statements is a moral tautology or necessary truth, though the fact that some act *is* immoral is good evidence that it should not be done.

What I wish to suggest is that one not engage in immoral acts blindly, or aid others to so act blindly. We all do immoral acts, perhaps we do them far more frequently than we imagine, but one step in avoiding immorality and one step in appreciating the precariousness of moral life is to recognize the seriousness of serious acts. They are more easily avoided in our future behavior by recognizing their occurrence in our past and present behavior, and becoming conscious of our mistakes and our moral self-compromises may tend to strengthen our resolve and our ability to be better human beings in the future.

It seems to me that the greatest danger created by the legalization of abortion is the support it has given to the view that human life can be weighed on some utilitarian scale or even a hedonic scale. Once we countenance killing lives-to-be, while thinking that we are either doing something right or at least not wrong, we are easily able to raise questions about killing actual persons by using the same set of criteria that permitted us to terminate pregnancies. The criteria of desire, convenience, pleasure, advantage, social improvement, and the like, once substituted for the absolute criterion of the sanctity of life, know no bounds. Once any of these criteria morally overbalance in some calculation life itself every immorality is possible.

Thus really all I have said is that the legality of abortion and its morality are two different matters. Except perhaps in rare instances, abortion is always immoral, whether or not it is legal. Everyone from the doctor and patient to the boyfriend, husband, relative, or state functionary who is pleased with, happy with, unsaddened by, or unashamed at the abortion of a human fetus is immoral. The moral complicity comes from approving such immoral acts as well as participating in them, for with that approval the foundations of a more callous society are built. The voyage from passionate killing to rational murder is an intellectual one — a voyage of the minds, not the behavior of human beings.

# CHAPTER 13

# No Right at All: An Interpretation of the Abortion Issue
*Lisa H. Newton*

*The abortion issue* is commonly presented as a conflict of rights: the fetus' right to life *versus* the woman's right to decide what shall happen in and to her own body (usually subsumed under the right of privacy) [9]. I believe such presentation is misleading. I think it can be shown that the abortion issue will not yield to any analysis in terms of rights, prima facie or otherwise; that the only moral resolution to any conflict on abortion will involve a careful, painful balancing of moral claims, with compassion the decisive value and no appeal to right at all.

## IRRECONCILABLE RIGHTS AND SOPHISTRY
Of the multitude of possible distinctions among the various sorts of rights, one very rough distinction will suffice for present purposes: the distinction between prima facie rights and absolute rights. Prima facie rights are rights you can exercise until that exercise runs into someone else's right, at which point you must stop. That point, or boundary, may depend for its definition on a long history of adjudication. (Your right of free speech ends where my right of privacy begins; but where is that? May I exclude you and your unwanted ideas from my restaurant as I may from my living room? What about my factory? May my neighbors and I exclude you from our theaters and our sidewalks? Only the Supreme Court can say for sure, which it has on a case by case basis.) It is entirely possible for two parties to have conflicting rights prima facie, until adjudication in the particular case fixes the boundary more precisely. It is *not* possible, within an orderly system of rights, for two parties to have conflicting rights absolutely. Absolute rights are the specific rights that particular parties may exercise to their fullest extent, rights fixed by adjudication or absolutely guaranteed. Among the absolute rights are rights that are by nature exclusive; if you and I are both awarded sole right of possession of a certain piece of property, there must be war between us unless one decision is reversed. Also among absolute rights are rights that are meaningless if only prima facie, and the example usually given of such a right is the right to life. Unlike the right of free speech, the right to life cannot be limited, suspended, or abrogated in a particular situation while remaining in force for others. Once dead, all your rights are permanently terminated; there is no appealing for a more favorable decision. Prima facie rights in conflict are regularly reconciled by adjudication; absolute rights in conflict are irreconcilable.

The abortion issue appears to be a case of an irreconcilable conflict of rights. If the fetus is to be granted the right to life, the right must, as above, be absolute. A *right* to life is not the sort of thing you can have contingent on someone else's decision to tolerate your existence; in such contingency, your life would be only an uncertain gift, or momentary privilege. (Nor can that right come into being only after a certain time has elapsed; your solemn assurances of my right to life effective

Thursday will do me little good if you plan to have open season on me all day Wednesday. That, of course, is what is wrong, on this account of rights, with the Supreme Court's abortion decision, *Roe* v. *Wade* [8].) And if the woman is to have, under any description, the right to decide what shall happen with her own body, surely that right, possibly not universally absolute, must be absolute in the case of pregnancy. Pregnancy and childbirth are not brief, or painless, or completely safe, nor do they entail only a minimum of medical humiliation and interference with other concerns, nor are they without effects on the lasting health and appearance of the body. Any right to control over the events of one's own body must include control over this enormous event, or be worth nothing at all. But then any case of pregnancy gives rise to two absolute rights in conflict: the right of the fetus to live (i.e., to live in the womb, since it can live only there) and the right of the woman to decide whether or not the fetus shall be allowed to live (and grow) in that womb.

The presence of "irreconcilable rights," which are no rights at all, inevitably gives rise to sophistry. In order to defend the right that is claimed, the conflicting right must be explained away, and any twist of logic or fact that negates it will be employed. Hence the antiabortion tradition of the last century explained that although in general persons have the right to control the events of their own bodies, the woman who now wants an abortion has none, for she has forfeited that right by knowingly engaging in (sinful) sexual intercourse; if intercourse resulted in pregnancy, she had only her own lewdness to blame and should suffer the burdens of pregnancy and childbirth as a proper lesson in continence. Echoes of this argument still appear in the "prolife" literature. But the sophistry of the proabortion side is at least as unsavory. Granted that persons have a right to life, they argue, the fetus does not have that right because it fails to satisfy the "criteria for personhood," criteria recently discovered by defenders of the practice of abortion. These criteria are many and various, differing from writer to writer, but tend to be grouped around consciousness, self-consciousness, the desire for life, and the ability to communicate sensibly with others [10]. As with the antiabortion sophists, the proabortion writers gladly toss in exceptions to their broad categories; no one really wants to claim that my "personhood" disappears every time I fall asleep. And they admit to genuine drawbacks in the "criteria": For example, no way seems available to distinguish the fetus (who, by this argument, may be disposed of at will) from a large number of the mentally ill and severely retarded; and no one really wants to license mass slaughter in the back wards of the state hospitals. But the "criteria for personhood" also persist in the literature, to justify a type of killing that without them would be murder.

But both arguments are thoroughly misguided. Neither in logic nor in fact, nor in any sane morality, is the pregnant woman in some state of sin redeemable only by a state-imposed sentence of nine months suffering and labor. And on the other side, the fetus' intrauterine development displays a most inconvenient continuity from diploid cell to 10-pound thumb-sucking baby, undergoing staggering changes in size and appearance (i.e., recognizability as a human), but not one *essential* change throughout the whole beautiful process. The whole individual is there in the cell; all we do is feed it and, after nine months or so when it starts getting cramped, move it into the outside air where it can squirm around more freely. If those of us who now, grown up, walk about in the outside air are in fact persons, the fetus is in fact a person [7]. Occasionally we long, I think, to believe in a pagan metaphysic like that propounded by Anaximenes, who thought the air was holy; then we could believe that the soul entered the body with the first breath of air. Then we could believe that fetuses had no souls and could be killed at will. But we hold to no such metaphysic and ethic; stuck with our facts, we are no more enlightened on the proper course of action in an unwanted pregnancy.

On the face of it, then — given the clear pleas of right on each side and the futility of the sophistry that attempts to explain away one set or the other — the abortion issue cannot be usefully presented as a case of "conflict of (prima facie) rights," where a "just solution" must be found somewhere between the two extremes; those claims of right are absolute and compelling, hence mutually destructive.

## LEGAL RIGHTS AND CONSTITUTIONAL PERSONS

It could be argued that the simplified account of "rights" given earlier is to blame for the logical impasse that has been reached; useful as such crude distinctions may be for some topics, more sophisticated accounts are available. One of the most promising moves is toward "legal positivism," the school of thought that treats of rights and assumes that rights are real only within the framework of a system of legal rules, promulgated by a sovereign and enforced by sanctions [1]. The vague and diffuse rights to life, to privacy — existing somehow independently of law, by nature perhaps — are treated by legal positivism not as rights but as *claims* to rights, felt needs of the people, desired and perhaps desirable, but lacking in legal force. A good positivist analysis of rights is presented by Bertram Bandman in his "Rights and Claims" [2]. Contra Feinberg [4], Bandman argues that claims are prior to rights. Claims are conative, expressions of desire, including desire for certain types of power. Claims, not rights, are often in conflict, often in potentially violent conflict. Arising naturally in any situation where contrary interests coexist (i.e., any situation in which two or more human beings are involved!), conflicting claims can destroy a society unless some mechanism is developed to keep *them* under control; this mechanism is the legal system. Thus there are no "rights" in the abortion issue at all until the conflicting claims of fetal life and maternal privacy have been resolved by the courts, and then there shall be only those rights that the courts assign. There is much to be said in favor of such an account. It solves immediately the problem of irreconcilable rights: The Supreme Court is perfectly free to grant a right of maternal privacy that extends only through the first one or two trimesters of pregnancy and a right of fetal life that springs suddenly into being after six months of open season, for it is bound by no concept of metaphysically prior, or preexisting, right in either case; what rights there ever will be are only the rights that the Court shall decide. If courts disagree on what rights shall be granted, a clear legal hierarchy of courts makes it possible to know immediately what grant is effective. Daily life and important decisions for living at least are taken out of the realm of philosophical debate and put on certain ground. I would be inclined to argue (as Bandman would

not) that legal rights are the only real rights, that "moral rights" are at best meta-
phorical extensions of the notion. When it comes to the point of practice, after all,
a "moral right" is like a "moral victory"; *you* feel good about it and take it seriously, and you know that all fine and decent people will see it as you do and respect
and honor it — but when you look at the scoreboard, it simply is not there. "Right"
is properly a legal term; "moral right" may be analyzed best as a talking point, a
manner of arguing for something desired — in short, a claim.

On a positivist analysis, then, it appears that we may be able to handle the abortion issue. All it requires is that a *legally* coherent, and practically workable, specifi-
cation of permissible action be handed down, and *Roe* v. *Wade* was designed to be
just that. The rule handed down by the Court in that decision is certainly clear
and workable: In the first trimester of pregnancy, the maternal claim to privacy is
granted the status of absolute right, and the fetal claim of life is rejected; in the
second trimester, the maternal claim gains the status of qualified right, limited by
the state's right to regulate the medical aspect of the abortion, while the fetal claim
is again set aside, in the third trimester, the maternal claim and the fetal claim are
both denied universal protection, and it is left to the decision of each state legislature whether the maternal or fetal claim shall prevail. No practical difficulties arise
in ordinary application of this rule (although some impediments might appear if the
state granted the maternal claim in the third trimester, resulting in significant numbers of live births). But when its legal coherence is investigated, the first appearance
turns out to be deceptive; in no way can *Roe* v. *Wade,* or any decision the Supreme
Court might come up with on the subject of abortion, be reconciled with the constitutional requirements that apply to the situation [6].

The crux of the problem is that the Court is bound to use the categories present
in the United States Constitution; it is not free to make up its own as it goes along.
Of the categories of recipients of constitutional protection (speech, religion, etc.),
the one that concerns us is the category of "person." The Constitution (in the
Fourteenth Amendment) forbids any state to deny to any persons in their boundaries "the equal protection of the laws" — no group may be singled out for worse
legal treatment than the others. It does not have this provision for any other class
of beings. If the Supreme Court determines that a fetus is *not* a person, it has no
right to extend it any protection at all, nor may it allow the states to extend it any,
where the fetal claim to protection conflicts with any already established right of
persons. The maternal claim to privacy is, as above, assigned a zero value in any
state law requiring a woman to go through pregnancy and childbirth. The woman is
certainly a person, and the provision that every citizen shall be given the equal protection of the laws guarantees her equal right to privacy. If the fetus is not a person,
no state can require her to carry such a thing around in her body if she does not
want it there. On the other hand, if the fetus *is* a person, the Court is bound to require each state to extend to it the same protections enjoyed by other persons in
the state; if the state makes the premeditated taking of human life (in the absence
of certain standard justifications) a felony, murder in the first degree, then the premeditated killing of a fetus must be so classified and punished accordingly. Perhaps
the standard plea of self-defense could be admitted to defeat the charge of first-

degree murder, when the pregnancy was a clear danger to the mother's life; but very few other exceptions would be allowed. Ordinarily, after all, I am not permitted to kill someone simply because three psychiatrists testify that I am not psychologically prepared to tolerate that person's existence. Nor should sophistry be permitted, of the sort that explains that abortion is not really *killing* the fetus, it is just evicting him from his mother's body, where he had no right to be if she did not want him, and it just happens that he cannot breathe in the air so he dies [9]. Ordinarily, I may not drown a stranger in my swimming pool merely because he is trespassing; my exercise of my rights in property has always been limited, in law, by my duty to protect human life. Only these two alternatives are open to the Court: to forbid the states to interfere at all with the maternal right of privacy, first trimester or last, if the fetus is *not* a person, or to insist that the state protect the fetus as it protects all others and treat abortions as murders in the first degree, if the fetus *is* a person. The trouble is, of course, not only that it would be politically impossible to enforce either decision, but also that the Court is nowhere empowered to decide what a "person" is. The Constitution wisely avoids that speculation on the nature of "personhood" that entangles the abortion issue in the philosophical literature; it leaves it open for the Court to recognize nonhuman artificial persons for certain legal purposes, but merely assumes without discussion that all human beings will fall into that category. Did the Founding Fathers contemplate the case of the unborn? The Justices of the Supreme Court have no way of knowing. It seems that the abortion issue simply cannot be handled in our law. But then it cannot be analyzed in terms of legal rights any more than it can in terms of absolute or moral rights. Some interpretation will have to be found that does not appeal to "right" at all.

## THE QUALITY OF MERCY

Sebastian DeGrazia concludes his study of the state as we will have to conclude this study of the abortion issue: "The Theologian is right. Why not admit it? More than anything else, the world needs Love" [3, p. 87]. No appeal to rights is going to solve the abortion issue; we are forced to meet the problem where it began, as an agonizing conflict of moral claims, to be resolved only case by case, in such a way as to maximize all human values for all the parties concerned. Individualized decisions can, and must, be guided by love (or "caring," the more fashionable word for the same quality [5]), by a deep personal concern for the human beings involved. Appeals to "justice," as if there were to be an adjudication between mother and infant, are worse than futile. At the end of a judicial process, the victorious party has every right to feel justified, blameless, "in the right." But at the end of an abortion, or at the birth of an unwanted baby reluctantly carried to term, no one has the victory. There has been a serious loss in either case, a diminution of humanity. The quality of the loss will vary from situation to situation. The loss of the fetus where the decision is to proceed with an abortion may be regretted very little in the case of the raped teenager, very much in the case of the older woman who has discovered that the fetus she is carrying is deformed; the suffering of an unwilling mother will also assume different weight in each case where abortion is ruled out. The essential point is that it is the attitude of the deciders toward these losses that really matters. There is no way of getting a right *result* in any situation where an abortion is desired and a decision must be made whether or not to proceed with it; there is only a right *manner* of approaching the decision, of weighing the factors, and of reaching a solution. That manner is one of compassion, of concern for the individuals, that recognizes that any decision must be made with at least some regret, that proceeds with sorrow for the value lost in either conclusion, and that acts to heal the inevitable injuries as quickly as possible.

## REFERENCES

1. Austin, J. *The Province of Jurisprudence Determined and the Uses of the Study of Jurisprudence* (introduction by H. L. A. Hart). London: Weidenfeld & Nicolson, 1954.
2. Bandman, B. Rights and claims. *Journal of Value Inquiry* 7:204, 1973.
3. De Grazia, S. *The Political Community: A Study of Anomie.* Chicago: University of Chicago Press, 1948.
4. Feinberg, J. The nature and value of rights. *Journal of Value Inquiry* 4:243, 1970.
5. Fletcher, J. *Situation Ethics.* Philadelphia: Westminster Press, 1966.
6. Newton, L. Abortion in the law: An essay on absurdity. *Ethics* 87:244, 1977.
7. Newton, L. Humans and persons. *Ethics* 85:332, 1975.
8. *Roe* v. *Wade.* 410 U.S. 113, 93 S. Ct. 705 (1973).
9. Thomson, J. J. A defense of abortion. *Philosophy and Public Affairs* 1:47, 1971.
10. Tooley, M. Abortion and infanticide. *Philosophy and Public Affairs* 2:37, 1972.

# Euthanasia and the Right to Terminate or Prolong Life: Reasons for and Against

# CHAPTER 14

# Karen Quinlan: Human Rights and Wrongful Killing
*Marvin Kohl*

## I

Few cases have received more publicity or stirred the interest of the general public more than that of Karen Ann Quinlan. The case is unusually complex, yet several things seem clear: First, Joseph Quinlan asked that the respirator be removed and that his daughter be allowed to die; second, Karen was quoted as saying, on three different occasions, that she never wanted to be kept alive by extraordinary means; third, none of her doctors claimed that Karen was in any technical sense dead; fourth, because there was evidence that she was comatose with irreversible damage to at least one part of her brain, there was neither any immediacy in terms of relieving pain nor a way of having her directly express her own preferences; and, finally, Judge Muir (in the lower court decision) denied the plaintiff's request, indicating that he did not find grounds for the so-called right to die in the Constitution. He also maintained that "such authorization would be homicide," and that it would be a violation of the right to life, presumably Karen's constitutional right [8, pp. 33, 41].

This decision was reversed by the Supreme Court of New Jersey. The Court essentially concluded that there is a right of privacy that might permit termination of treatment and that, even if the acceleration of Karen's death were to be regarded as homicide, it would not be unlawful or criminal homicide.

I wish to consider, not the legal question, but what is perhaps the most difficult of the moral issues. Suppose the act of removing Karen Quinlan from the respirator was an actual homicide and, at that, the killing of a person. Would it be a morally wrongful act in the sense of being an injury to Miss Quinlan?

The question is a fundamental one. For if it can be shown that the act of removing Karen from the respirator (hereafter referred to as "the act in question" or simply as "the act") is a wrongful one, then it is not clear just how anyone in essentially similar circumstances has a right to be so treated. An example should serve to illustrate this point. If it is morally wrong to shout "fire" in a crowded theater that is not on fire, then we cannot argue with any force that one has an actual right to shout fire. On the other hand, if it can be shown that the Quinlan act is not a wrongful one, then — whatever the Constitution may not directly say about the right to die — at least we have a way of bringing the latter question to a precise issue. One way of doing so is to appeal to the right of privacy and argue that both common law and the Constitution secure "to each individual the right of determining, ordinarily, to what extent his thoughts, sentiments, and emotions shall be communicated to others" and also secure "the more general right of the individual to be let alone" [2, pp. 127, 134].

Another perhaps more interesting way is to appeal to the Ninth Amendment. This amendment tells us that the enumeration of certain rights in the Constitution "shall not be construed to deny or disparage others retained by the people." So that, even if the right to die is not listed as one of the rights in the Constitution, if it can be established in certain carefully specified situations that people have and retain the right to die how, when, and where they choose to, then we have established both the moral and constitutional grounds for that right. Nor, I suggest, would it be necessary to show that the people possess the full right to die. All that would be necessary would be to establish that each individual has a certain negative right, the right not to be forced to live a completely meaningless life, if that state of affairs is irreparable and irreversible.

I shall not argue for the existence of these rights. Although I shall have something to say about the notion of having a meaningful life, I will be content to show that, even if the act in question were an actual homicide, it would not have been a morally wrongful one because it would not have been an injury to Miss Quinlan.

## II

Concerning the question of possible killing, one can argue (as Joseph Quinlan has) that removing Karen from the respirator would not be an act of killing but merely an act that would allow her to die with some semblance of dignity. Or in somewhat different language, one can say that the act in question permits death to occur but does not cause death. The difficulty here is as follows: Is there not a difference, a significant conceptual difference, between allowing nature to take its course by doing nothing and, so to speak, expediting the course of nature? If I pass a drowning man and do nothing, it may perhaps be said that I merely allowed him to drown. But can we make exactly the same judgment when I pass a nonswimmer and remove his life preserver?

It is one thing not to place a patient on the machine and another to remove the machine. Not placing a person on a machine is roughly analogous to not throwing a drowning person a life preserver. But the removal of the patient from the machine seems to be sufficiently like that of removing the life preserver to warrant our saying that it is an act of killing.

Notice that I am not denying the general validity of the distinction between "allowing to die" and "killing." Nor am I denying that clear-cut cases of each kind exist. What I wish to suggest is that there is another class of acts, acts that expedite the course of nature, and that within this class some are rightfully called acts of killing and some are not. Few of us would say that giving another cigarette to an average smoker is an act of killing, though it may indeed be expediting a certain course of nature. But it would be hard to deny that the turning off of a heart pacer of someone who vitally needs it is expediting the course of nature and is an act of killing. What is unclear is how, in respect to killing qua killing, this case or the one of removing the life preserver differs significantly from that of the Quinlan case.

Another way of denying that the act in question is the wrongful killing of a person is to deny that Karen Quinlan is, in fact, a person. There are a variety of definitions of personhood, ranging from being a member of the biological species *Homo sapiens* to being a person only if the organism is a self-actualized human being. I do not consider these extreme characterizations worthy of serious attention for three reasons. First, in the former, all human beings become persons while, given some standard interpretations of the latter, almost no human would be a person.

Second, if the term "person" is defined so that it is perfectly synonymous with "human," then it becomes a needless redundancy. And, finally, to insist that an individual is a person only if he or she is a self-actualized human being seems to confuse personhood with being a self-fulfilled or excellent person.

One of the more helpful ways of remedying this situation is to follow a suggestion made by Eike-Henner Kluge. Professor Kluge maintains that "a person is an entity that is a rational being" [6, p. 91]. He then adds that the concept of a rational being "is tied not to actual behavior, but to inherent and constitutive potential," and that "all and only these entities are rational beings whose neurological activity or relevant analogue thereof has a mathematically analyzable structure that is at least as complex as that of a human being, or whose brain or relevant analogue thereof has a structural and functional similarity to that of a human being, particularly with respect to those substructures that are the relevant analogues of the non-limbic cortex" [6, pp. 90, 94]. This analysis has the merit of telling us that an examination of behavior in terms of mere body movement is not a sufficient criterion, that we have to determine whether or not certain physical correlates exist, and that when an individual permanently lacks the relevant analogues of the nonlimbic cortex, they are not persons. What the analysis, however, fails to tell us, aside presumably from the criterion of brain death or being born with no cerebral hemisphere at all, is what it means to lack the relevant analogues of the nonlimbic cortex.

There is consensus that Karen is not brain dead. It is generally agreed that she is in "a chronic persistent vegetative state," that she is comatose with irreversible damage to at least one part of her brain, probably a lesion in the high brain stem. But some have taken this diagnosis to imply that since an individual lacking the capability of conceptualizing a continuing self and having other mental states is not a person because he or she lacks the neurological analogue that makes those things possible, and since Miss Quinlan lacks that analogue, it follows that she is not a person. If this is so, then to the extent that we may be killing, we are not ending the life of a person.

I do not find this argument convincing. My hesitation begins with the use of a definition of "person" that purports to resolve a moral issue without reference to acts of preference and decision. It increases when we consider the medical testimony and realize that there is an important difference between knowing that Karen no longer functions as a person and knowing that she no longer is a person. The former may be granted; the latter, however, is dubious. The type of definition of "person" we appear to be using is being made to do more than it is really capable of doing. Strictly speaking, it cannot tell us or help us find out whether Miss Quinlan is a person or not. Miss Quinlan is neither brain dead nor apparently lacking a cerebral hemisphere as an anencephalic child would be, for example. Furthermore, tests that would confirm the extent of injury (pneumocephalogram or computerized tomography) were not performed, so that inferences as to the extent of injury are largely based on other cases and the principle of induction. Finally, it should be observed that the principle of induction seems to be of little help in this case. So that even if one argues that the greater the number of cases in which the lack of relevant neurological analogues of this sort has been found associated with cases of Karen's sort, the more probable it is (if no cases of failure of association are known) that this correlation exists, we still would not get very far, since all medical parties to the dispute admit that Karen's case has unique features and that these may indeed be relevant. The view to which I find myself driven, in the attempt to avoid these objections, is that there is a difference between knowing Karen no longer *functions* as a person and knowing that she no longer *is* a person, and since only the former appears to be true, I may assume she may still be a person.

## III

It is not unusual when approaching the question of wrongful killing to begin by appealing to existing law. Undoubtedly, as Judge Muir noted in the Quinlan case, the intentional taking of another's life regardless of motives is likely to be regarded as sufficient ground for conviction of some form of homicide [8, pp. 33, 41]. But upon reflection, what should we make of this? Does it not suggest that the problem may well reside in an inadequate characterization of "wrongful killing"? Does not the Quinlan case, at its very heart, suggest that "wrongful killing" ought not be considered simply as "the intentional taking of another's life"?

Richard Brandt suggests the prima facie duty not to kill is perhaps derivative from two more basic duties: (1) not to cause injury and (2) to respect the rational wishes of others, and hence that the duty not to kill simply does not apply where killing is wanted or desired and is not an injury [3, pp. 106–114]. Thus he writes that "there are two things that are decisive for the morality of terminating a person's life: whether so doing would be an *injury* and whether it conforms to what is known of his preferences" [3, p. 113]. "If I am right in all this," he concludes, "then it appears that killing a person is not something that is just prima facie wrong *in itself*; it is wrong roughly only if and because it is an *injury* of someone, or if and because it is contrary to the *known preferences* of someone" [3, p. 114].

Brandt's analysis in large part provides the grounds for explaining why most of us hold the discontinuance of certain medical procedures in cases like that of the Quinlans not to be wrongful killing. What is probably not as clear as one might like is what constitutes an injury and why the act in question is not or does not constitute an injury. The former question is, however, very difficult and is differently understood by various thinkers.

In ordinary discourse the term "injure" is used as a synonym for almost any kind of harmful, hurtful, detrimental act or act of impairment. This in part explains why many thinkers believe killing, or even death itself, is always an injury. They hold harm to be the violation of an individual's interests, and in turn they view death always to be a harm. It has been suggested elsewhere [7, pp. 139–140] that, according to this interpretation of what it means to have an interest, even predominantly helpful acts of killing are harmful in part. The reason being that, even in acts such as beneficent euthanasia, we violate an interest (presumably an extremely weak interest), namely, to remain alive only if one's life could be radically different.

Almost everyone views death as a definite and irreversible kind of impairment. So that if by "injury" is only meant "impairment," then it is true that the act in question is an injury. But that only means that an act of ending life diminishes the quantity of life which, though true, appears to be tautological and trivial. If, on the other hand, by "injury" is meant "harm in terms of infliction or the feeling of pain," then it is not true to say all death is an injury. We need not, I hope, belabor the point that some people do not feel pain when they die or that in normal conditions in modern countries most patients need not suffer pain, especially unbearable pain, when they die.

We have seen that the word "injury" has acquired, in the course of time, some associations that are apt to be misleading or needlessly problematic. We shall therefore use "harm" to refer to any violation of interests and use "injury" only to refer to a very specific kind of violation of interests. We shall say provisionally that we injure a man only when we deny or deprive him of some important or vital need he would prefer to have. In this sense, a person who removes a minor part of another's toenail without permission while the latter is asleep may or may not be harming him, but it is difficult to see how this act could constitute injury. On the other hand, to remove the entire limb would normally constitute an injury unless, of course, it was gangrenous.

It will serve to make the point clearer if we first apply it to the heroic struggle of Charles Wertenbaker. His wife, Lael, in *Death of a Man* has written of their last 60 days together, of his struggle with cancer, and their decision to end his life when he decided the time had come [9]. In essence, Mr. Wertenbaker, upon learning that he was terminally ill, decided to bear the test of pain and live as full a life as possible as long as it was a meaningful one. In the end, he takes his own life in the company of, and assisted by, his wife. I shall not attempt to describe the beauty or the agony of their last moments together, of a wife saying, "I love you, please die."

We agreed provisionally that we injure a man only when we deny or deprive him of some important or vital needs he would prefer to have. So that our question comes down to the question: What did death deprive Charles Wertenbaker of? If Lael's description is accurate, and we have no reason to believe it is not, what made life meaningful to her husband during his last 60 days was his ability to eat, drink, read, listen to music, write, and relate sensitively with his children and wife. However at the time he took his life, he could not do any of these things. He had no control over his functions. He could not hear music or even drink tea. Moreover, there was no reasonable possibility that these conditions could be reversed or that he would recover. Now if this is true, and we can discount the fact that he still enjoyed seeing the face of his wife, then where is the injury Mr. Wertenbaker allegedly inflicted upon himself?

We are now in a position to understand why the argument for noninjury in the Quinlan case in a most significant respect is like the Wertenbaker case. For this purpose we shall follow a common characterization of meaningful life. In judging a span of life as meaningful many of us seem to mean, first, that during the period in question the individual has some overall dominant goal or goals that give direction to those parts of his total life pattern and circumstances that he thinks are important; second, that the individual believes there is some genuine possibility that he will attain these goals; third, that the having of the goal or goals is sufficient to thwart chronic depression or melancholy and in more optimistic situations adds a special zest to life [cf. 4; 5, pp. 93–104; 10, pp. 148–150]. On this view, a span of life becomes devoid of meaning roughly when, or to the extent to which, an individual believes he cannot possess goals or when, if he can and does have goals, they are impossible of being achieved.

It would obviously take us far afield to examine the merits of this point of view, but it may be useful, in order to avoid serious misunderstanding, to stress several points. Advocates of this view are maintaining neither that everyone talks this way nor that everyone believes life is meaningful in this, or merely in this, sense. They realize that there are different visions of meaningfulness, that many people hold life to be intrinsically valuable, and that others would maintain that life is always meaningful because even the greatest suffering has a role in the fulfillment of God's purpose. Nor are they maintaining that, because a life is no longer worth living, it somehow follows that that life, in part or as a whole, does not have worth, even great worth. This confusion is often perpetrated by those who, because they believe voluntary dying or suicide would repudiate the meaningfulness and worth of their own lives, mistakenly transfer their values to others, and oddly enough then conclude that these persons (usually their critics) are guilty of such repudiations. Finally, and this closely relates to the last point, it is one thing for someone to have a teleological vision of the meaning of life that includes the conviction that suffering is sacred because it confers upon those whom it rends the most intimate resemblance to Christ, and perhaps because of this insists that, given his interests, the ending of his life would be an injury; but it is another to show that where there is a different vision and different major interests that the ending of such a life is, in all circumstances, an injury.

Returning to the cases before us, we may say the following. The judgment of Mr. Wertenbaker to end his life can be interpreted to be approximately synonymous with the judgment that his present mode of existence was almost totally devoid of meaning, was becoming increasingly worse, and that this process was irreparable and irreversible. The judgment of Mr. Quinlan to end, or allow, his daughter's life to end may be similarly interpreted. There are, however, two important differences. Because Mr. Wertenbaker was aware of the fact that it was highly probable that his life was going to change from that of having a meaningful existence to that of having a meaningless one, he explicitly requested to be allowed to take his own life if and when that occurred. Aside from courage, his major problem was that of having to draw a line, and to draw it before it was too late for him to help himself. Unlike Mr. Wertenbaker, Miss Quinlan can neither directly state her preference nor take her own life. Yet her present mode of existence is not one in which goals can be met, and therefore, if her condition is irreparable and irreversible, as it seems to be, then to the extent she remains alive she is doomed to a completely meaningless life.

According to Paul Armstrong, the Quinlan's attorney, "Karen's mother, Julia Quinlan, her sister, Mary Ellen, and her friend and confidante, Lori Gaffney, testified that Karen has expressed to each of them that should fate find her in the tragic circumstances which bring her father before this Court, she would elect and request the discontinuance of the futile medical measures presently employed" [1, pp. 10–

11]. On this basis and given certain legal precedents, Mr. Armstrong, I submit, correctly concludes that there is evidence "of sufficient probative weight to compel the conclusion that she would elect to remove the futile measures presently being administered to her" [1, p. 9]. But what if, we may ask, there was no such evidence? What if Miss Quinlan was much younger or had never stated such a preference or both of these things? Would we then be willing to say that to remove her from the respirator is to injure her? I think not. Of course in doing so we are tacitly making an additional claim. The assumption or claim is that while a rational man might prefer a predominantly meaningless existence over death, he would not prefer to live a completely meaningless life. This factor, I believe, is not being advanced as an a priori truth. The claim appears to be an empirical one, namely, that as a matter of fact it is highly improbable that a fully informed man would choose to live a completely meaningless life, in the sense in which that is here understood. If this claim is warranted, and the above account correct, then not only is the act in question not wrongful to Miss Quinlan because it is not an injury to her, but it would also seem to follow more generally that we do not injure human beings when we allow their lives to end or put them to death even if they cannot, or have failed to, state their preferences if they are living a completely meaningless life, if the state of affairs that warrants that judgment is irreparable and irreversible, and if there is no or insufficient evidence to indicate they would not accept the judgment that their lives were meaningless.

REFERENCES
1. Armstrong, P. W.   Brief and Appendix for Appellant, In the Matter of Karen Quinlan: An Alleged Incompetent. N.J., A-116 (Dec. 16, 1975). Opinion is reported at 137 N.J. Super. 227 (1975).
2. Brandeis, L. D.   The Right to Privacy. In R. Pound (ed.), *Selected Essays on the Law of Torts*. Cambridge, Mass.: Harvard Law Review Association, 1924.
3. Brandt, R.   A Moral Principle About Killing. In M. Kohl (ed.), *Beneficent Euthanasia*. Buffalo, N.Y.: Prometheus, 1975.
4. Edwards, P.   Life, Meaning and Value of. In P. Edwards (ed.), *The Encyclopedia of Philosophy*. New York: Macmillan, 1967.
5. Joske, W. D.   Philosophy and the meaning of life. *Australasian Journal of Philosophy* 52:2, 1974.
6. Kluge, E-H. W.   *The Practice of Death*. New Haven: Yale University Press, 1975.
7. Kohl, M.   Voluntary Beneficent Euthanasia. In M. Kohl (ed.), *Beneficent Euthanasia*. Buffalo, N.Y.: Prometheus, 1975.
8. Muir, R., Jr.   In the Opinion of Robert Muir Jr., In the Matter of Karen Quinlan: An Alleged Incompetent. Super. Ct. N.J., Chancery Div., Morris Co., C-201-75 (Nov. 10, 1975).
9. Wertenbaker, L. T.   *Death of a Man*. Boston: Beacon Press, 1957.
10. White, F. C.   The meaning of life. *Australasian Journal of Philosophy* 53:2, 1975.

# CHAPTER 15

# To Kill or Let Die
*James F. Childress*

*In recent years,* several critics have contended that the distinction between killing and allowing to die lacks "moral bite" and is a mere "moral quibble" [7]. In response, one theologian has held that it has moral significance, not as an abstract proposition, but within "the religious context out of which it grew" [6]. But we can defend the distinction without reducing it to an abstract proposition or to a claim about intrinsic differences and without reviving the religious story within which it originated. Even in a secular context, the relationship of care and trust can support an argument that we ought to (continue to) prohibit killing but accept some instances of allowing to die.

Because the distinction between killing and letting die is inextricably tied up with our understanding of medical care, we cannot remove it without tearing the whole fabric. For the community or the medical profession to authorize physicians actively to kill patients would so alter the moral ethos of medicine as to necessitate a new basis for trust. Trust is the expectation that others will respect moral limits. When we trust others, we have confidence in or rely on them to act within certain limits and boundaries toward us, for example, not to harm us or lie to us. Trust in the context of medical care involves the expectation that medical practitioners will work for our health and life, will provide "personal care," and will do us no harm [2, 8].

This trust is not to be confused with the dependence of the "sick role," although the patient's vulnerability and dependence certainly enhance the value of a relationship of trust. Furthermore, much medical care depends on the patient's trusting response to the physician. While relationships of trust are valued as conditions for other relationships and as instruments or means to accomplish other ends, such as successful medical care, they are also valued as ends or goods in themselves. One Harris poll in late 1973 discovered that only 2 of 22 institutions were deemed trustworthy by a majority of those questioned. Fifty-seven percent trusted the medical profession, and 52 percent had confidence in local trash collectors! The other institutions, including the police, press, religion, Congress, and so forth, evoked less confidence. Trust is fragile and, once weakened, is difficult to restore.

David Louisell [4] contends that "Euthanasia would threaten the patient-physician relationship: confidence might give way to suspicion . . . . Can the physician, historic battler for life, become an affirmative agent of death without jeopardizing the trust of his dependents?" It is implausible to hold that trust within medical care is impossible without a prohibition of all killing. But the trust that now characterizes the relationship between patient and physician is based on the medical profession's implicit and explicit commitments to foster life and health. Those commitments appear to be in tension with and perhaps would be weakened by a policy of selective killing. The distinction between killing and letting die is one important *expression* of the ethos that directs medicine to the patient's life and health in the form of personal care. But it also appears to be *instrumentally* as well as symbolically important. To remove it, Hughes [3] contends, would weaken a "climate, both

moral and legal, which we are not able to do without." At stake is the set of commitments and values that undergird the medical profession and provide the grounds
for trust in medical practitioners.

Is my argument simply another version of the wedge or slippery slope argument?
The wedge argument may take different forms. First, it may focus on the moral
reasoning for rules or acts, holding that if the reasoning is bad, there will be no line
between the conduct that is considered legitimate and the conduct that is deemed
reprehensible. Such an argument is really concerned with the hammer back of the
wedge, for justification of one act may be justification of another. Second, a wedge
argument may focus more on what the wedge is driven into. Instead of being concerned with the hammer, the logic of moral reasoning and the possibility of logically
distinguishing between two acts, it may examine the forces in society that, at least
in part, determine the impact of alterations in rules. It may hold that a moral decline is probable or perhaps even inevitable if certain restraints are removed, for we
cannot count on people, because of certain psychological or social forces, to draw
the important distinctions even if those distinctions are clear and defensible.

Although my argument does not exactly fit either pattern, it is closer to the
second, for it holds that the distinction between killing and allowing to die — the
prohibition of the former and authorization of the latter under some circumstances
— expresses and may help preserve some important moral commitments by the
medical profession. Such commitments are the basis of the valued trust between
patient and physician.

One of the strongest arguments for allowing physicians to kill some patients is
the relief of unbearable and uncontrollable pain and suffering. No one would deny
that pain and suffering can so ravage and dehumanize patients that death appears to
be in their best interests. Prolonging life or even refusing to kill in such circumstances
may appear to be a cruel infliction of unnecessary suffering. Often critics of the distinction between killing and letting die appeal to situations outside the medical context to show that direct killing may be more humane and compassionate than letting die: For example, a truck driver inextricably trapped in a burning wreck cries
out for "mercy," that is, to be killed. No doubt we are reluctant to say that in such
tragic situations those who kill out of mercy on behalf and at the behest of the victim act wrongly. Furthermore, juries may even find persons who have killed a suffering relative not guilty by reason of temporary insanity. There are, nevertheless,
serious objections to building into medical practice an explicit exception: Physicians
may kill their patients in order to relieve great pain and suffering that cannot be
controlled or managed by other means.

First, there may be few if any cases in medical practice that are really parallel to
the truck driver trapped in a burning wreck, for the physician may have the means
to relieve pain and suffering short of killing that may not be available to a bystander
at the scene of an accident.

Second, even if medical practitioners confront some cases of uncontrollable and
unmanageable pain and suffering, it is not clear [5] that we should build a social
ethic, a professional ethic, on the *Grenzfall,* the boundary case. An emergency ethic
is just that, and it should not be taken to provide the ethos for normal medical practice. Hard cases may make bad social and professional ethics as well as bad law. Put-

ting society and the medical profession on record in favor of killing in such cases may not be necessary or desirable.

Third, some critics of the distinction argue that limiting killing to cases of *voluntary euthanasia,* or assisted suicide, would enable us to relieve many sufferers who desire to die while retaining adequate controls (although we might want to establish procedures and safeguards in order to ensure voluntary requests). But such a strategy is not without serious drawbacks. As William May has suggested, patients often assert a right to be dead rather than a right to die. And they conceive the right to be dead as the right to have the physician guarantee this death by being its agent. A patient's request to be killed does not alter the physician's role in its execution, although it may attest to the purity of his motive. It calls for acts that are in tension with the ethos of the profession and society's vision of the profession. The physician becomes an agent or even an instrument in a practice that threatens the ethos of medical care. A formal policy change to permit such actions would be extremely risky.

Do the proponents or the opponents of such a policy change have the burden of proof? Antony Flew [1] argues that maintaining the existing practice of prohibiting voluntary euthanasia requires justification because it violates the principle of liberty by refusing to respect individual wishes. But supporters of voluntary euthanasia do not merely want suicide or refusal of treatment or allowing a patient to die. They want the patient dead when he wants to be dead, and they want this accomplished through the physician's agency. This policy would involve such a change in the orientation of the medical profession as to shift the burden of proof to the proponents of change. The prohibition of active euthanasia (even when it is voluntary) has not been arbitrary. It expresses some important moral commitments whose loss would be very serious. Since this policy embodied in the medical profession and the law has served us well, although not perfectly, its alteration should be undertaken only with the utmost caution. While lines are not easy to draw, our society has been able to maintain the line between letting die and killing in medical practice with considerable success. Therefore, before we implement changes that might have serious negative consequences, we need evidence that major benefits will result and that they will outweigh the bad effects.

Another argument against respecting the distinction between killing and letting die is that decisions about treatment will be made on "irrelevant grounds." While Rachels [7] develops this argument with reference to defective newborns, it applies to other cases as well. For him the basic question is whether life or death is in a particular patient's interests, and the answer to that question should govern the physician's actions. If a physician determines that a patient would be better off dead, why should he violate that patient's interests because of the "irrelevant" factor that the patient will not die when artificial treatment is discontinued? Why should the means of meeting the patient's interests be limited in this way? The main objection appears to be that some patients will continue to live when their own interests would dictate death.

This objection is outweighed by the importance of preserving the medical ethos. While it is true that some patients may continue to live when it is not in their interests to do so, our society should tolerate that "cost" in order to preserve the primary commitment of the medical profession to care for life and health.

I have argued for a policy that excludes killing but accepts letting die under some conditions. In order to protect the patient's interests, it is sometimes necessary to relinquish efforts to prolong that patient's life. Such an entrusting of the patient to "nature" or "God" is difficult for a society that finds those terms meaningless and that views death as the implacable enemy and its triumph as the failure of medicine. Nevertheless, even in a secular age dominated by a technological imperative, the obligation to care does not require that physicians engage in hopeless efforts or needlessly prolong the dying process when it appears to be irreversible. Care involves easing pain and suffering even when it hastens death, providing comfort, and so on, but futile acts to prolong life may only be rituals to affirm the power of medicine or to lessen the guilt of families. Optimal care should not be confused with maximal treatment.

A decision to allow a patient to die may be as reprehensible as killing the patient. Thus a satisfactory policy of letting some patients die would require attention to the conditions under which such decisions would be justified. While I would argue for concentration on the patient's interests in contrast to familial or social considerations, attention would also have to be given to such matters as criteria of the quality of life, living wills, and procedures for making sure that the patient's interests and wishes are respected. Although a policy of selective killing cannot easily express or preserve a satisfactory moral ethos in medicine, a poorly developed and articulated policy of letting die may also weaken this ethos.

## REFERENCES

1. Flew, A. The Principle of Euthanasia. In A. B. Downing (ed.), *Euthanasia and the Right to Death: The Case for Voluntary Euthanasia.* London: Peter Owen, 1969.
2. Fried, C. *Medical Experimentation: Personal Integrity and Social Policy.* New York: American Elsevier, 1974.
3. Hughes, G. J., S.J. Killing and letting die. *Month* 236:42, 1975.
4. Louisell, D. W. Euthanasia and biathanasia: On dying and killing. *Linacre Quarterly* 40:234, 1973.
5. McCormick, R. A., S.J. Notes on moral theology. *Theological Studies* 37:70, 1976.
6. Meilaender, G. The distinction between killing and allowing to die. *Theological Studies* 37:467, 1976.
7. Rachels, J. Active and passive euthanasia. *N. Engl. J. Med.* 292:78, 1975.
8. Ramsey, P. *The Patient as Person.* New Haven: Yale University Press, 1970.

# CHAPTER 16

# Living Wills and Mercy Killing: An Ethical Assessment
*Arthur J. Dyck*

*Modern medicine* has greatly increased skills in the service of life. Mortality has been considerably decreased, and it is possible to maintain and extend the life span of people with serious conditions, including patients with severe brain damage who can be kept on respirators in a comatose state, sometimes for months. This greatly augmented ability to extend life under medically devastating conditions has for many raised the question more sharply as to whether there are values that outweigh the value of life itself. In short are there conditions spawned by some of the interventions of contemporary medicine that would justify loosening our usual constraints against taking life?

This question is not academic. The pressures to deal with this question are revealed in at least two ways: the surge for living wills and their legislative sanction; the increasing failure to distinguish clearly in word and deed between allowing to die and direct killing.

## LIVING WILLS

There are three major arguments for making living wills and for having them legally sanctioned. The first argument is that there is the desire to minimize pain and suffering and, for many, to minimize the time spent in certain states of being that are considered to be undignified.

A second argument is rooted in a concern for others. Health professionals need legal protection against malpractice suits and also some assistance in knowing what patients wish to do under difficult and morally ambiguous circumstances. In addition, if one refuses to have scarce and expensive resources used to prolong dying or in a situation like an irreversible comatose state, one will thereby free these resources for others.

The third and most basic reason behind the drive to legislate living wills is the desire to extend the right to choose how one is treated as a dying or severely handicapped individual in situations where the patient's capacity to make reasonable judgments has been lost. By writing down one's wishes, persons hope to retain and exercise their highly treasured rational autonomy by rejecting medical intervention they do not wish or consider morally unwarranted.

Although the arguments for the living will are very plausible and in many ways attractive, there are serious reasons why it would be unwise to take living wills out of the private sphere and make them legally binding, as in the California Natural Death Act [1]. First, with respect to minimizing pain and suffering, it should be recognized that pain and suffering can be very well controlled, and health professionals, as well as relatives, are inclined to make dying patients as comfortable as possible unless the patient is in a position to object to increasing pain control. But more important still is the fact that living wills add to the risk of being considered terminal or irreparably and severely handicapped in situations where there may still be some hope for life and rehabilitation of some kind. In short, the living will influences health professionals to act against the usual dictum to "err on the side of life."

The concern for others is extremely laudable, and it may well be the case that, under present circumstances, living wills may somewhat ease the burdens of decision making and also free some resources. However, it would pressure health professionals to be less aggressive to save those with a living will as over against those who have no will. It does not press the claims of justice. From the standpoint of society, resources should be available on an equitable basis. As a general rule, patients similarly at risk of dying should be treated on a first-come, first-served basis, and when this is unworkable, by means of a lottery [6]. Where possible, the resources needed for maintaining the health and life of a country's citizens should be provided. Living wills, particularly where health needs are met by government money, would provide a way of saving money and would make moving toward compulsory living wills attractive. This idea should not be dismissed as a mere hypothetical possibility. A recent memo within the Department of Health, Education and Welfare, described as an idea paper to guide preparation of the agency's budget, explicitly encourages the adoption of the living will concept. The memo estimates that doing this would result in an estimated medical savings of 1.2 billion dollars in fiscal year 1978 [3]. If there are certain health needs that cannot be met for everyone, difficult decisions have to be made as to whether this need will be met at all or whether it will be meted out on the basis of some type of lottery system. There is no space here to clarify the role of lotteries; this concept has been discussed elsewhere [2, 8].

I have no quarrel with the attempt to extend one's right to consent as a patient. However, the freedom achieved through a living will is not as great as one might imagine. One cannot describe individual cases so accurately in advance that they are likely to fit one's own case. In any event, patients are still very dependent on the judgments of health professionals as to whether their conditions are actually terminal or irreversible. It would appear that the freedom of health professionals is curtailed by living wills. But health professionals can appeal to the finitude of their judgments. Whether a living will is being followed or not is, in the end, determined by the judgments of health professionals, and persons who dispute these judgments will be hard pressed to win their arguments. There is no substitute for good judgment and the integrity that breeds trust. But the living will can undermine trust in the health professional's desire to restore health and life. Allowing to die under the conditions of a legally sanctioned living will may merge into what amounts to selective killing, depending upon the dominant moral formation of the health profession. As James Childress has argued in this volume, selective killing would entail a serious change in the ethos of the medical profession and be a source of distrust.

There is still another consideration that weighs against the legalization of living wills. To legalize living wills is to bring the government into the business of assuring death rather than protecting life. As Hobbes contended, the purpose of government, the justification for submitting to its power, is that our lives are thus better protected than they would otherwise be. The full effect of the living will is to give government a new function, which is to assure the death of persons who are not criminally liable. Even capital punishment, which I would not attempt to justify, derives its alleged justification from the protection of innocent lives against aggressors.

## MERCY KILLING

What we have said against the legalization of living wills would not touch those who are asserting a right to commit suicide or a right to have death induced, which would amount to assistance in a suicide. Such persons may precisely be seeking the protection of government for what has been called a right to die but which includes a right to have death induced. Legislative proposals of this kind have been introduced in some states [5, 9]. Childress has already argued in this volume that the health profession would be undermined by the recognition of such a right and the obligation of the health profession to implement it. But some have argued that there is no morally justifiable distinction between allowing to die and directly inducing death, and that if one can justify the former, as health professionals generally can, one can justify the latter. James Rachels has recently argued this point quite forcefully.

To make his case that a failure to prolong life or any efforts to shorten it that are entailed by the practice of "allowing to die" is akin to killing, Rachels [7, p. 79] offers us the following examples:

In the first, Smith stands to gain a large inheritance if anything should happen to his six-year-old cousin. One evening while the child is taking his bath, Smith sneaks into the bathroom and drowns the child, and then arranges things so that it will look like an accident.

In the second, Jones also stands to gain if anything should happen to his six-year-old cousin. Like Smith, Jones sneaks in planning to drown the child in his bath. However, just as he enters the bathroom Jones sees the child slip and hit his head, and fall face down in the water. Jones is delighted; he stands by, ready to push the child's head back under if it is necessary, but it is not necessary. With only a little thrashing about, the child drowns all by himself, "accidentally," as Jones watches and does nothing.

From these examples, Rachels argues that Smith killed the child, whereas Jones allowed the child to die. That is the only difference between them. Is there any moral difference between killing and letting die in these two instances? Both men had the same result in view and achieved the same end when they acted or failed to act.

Now interestingly enough, Rachels has made a very good point with respect to certain cases in which the distinction between allowing to die and killing has eroded within the medical profession. Thus he cites just such instances in which physicians have decided, with the concurrence of parents, not to save the lives of certain infants whose lives could be saved, but who are diagnosed as having some irreversible handicap that would not be ameliorated by lifesaving interventions. Rachels has made an important point. There is no moral difference between these particular instances of allowing to die and directly taking life. In fact, Rachels is quite right to point out that the device of not feeding these infants and letting them linger is less merciful than the direct inducement of death. What I wish to argue, however, is that it is as morally wrong to let those die who can be saved as it is directly to kill them, and at the same time to argue that there are instances of allowing to die that are morally justified.

To understand the situation of the dying as well as the situation of health professionals in the face of those who are surely and imminently dying, we need somewhat different examples from those that Rachels gives. His examples are sufficient for some cases with which he is concerned but not the great bulk of cases in which decisions to allow people to die are being made. Imagine that an individual X is caught on a small rocky ledge in the midst of rising flood waters. The area is quite inaccessible, it would take days to summon help, and the flood waters will inevitably rise over X and sweep him away well before help could be summoned and brought back. Nor can he be rescued from the shore, even though it is close enough to allow conversations to be heard between X and someone on the shore.

In the first scenario, let us imagine that an individual Y is standing on the shore. X requests that Y go and summon X's child so that X can talk to his child before the flood waters overtake him. Y believes that this is sheer misery and that talking to his child would be torture for X and for his child, and so he shoots X to put him out of his misery.

Let us imagine another scenario with Y. X requests Y to shoot him so as to avoid suffering. Y honors the request. Now imagine, also, that Y in this instance, as well as in the other instance described in the previous paragraph, possesses a pellet that he is capable of shooting into X to relieve him of any pain or suffering and to render him unconscious until the time that the flood waters take him. This makes Y's position comparable to that of health professionals. Why would one shoot X with a pellet rather than with a lethal weapon? Well, as in virtually all cases of persons judged to be terminally ill, there is still some chance that X will be rescued — for example, a helicopter might fly over the flood waters, see X, put out a ladder that allows someone to descend, and rescue X. In any event, the scenarios with Y raise the question as to whether killing or asking to be killed under those circumstances are morally justifiable for Y or X.

Imagine that an individual Z is on the shore and Z is a person who will not honor any direct request by X that Z kill him. Consider a situation in which X requests Z to bring X's child to him for a last conversation before the flood waters take him, but Z refuses. Z thinks it would be cruel for the child, or he simply has no inclination to bother with any request on the part of X. Here we might think of Z as a scoundrel, as morally reprehensible, but Z would not be charged with the murder of X nor would he be held in any way responsible for X's death. We can have thoughts of Z, therefore, as a morally reprehensible person doing nothing for X, but in no way being the cause of X's death and in no way being responsible for that death. Allowing to die in this situation does not in any way make Z morally culpable for the death of X, whereas Y in both instances of killing X is morally culpable for the death of X, even though X's death is imminent and inevitable. If X's child should see Y killing X, he can serve as a witness to the killing and Y could be indicted for murder. The fact that X was imminently dying would not clear Y of guilt, but the fact that X is imminently dying is what clears Z of guilt because the situation as we have set it up does not allow Y or Z any chance of rescuing X.

Let us imagine another variation on this story involving Z. Suppose Z is asked by X to throw him a rope, too short for rescue but useful for another purpose. By means of this rope, X will be able to lower himself from a jagged rock to a flat piece of rock, so that he will be more comfortable even though he will be caught by the flood somewhat sooner. If Z honors this request, he will be a party to allowing X to shorten his life for the sake of being more comfortable. Z does it after carefully scanning the sky and land for any imminent sources of rescue. Is Z in any sense responsible for the death of X? No. But what if one argues that he is an accomplice to the shortening of X's life? Is that tantamount to killing X?

The reason why we would not accuse Z of killing X, even though in providing him some means to be more comfortable, Z has thereby shortened X's potential life span, is that there are other considerations that enter into the choices we make with regard to how long we will live. For example, we make decisions to work hard. These decisions could shorten our lives. If, however, this work is important and justified on various grounds, and if, as is usually the case, we cannot be certain of the direct effect of this work on the shortening of our lives, we do not consider such decisions as tantamount to suicide. Similarly, we permit vacations. Vacations may contribute to health, but vacations may also be very risky. Some vacations involve greater exposure to potential risk of loss of life and loss of health, but we would not generally deny people such vacations nor consider them to be morally unjustified. Why, then, would we deem it morally unjustified for this struggling, imminently dying man out on a rock to ease his pain on this jagged rock by lowering himself to a more comfortable position for his last hours? Is this man not entitled to a vacation, even if it involves the risk that he will die sooner? Now people who are dying may not wish to have their lives shortened by being made more comfortable, but if they do, there is nothing morally reprehensible about requesting what amounts to a vacation from pain and discomfort.

The question that remains, then, is whether one should honor a request to kill or whether one should kill oneself under certain circumstances. In other words, what are the arguments that can be offered for justifying suicide and assisting in a suicide?

Recently Karen Lebacqz and H. Tristram Engelhardt have reviewed a whole series of traditional arguments against suicide and have found them wanting. They argue that "Persons should be permitted to take their own lives when they have chosen to do so freely and rationally and when there are no other duties which would override this freedom" [4, p. 689]. However, they contend that suicide is usually wrong because it does violate our "covenantal obligations to others," such as promise keeping, gratitude, reparation, and the like. Nevertheless, they believe there are three circumstances that, when they arise, justify suicide [4, p. 695]:

(1) when it is impossible to discharge one's duties, (2) when suicide supports rather than violates obligations of covenant-fidelity, and (3) when the intention of the suicide is to support the general good of persons and the conditions which make covenant-fidelity possible.

The first set of circumstances involves the issue of voluntary euthanasia. Lebacqz and Engelhardt argue that there are certain life circumstances, as in terminal illness accompanied by great pain, where it may be impossible to fulfill one's usual obligations. This being so, the right to suicide, which they claim is an extension of the general right to control over one's own body, is morally permissible.

This argument does not take account of important alternatives that remain for the terminally ill. Dying persons retain the moral option of having their pain reduced. It is certainly morally justifiable to provide for a patient what we have considered to be the moral equivalent of a peaceful rest or vacation and to withdraw from interventions that prolong the dying process. Relief of suffering that foreshortens the dying process does not require killing. As we noted above, it is possible to distinguish shortening human life by killing and shortening human life by withholding medical interventions. One of the time-honored criteria in deciding whether killing can be morally justified is to ask whether it is a last resort. In the case of terminal illness, killing is not the only option and not the last resort.

With respect to what Lebacqz and Engelhardt call covenant-affirming suicide, they have in mind either "suicide pacts" or self-sacrificial suicides in which people would rather die than remain a burden to their families or to others. Here we see a rather narrow interpretation of the notion of covenant. The fundamental basis of community is our agreement to live together in peaceful, nonviolent cooperation. We are all dependent on one another and here at one another's mercy, which is one reason among others why the injunction against killing is recognized as one of our most stringent obligations. An act of suicide, whatever agreements it might involve with some part of the community, is a violation of the covenant on which such communities are based. It is not strictly true to argue, as Lebacqz and Engelhardt do, that there are circumstances under which one no longer has any obligations to anyone. The obligations not to kill and not to engage in actions destructive of this principle remain to the very end of our lives. That is why it is important to bear in mind that the relief of pain does not require direct killing, and so does not require us to violate our obligation not to kill.

The third argument for suicide is what Lebacqz and Engelhardt call symbolic protest, or supporting the general good. Since they offer no specific illustrations, it is difficult to imagine what kinds of suicide Lebacqz and Engelhardt would wish to justify as supporting the general good of persons or as symbolic protests. Among our most revered and cherished saints, heroes, and heroines are those who have given their lives so that others might live. We do not, however, think of these actions as suicide. They represent circumstances in which people place their lives in jeopardy for the sake of others. We can think of all kinds of instances, like smothering a bomb that is about to explode, stepping in front of an assailant's bullet, or being killed rather than reveal a secret, in which something is done so that others may live.

To justify dying under these circumstances, however, depends, as Lebacqz and Engelhardt rightly discern, upon what our act symbolizes. Killing oneself symbolizes a direct repudiation of the general injunction against killing and of the covenant on which our communities are based. If, however, we risk death for the sake of saving someone's life, this action would symbolize the same high value placed on human life that the constraint against killing is designed to foster. Where "symbolic protest" involves a high risk of death or virtual certainty that death may result from the action in question, it may be morally justified if it symbolizes the affirmation of life.

Consider the doubtless well-intentioned persons who burned themselves to death protesting the war in Vietnam as an example of what Lebacqz and Engelhardt may wish to condone. This action was certainly not a last resort. I see no reason why these same persons could not have made their symbolic protest by entering villages under bombardment and using their bodies, for example, to smother the flames from the clothes of napalmed children. The difference between this way of dying or risking death as against simply burning oneself is that throwing oneself on the body of a burning child precisely honors the distinction between taking one's life and giving one's life, and it entails the actual and symbolic hope that thereby some lives now and many lives in the future may possibly be saved.

Ultimately each of us has an obligation to encourage respect for the preciousness of life, the fostering of which has made our own lives possible until our dying day. The only circumstances under which we could feel confident that it is right to give or yield our lives would be those in which we have a clear opportunity through our death to save, affirm, and protect life. People who are dying are not expected to be heroic either by extending their lives or by giving it up for others. What we do expect is that they refrain from acts that diminish rather than enhance the value we place on life.

## REFERENCES

1. California Natural Death Act. In S. J. Reiser, A. J. Dyck, and W. J. Curran (eds.), *Ethics in Medicine*. Cambridge, Mass.: MIT Press, 1977. Pp. 665–666.
2. Dyck, A. J. *On Human Care: An Introduction to Ethics*. Nashville: Abingdon Press, 1977. Pp. 162–166.
3. *Hosp. Week* 13(25), June 24, 1977.
4. Lebacqz, K., and Engelhardt, H. T. Suicide. In D. J. Horan and D. Mall (eds.), *Death, Dying and Euthanasia*. Washington, D.C.: University Publications of America, 1977. Pp. 669–700.
5. Montana House Bill No. 256. Montana Self-Determination of Death Act of 1975.
6. Outka, G. Social justice and equal access to health care. *The Journal of Religious Ethics* 2:11, 1974.
7. Rachels, J. Active and passive euthanasia. *N. Engl. J. Med.* 292:78, 1975.
8. Ramsey, P. *The Patient as Person*. New Haven: Yale University Press, 1970. Chap. 7.
9. Wisconsin Assembly Bill 1207. Introduced by Rep. Barbee, October 22, 1975.

# CHAPTER 17

# The Right to Death and the Right to Euthanasia
*Natalie Abrams*

*Discussions of euthanasia* frequently focus on two aspects of the problem — whether a person can be said to have a so-called "right to death" and whether, if he does, there is any morally significant difference between active and passive euthanasia. In the following brief discussion, I suggest a possible relationship between these two issues, rather than discuss the issues themselves. I shall try to support the claim that if people can be said to have a "right to death," then this claim has implications for the active-passive debate.

A central question in discussions about a right to life or death is what would actually be entailed by such claims. Here I believe Philippa Foot's distinction between doing harm and bringing aid is relevant [1]. She claims that we ordinarily believe a person has a greater duty not to do harm than he has a duty to do good or bring aid. In other words, a "right to life" should not be seen to entail a right to be given whatever is necessary to maintain that life, but rather should be seen to entail a right not to be killed. This concept, of course, is working on the assumption that in order for a right to be meaningful, it is necessary that some duty or obligation be correlated with that right. To correlate a positive duty in relation to the right to life would impose enormous and unreasonable positive obligations for other people to supply aid.

This view would also be consistent with Judith Thomson's argument [2] that simply establishing the "right to life" of a fetus does not automatically imply the impermissibility of abortion. Certainly many other considerations are relevant to the abortion debate. What is being claimed here is simply the fact that even if the fetus is assumed to have a "right to life," this right cannot be seen to automatically imply the positive duty on the part of the mother to sustain that life. Abortion may still be impermissible, in many situations, but not simply as a result of the fact that the fetus may have a "right to life."

The central question in the euthanasia controversy, which would be analogous to the issue presented by the abortion situation, is what exactly is entailed by the claim that a person has a "right to death." Does a "right to death" entail the duty of others to bring about one's death?

Just as the establishment of the "right to life" of the fetus does not automatically imply the impermissibility of abortion, nor solve the abortion problem, similarly, the establishment of a "right to death" does not eliminate the problems surrounding the practice of euthanasia. As the "right to life" must be seen to imply the negative duty of others to refrain from killing, so the "right to death" should be seen to imply the negative duty of others to refrain from prolonging one's life. Neither the "right to life" nor the "right to death" entails the positive duty of others either to sustain one's life or to bring about one's death.

If one were to interpret the "right to death" in this way, that is, to correlate it with the negative duty not to prolong one's life, then an individual should be seen to have a right to passive but not active euthanasia. An individual would not have the right to require that others do whatever is necessary to end his life, although he would have the right to require that others not prolong his life. On this interpretation, establishing a "right to death" can, at most, entail the obligation of allowing patients to die. Passive but not active euthanasia can therefore be seen as obligatory in the sense that it emanates from the patient's right to death.

It can also be seen as obligatory from the physician's point of view. Knowledge of any role must entail awareness of the situations in which one's role should and can be performed. Knowing when his skills are not applicable, that is, when it is not possible to restore the patient to health, is an essential part of the physician's role.

Passive euthanasia can therefore be seen as obligatory from two perspectives, that of the patient's right to death and that of the physician's role as healer.

REFERENCES
1. Foot, P.   Abortion and the doctrine of the double effect. *Oxford Review* 5:5, 1967.
2. Thomson, J. J.   A defense of abortion. *Philosophy and Public Affairs* 1:44, 1971.

# CHAPTER 18

# Rights and the Alleged Right of Innocents to Be Killed
*Peter Williams*

*A number* of moral positions can be taken about the propriety of killing or helping to kill a person who wishes to die. One position is that in all circumstances we are morally required to make every effort to save lives, our own or others; another is that each of us has an unfettered right to choose the time and manner of death and, if we need help to carry out the act, we may insist that others assist us.

I believe we have no right to be killed. Three features of rights serve to organize my argument. First, the relation between rights and duties shows that if one person has a right to be killed, then another has a duty to kill him. Second, analysis of the moral force of claims of right demonstrates that we cannot be duty-bound to kill innocents. Finally, the limited domain of rights leaves open the possibility that killing an innocent person who so requests may, nonetheless, be morally proper and even laudable.

## I. RIGHTS AND DUTIES
A person's right to be killed gives rise to someone's (or everyone's) duty toward that person. If anyone can be said to have a right to be killed, someone else must have a duty to cooperate in the killing. If there were a right to be killed, someone else would be bound, if and when demanded, to bring about or help bring about the death of the right-holder. The killing might be done in any of a number of ways. The important thing is that someone — a doctor, a nurse, a candystriper, a relative — intervene actively or passively to end the right-holder's life.

## II. THE DEMAND QUALITY OF RIGHTS
There is an important distinction between claims of right and seeking something as a matter of privilege, favor or permission. Rights are the basis for the legitimate demands that we can make on others. For example, when we have a right to X we may *require*, not merely request X, *insist* on, not merely invite receipt of X, *stipulate,* not merely supplicate that we receive X. I call this feature of rights their demand quality, and this feature is crucial in deciding whether or not we have a right to be killed or to die. If we have such rights, then we can demand that another kill us or demand that others not interfere with our dying.

A demand quality of rights makes them essentially and importantly different from requests or pleas. When a right is asserted, a duty-holder is not allowed to weigh consequences as though there were no right. If a person has a right to be killed, someone else has a duty to cooperate in killing the claimant. Is there ever

Excerpted from Rights and the Alleged Right of Innocents to Be Killed, in *Ethics* 87:383, July 1977. Copyright 1977 by the University of Chicago Press, Chicago, Ill. Reprinted with permission of the author and publisher.

such a duty to kill an innocent person? What has been said about the demand quality of rights makes it possible to reword this question. Is an innocent's demand that B kill him ever a sufficient reason for B to kill him?

Since the issue of the right to be killed occurs in hospitals, various health care professionals are possible duty-holders; so are family members, friends, religious advisors, etc. The argument does not turn on who the B is but, since humane killing would probably require some, though not much, medical knowledge, the assumption will be made that a physician is the supposed duty-holder.

The question does not ask whether someone's death would be good or bad or the killing of him good or bad; nor is the issue one of risking a life. The question is whether a physician (or someone else) is duty-bound to intentionally kill a particular patient when and because that patient so demands.

The first argument that such a duty does not exist stems from the fact that the person asked to do the killing would normally be expected, if not morally required, to process the demand in an "unrights-lie" way. The prospective killer would be expected to ask whether or not the victim's reasons for wanting to die were good ones. As opposed to contexts in which genuine rights exist, one may legitimately refuse to kill another when it seems that the death would be morally worse than continued life. The potential killer need sustain only the burden of proof typical of non-rights contexts. "I don't think it is prudent of you to die," is sufficient to excuse the reluctant executioner. The decision doesn't belong to the potential victim in the way characteristic of rights.

Also a physician, or another potential killer, may excuse himself by saying "I simply don't do that kind of thing." Professional and personal codes of ethics prohibiting killing always justify a refusal to kill innocents. One cannot give a priest a hand gun and *demand* that the priest use it. One might ask, or beg, or plead to get shot with a gun or hypodermic but not as an assertion of right. But, when genuine rights are asserted, the one who has the relevant duty does not have any kind of blanket excuse. Even those who decide to honor another's request for help in dying may choose to terminate their participation or refuse to cooperate in specific killings. Such latitude in being excused from performing one's duty is, at best, unusual. Rights can be demanded urgently, preemptorily or insistently. Killing another is so serious that we cannot surrender the responsibility over or control of the decision to another. Others cannot *insist* that we kill them.

Finally, a physician's failure to kill a patient does not call for indignation. We might believe the physician should kill the patient, but refusing to do so does not call for the opprobrium characteristic of a violation of rights. Also a physician administering a painless and lethal drug to a terminally ill patient who had requested it might be appropriately thanked. Here, too, killing another at their request lies outside the domain of rights. The moral control a potential killer has over his decision to kill or not indicates that the victim has no *right* to be killed.

People cannot preemptorily demand, insist, or require that others kill them or let them die. Because death is irreversible, our privilege and our responsibility for making an independent judgment of the moral propriety of the death cannot be limited. What, then, of those facing an indeterminate existence of suffering and/or dependence?

## III. ACTING ON RIGHTS AND ACTING RIGHTLY

What we ought to do cannot be determined merely by knowing what someone's rights are. More important, there are things we ought to do that are not governed by rights. Killing others may be the right thing to do without being a response to a right of the victim. Gracious, loving, charitable, sacrificial, heroic, or saintly acts of killing are of enormous importance and value morally. Though rights are crucially important to our moral universe, there are other constellations. Acts of euthanasia — whether positive or negative — probably fall within these other categories. Administering a coup de grace can be an act of graciousness and of love instead of duty. To view the morality of euthanasia in terms of rights and their correlative duties is not only incorrect, but may well lessen its significance as an expression of truly humane interaction.

# CHAPTER 19

# The Sanctity of the Human Body
*Burton M. Leiser*

*The common law* and the statutes of most states declare that indecent treatment of a dead body, failure to bury it, selling it, mutilating it, or disinterring it, except under certain exceptional conditions, is a criminal offense [1, p. 526]. Such laws have undoubtedly arisen out of recognition both of the personal anxiety regarding the disposition of one's own body that so many people feel and the extreme distress that survivors may suffer upon learning of the mutilation or other mistreatment of a loved one's body. Vestiges of such ancient religious beliefs as those concerning the survival of the soul and the likelihood of bodily resurrection have undoubtedly played a role in the development of the law and custom of burial practices. But it is altogether possible that those beliefs themselves arose out of a need to explain or justify traditional practices, that they helped to rationalize venerable customs that seemed to need some explanation. The customs themselves, however, may have arisen out of profound anxieties and fears that all or most persons experience in anticipation of death or on the occasion of the death of a loved one. If I am right about this, then acceptance of the proposition that men are mortal and that bodily resurrection is either impossible or unlikely to be directly related to the condition in which one's body is committed to the ground would be completely irrelevant to issues surrounding the treatment to be accorded dead bodies. The facts of human psychology do not change. The rejection of archaic rationalizations born of nearly universal fears and anxieties has no logical bearing upon the question of the extent to which those fears and anxieties affect the lives of persons who are presently living. We should carefully distinguish between fears based merely upon false beliefs and beliefs designed to satisfy the desire for an explanation of unavoidable apprehensions. Rational persons may be relieved of their anxieties by being set right about the facts in cases of the first type, but no amount of scientific explanation will relieve those who suffer fears of the second type.

## THE RIGHTS OF THE DECEASED

Under the law in most American jurisdictions, every person has the right to determine the disposition of his body after his death, and in some states, where the right has been conferred by statute, the decedent's wishes are regarded as paramount [1, pp. 493–494]. As one court put it, "It is a universally recognized principle that there is a duty owing both to society and to decedent that the body of decedent shall be decently buried without unnecessary delay" [5]. Lack of space prevents an analysis of what it might mean to speak of a present duty owed to a person who is no longer alive, but it should be observed that any difficulties there may be about the matter are equally problematic in regard to such other well-recognized duties as those of carrying out a person's last requests, fulfilling the terms of his will, fulfilling promises made to him, and the like.

## GOOD INTENTIONS ARE NO EXCUSE

One might argue, however, that good intentions and worthy projects might relieve one of such duties, since the decedent is not likely to be harmed by whatever one might do anyway. Attorneys for Paul Hartzler, spiritual leader of a Bible study group in New Mexico, offered such an argument in their client's defense when he was convicted of indecent handling of a dead body. When the 19-year-old daughter of one of his followers died, he persuaded her father to keep her body in a shed, since he and other members of his religious group had reason to believe that she would return to life by act of God and that God would need her body for reincarnation. After citing numerous authorities, the court concluded that, despite the fact that New Mexico had no statute specifically covering the matter, under the common law there is "a well-established and known standard of decency and morality with respect to the disposition and treatment of dead bodies," and behavior that contravenes "the established and known public standards of decency and morals, relative to the care, treatment or disposition of a dead human body, is punishable as an act of indecency" [6, pp. 234–235]. That the defendant acted out of "sincere motives and with no evil intent," the court said, is immaterial. "[I]f a man deem that to be right which the law pronounces wrong, the mistake does not free him from guilt" [6, p. 235]. Hartzler's good intentions were not sufficient to free him of the duty to treat the dead in accordance with prevailing standards of decency. Notice that the court said nothing about the truth or falsity of Hartzler's beliefs. No judgment of that matter was necessary to pass judgment on his actions. In this respect, the case of a physician who sincerely and truly believes that medical science will be furthered by autopsies on certain deceased patients is identical with that of Hartzler. Neither his good intentions nor his beliefs would be enough to save him from legal sanctions if he took it upon himself to perform an unauthorized autopsy. Nor should the law be liberalized to permit him to do so.

There is no doubt that autopsies have been an important factor in making significant discoveries and in saving the lives of future patients. From a purely utilitarian point of view, a very strong case could be made for granting blanket permission to physicians to use their best judgment and order autopsies as they see fit. But this practice would be small comfort to many patients and their families, to whom the notion of being cut up, or having their loved ones cut up, their organs removed, handled, and examined, is gruesome, frightful, even horrifying. Their rights and their feelings deserve to be protected. In one case, the widow of a cancer patient, kneeling at the bedside a few moments after her husband's death, refused to sign a consent form for an autopsy, saying that her husband had "gone through enough." Ten days after the funeral, she learned that an autopsy had been performed anyway — evidently because the permission slip, prematurely signed by two "witnesses" to her signature, was not carefully examined by the pathologist. She suffered from sleeplessness and worry over the possibility that her husband might not have been dead when they dissected his body. The appeals court *raised* the damages awarded her by the trial jury, arguing that "her understandable sense of outrage and frustration at being deprived of her intention to fulfill her husband's wishes, despite her efforts to do so," justified the increased award [2]. This act involved negligence.

Where the acts are intentional, the courts are likely to be even harsher, and rightfully so [4].

Besides such feelings of shock and outrage, strong religious convictions may come into play. The physician's professional interest in learning more about his patient ought to be subordinated to the intense personal interests of the family most immediately concerned [3].

THE HUMAN BODY AND SOCIAL WELFARE

While autopsies and dissections can be valuable both for the training of future physicians and for the discovery of facts that may help to save human lives in the future, no person ought to be treated merely as a means for the advancement of medical research or the well-being of society. Nor should anyone's body be appropriated for such purposes without consent. There could be no moral objection to encouraging people, through educational campaigns, to see their bodies and the bodies of their loved ones as instruments for the healing of the sick and the relief of human suffering. But so long as the deceased person and his family withhold their consent because they see matters differently, unless an overwhelming public interest is at stake, no autopsy should be performed. When autopsies are either necessary or consented to, they should be performed with a deep sense of the sanctity of the human body.

REFERENCES
1. *Corpus Juris Secundum, "Dead Bodies".* Vol. 25A, esp. §§3 and 10. Brooklyn: American Law Book Co., 1966.
2. *French* v. *Ochsner Clinic.* 200 So. 2d 371 (La. 1967).
3. *Lott* v. *State.* 225 N.Y.S. 2d 434, 32 Misc. 2d 296 (1962), in which a state hospital was held liable to pay damages to two families as a result of its negligence in mistakenly sending the bodies of two elderly women to the wrong funeral homes.
4. *Phillips* v. *Newport.* 187 S.W. 2d 965 (Tenn. 1945), in which an unauthorized autopsy on a two-year-old was discovered by the child's mother when she went to the funeral parlor to dress the body prior to the funeral.
5. *State* v. *Bradbury.* 9 A. 2d 657 (Me. 1940).
6. *State* v. *Hartzler.* 433 P. 2d 231 (N.M. 1967).

# The Right to Live as Persons and the Responsibilities for Changing Behavior

## COERCION VERSUS FREEDOM

Issues of coercion and freedom arise in deciding what it means to live as a person and to exercise control over one's behavior and that of others. In this connection, Plato pointed out that human beings are not independent:

A state comes into existence because no individual is self-sufficing; we all have many needs. Having all these needs we call in one another's help to satisfy our various requirements and when we have collected . . . helpers and associates to live together in one place we call that settlement a state [13, p. 56].

A state makes human life possible, but only by achieving a balance between requiring people to function in one another's interest, in doing the world's work, in minimizing harm and injury between people, and in maximizing every person's freedom to live well. For Socrates the highest priority was survival of the state, even when he was condemned to death by the Athenians for questioning its laws. His devoted followers urged him to escape what they considered to be an unjust and harsh verdict. He refused. Socrates believed in the finality of the community's laws and his need to obey them even at the cost of his life.

Socrates believed that he owed the fact that he was nurtured to the existence of society, and he believed that he must uphold that society, no matter how wrong or unjust he believed its verdict to be. He believed in the right to be subversive of injustice; to question authority in dialogue and free interchange of ideas; to follow the argument to the truth and to the right, without exception; but not to overthrow or to revolt against the essential structure of society itself. He thought bad rule was better than no rule, for he believed that the bad could be corrected by the good.

Thus the claims of coercion and authority point to the demands of society for survival. The right of the mature members of society in positions of authority and power to limit the rights of others in the interests of preserving a justly lawful and orderly relation is what Willard Gaylin calls "the right to limit rights" [7]. Toilet training illustrates the right of parents to coerce their children in the service of society's need for sanitary and aesthetic surroundings. Society rewards functioning members of the community. Nonfunctioning members are sometimes treated much less well, and malfunctioning members, those who grossly offend or who injure themselves or others, are committed to mental institutions or to prisons.

Aristotle, long before Freud and Bentham, held that "pleasure is considered to be deeply ingrained in the human race and that is why in educating the young we use pleasure and pain as rudders with which to steer them straight" [1, p. 273].

However, in the process of "steering them straight," excesses and exclusions occur to those who cannot or will not function. Karl Meninger has called this treatment "the crime of punishment" [10]. It includes victimizing persons who do no harm, such as the neglected aged in nursing homes or the mentally retarded in custodial institutions, or denying persons the right to exercise their rights, such as mental patients who wish to refuse psychosurgery or electroshock therapy.

Opposed to the demands of authority, coercion, and control are an individual's claims to self-determination. Dostoevski's often repeated remark that a measure of a civilization is the way a society treats its prisoners applies here [4]. One could extend Dostoevski's test to other segments of a population, including the very young, the sick, the mentally ill, and the aging. They are all vulnerable and dependent. Some are simply unwanted as undesirable persons.

A society decides the position and status of its less productive and dependent members. In the Eskimo culture, for example, a reportedly common practice is for the aged, nonfunctional member to be put on an ice floe or to wander off into the icy wastes to die.

The precarious balance is between "the common good" and the simultaneous capacity of each individual to live a fulfilling life. Accordingly, there are two major views, the first on behalf of coercion, law, and control of behavior regarded as essential to survival [6, 7, 13]; and the second on behalf of the fulfillment of the individual. According to this first view, people in society have the right and the power to reward merit and function and to control, limit, or even punish nonfunctioning, dysfunctioning, and malfunctioning members.

The second view holds that society is partly or largely to blame for its dregs, outcasts, and criminals through its unjust distribution of goods and opportunities. Moreover, the inhumanity of society is evident in "cruel and unusual punishment" through rejection of those who either do no harm to others or those whose "crimes" are idiosyncratic acts that depart from the customary norms of the community. Classic novels of the 1960s, such as *Easy Rider* and *Midnight Cowboy,* portray this victimization of deviant persons. They portray the plight of the powerless without rights when faced with the unqualified disapproval of the majority. Hence authority is imposed on behalf of the most urgent needs of society, those of survival.

The chapters for this topic reflect a reformist approach to the plight of such powerless groups as children, the aged, the mentally ill, and prisoners. An attempt is made to find reasonable criteria for rationally deciding how to resolve conflicts of rights between persons, for example, in the child-parent relationship. Some argue that a child's rights come first [10, 15]. On the other hand, Ruddick argues that the rights approach in certain parent-child relationships is inappropriate [14].

A third position is for an adolescent to assume complete responsibility for his or her own health care with or without parental guidance as provided by law. Still another position is that the parent-child relationship has strong moral features that are unsuitable for either the legal or the moral rights sphere, for example, decisions about kidney donation from one identical twin child to the other; it is believed that these questions must ultimately be settled by parents.

The issues in relation to aged persons are similar to the questions about children regarding autonomous rights. Understanding of the aging process is imprecise, with a wide range of individual variation. However, the older adult tends to be stereotyped as prematurely dependent and incapable of autonomous function, which results in the loss of rights. Issues in the care of the aged include society's obligations versus its ability or willingness perhaps to maintain the aged in their customary habitat with adequate supporting services. In contrast, many in the aged population have been relegated to nursing homes, with a loss of most of their basic human rights, such as the right not to be drugged into a stuporous, nonmobile state for the convenience of the staff.

Among the least powerful of the dependent groups in a community are the mentally ill. They are the stigmatized, which Goffman, a sociologist, characterizes as "the situation of the individual who is disqualified from full social acceptance" [8]. These persons are thought to be less than human, and "we exercise varieties of discrimination through which we effectively, if often unthinkingly, reduce his life chances" [8, p. 15]. The tendency is to expand the person's list of imperfections and view defensive responses to situations of stigma as still another expression of incapacity, thereby justifying the low esteem in which such persons are held [8].

The stigmatized individual, whether because of mental illness or imprisonment, presents an incomplete, precarious self, which is subject to being discredited and abused [8]. To be treated as a disadvantaged person is to suffer loss of autonomy and the ability to participate in selecting one's course of actions.

To have rights, however, is to be in a position to assert personhood and to claim one's proper due. It eliminates the need for supplication to authority on the basis of abject need and powerlessness. The use of rights elevates the individual to someone who counts as a person.

The extension of patients' rights to the mentally ill are regarded by some [2, 12] as simply a restoration of human rights to this underserved and undervalued patient group. The right to dignified and respectful treatment, the right to full disclosure with consideration of the risks and the alternatives, informed consent, and the right to refuse treatment are no less important for the mental patient than for the medical patient. Yet the medical patient's refusal of the suggested surgery, treatment, medication, or diagnostic procedure is both commonplace and institutionally recognized, as for example, in the form of signing oneself out of the hospital against medical advice.

The controversy surrounding refusal of treatment by the mental patient is charged with assumptions of patient incompetence to make appropriate decisions. Critics of this view contend that even the patient who is involuntarily admitted should be granted the right to refuse treatment in the absence of an emergency. If a declaration of incompetence is deemed necessary to institute treatment, it should be sought through proceedings in which due process and review procedures are used, such as interdisciplinary committees and outside consultation. These safeguards have effectively reduced the frequency of psychosurgery and electroshock therapy [2].

Another form of deviance is seen in persons who openly decline the social position given them. Among these persons, Goffman lists prostitutes, drug addicts, criminals, delinquents, and the "urban unrepentant poor" [8, p. 144]. All of these persons are viewed as being engaged in "collective denial of the social order . . . as failing to use available opportunity for advancement . . .; they represent failures in the motivational schemes of society" [9, p. 144].

Such persons are judged to be destructive of the very fiber of society, and so they meet with scorn, rejection, and punishment. Prisoners and the mentally ill alike are seen as lacking in a determination to do better. In R. D. Laing's words, they are "mad or bad." They are therefore perceived as less than full human beings, and issues of rights thus seem inappropriate. Prisoners have been the primary group of subjects for experimentation, for unduly restrictive methods of behavior control, for work paid at substandard wages, and for a long list of grievances in which the prisoner is unjustly deprived of human rights.

In the larger social context, the function of prisons is viewed from the perspectives of incarceration, retribution, deterrence, rehabilitation, and punishment. Retribution holds that a criminal owes a debt to society to obey its laws and is punished when he violates that commitment. The rehabilitation model assumes that if prisoners are treated as if criminal behavior were a form of psychosocial illness, criminals would be responsive to appropriate humanistic methods of behavior control for an indeterminate period of time [11].

Punishment viewed as the prisoner receiving what he deserves is seen by others as necessary for the ultimate protection of the community and, if conducted with kindness and compassion, possibly as the most effective penal system [7]. Critics of this view point to the commission of more crimes despite or as a consequence of punishment and ask what worthwhile purpose is served [5]. The poorest and the least advantaged persons are probably the most punished through deprivation of those rights necessary to a full human life, namely, adequate food, shelter, nurturance, education, and medical care [5]. Critics point out the fundamental injustice of retribution, since the same crimes are not punished equally. Perhaps the most glaring violation of rights occurs when a prisoner is discharged and must then face the stigma of being a former convict.

The chapters within this topic are oriented to the plight of the vulnerable and powerless and for the need to recognize these people's human rights, although little or nothing is said about the cost and burden to society for two reasons, one logical and the other social. The logical reason is that there can be no rights without duties on others. If all the groups mentioned have human rights, there are corresponding obligations and responsibilities on the part of all health professionals and members of other appropriate disciplines to provide for the exercise of the human rights of these vulnerable, undervalued, or even unwanted persons.

The social reason is that some people may be unable or unwilling to incur the burden implied by such obligations. Moreover, does one spend limited resources on the prevention or the cure of illness or both, or does one meet human needs? If human needs are to be met, what is the fair share for the aged poor population, for example?

Those who abuse other persons deny Martin Buber's I-Thou relationship [3] that ennobles human life. Some people have little sympathy for the feelings of others, such as the aged poor and the mentally ill. Perhaps this is the reason rights are vital. Rights assure a baseline of decency, consideration, and respectful treatment. Care and concern are manifest when other values like love and kindness fail and cannot be counted on. A world with rights partly makes up for the inability to feel and act on behalf of the feelings of others. Finally, rights provide a rational and just basis of authority and for the more equitable distribution of powers.

REFERENCES

1. Aristotle. *Nicomachean Ethics* (translated by M. Ostwald). Indianapolis: Bobbs-Merrill, 1962. P. 273.

2. Bandman, E. The Rights of Nurses and Patients: A Case for Advocacy. This volume, Chap. 48.
3. Buber, M. *I and Thou*. New York: Scribner, 1958.
4. Dostoevski, F. *The House of the Dead*. New York: Dutton, 1912.
5. Ezorsky, G. The Right to Punish. This volume, Chap. 37.
6. Freud, S. *Civilization and Its Discontents*. London: Hogarth, 1953.
7. Gaylin, W. The Functions of Prisons and the Rights of Prisoners. Paper presented at Conference on Bioethics and Human Rights, Long Island University, April 8, 1976.
8. Goffman, E. *Asylums*. Garden City, N.Y.: Anchor Books, 1961.
9. Goffman, E. *Stigma*. Englewood Cliffs, N.J.: Prentice-Hall, 1963.
10. Lieberman, F. Special Children. This volume, Chap. 23.
11. Meninger, K. *The Crime of Punishment*. New York: Viking Press, 1966.
12. Peplau, H. The Right to Change Behavior: Rights of the Mentally Ill. This volume, Chap. 29.
13. Plato. *The Republic* (edited by F. M. Cornford). New York: Oxford University Press, 1945.
14. Ruddick, W. Parents, Children, and Medical Decisions. This volume, Chap. 22.
15. Webb, S. The Rights of Parents and Children in a Biomedical and Social Context. This volume, Chap. 24.

# The Rights of Children and Parents in a Biomedical and Social Context

# CHAPTER 20

# Human Rights and Ethical Decision Making in the Newborn Nursery

*Natalie Abrams and Lois Lyon Neumann*

*The issue of human rights is discussed in relation to ethical dilemmas posed in the newborn nursery. Because of recent advances in neonatology, many infants are now able to survive who previously would have died. Some of these infants, however, may survive with serious handicaps. The problem arises, therefore, as to whether or not the lives of all infants should be maintained, regardless of the prognosis for the infant's physical or mental development. In this chapter, we discuss three basic issues concerning human rights in the neonatal situation: (1) to whom rights should be ascribed, (2) what interests should count as rights, and (3) how conflicts among rights can be resolved. Recognition of rights is necessary to assure that the interests of neonates, who cannot speak for themselves, are adequately considered.*

In the past there has seemed little need to delineate the respective interests and rights of those concerned in the care of sick and damaged newborn infants. Maintenance of the life of the infant has generally been considered the common goal of all. New complexities in decision making have been introduced, however, with the rapid development of intensive care for neonates in the years since 1960. Death rates have been lowered greatly for small premature infants and for infants with a variety of serious conditions [1, 7, 8]. Improved survival has been accompanied by a lowered incidence of neurological and other handicaps among the survivors [6, 9]. At the same time, however, some infants with serious congenital abnormalities, such as meningomyelocele and hydrocephalus [2], are now enabled to live. Others may be permanently injured by the very measures implemented to save their lives. For example, modern techniques have at least doubled the percentage of premature infants surviving hyaline membrane disease. However, a few of these infants develop a severe chronic lung disorder, which is, in part, caused by the respirators that supported them through their illness [4]. In addition, it is now possible to maintain some infants on life-support systems for extended periods, in spite of conditions incompatible with independent existence. Advanced technology has thus brought new problems. Should the lives of all infants be maintained, regardless of the prognosis for physical or mental development? This issue is a moral question and can be approached only by considering the respective rights, duties, and values of the concerned parties.

In our society, the "right to life," for example, is held to be both a moral and a legal right. The term "right" is generally used when it is thought that an individual has a just claim to something. When that claim is supported and enforced by the laws of the society, the individual is said to have a legal right; when the claim is thought to be based in some way on the fact of the person's humanity but is dependent on either individual or social conscience for satisfaction, rather than the legal system, the individual is said to have a moral right. The dilemma posed in the newborn nursery can be seen as whether or not all newborns have a moral right to life and, if they do, whether this right can ever be overridden by other rights or interests, namely, those of parents or society.

The most troubling issues surrounding an assertion about rights are: (1) to whom they should be ascribed, (2) what interests should be designated as rights, and (3) how conflicts among human rights can be resolved.

The first issue surrounding the question of rights is, if there are any human rights, to whom they should be ascribed. This question has two parts. First, who is to count as a "human being" or "person"? Second, should these rights be universally ascribed to all people or only to select populations, based upon specific criteria? Various attempts have been made to assign rights on the basis of specific characteristics. In one such attempt, the definition of a "person" is said to depend on the capacity for rationality. Others argue that rationality is not the basis, but rather, another capacity, such as the ability to sense pleasure or pain, to have a sense of self-consciousness, to remember, or to form attachments to other people. None of these capacities is possessed equally by all human beings. The application of any of these criteria would imply that certain "human beings" are not "people" and therefore do not possess rights or at least not equally. No definition of personhood carries with it the inherent possession of rights. The issue of the ascription of rights raises the question of whether or not infants (normal or defective) should be considered "persons" with protectable human rights. This question is similar to the one raised in the abortion controversy concerning the functions, cognitive or otherwise, essential to being a "person." The various positions concerning when "personhood" begins are not factual decisions. They are based instead on alternative conceptions about what is morally justifiable action at different stages. Therefore deciding if newborn infants are "persons" does not appear to be a worthwhile approach to the problem. It still must be decided what actions are appropriate toward what kinds of beings, whether they are "people" or not. The assignment of rights depends on prior value judgments and human decisions. "Assertions about natural rights are assertions of what ought to be as a result of human choice" [5, p. 49].

The second issue is the choice of which values or interests should be considered as "rights." We all have numerous interests, for example, obtaining a certain job, earning a particular income, or living to a certain age, which we would not necessarily consider as rights. Ultimately, decisions implicitly or explicitly made within the context of a given society determine which interests are designated "rights," as well as to whom they are ascribed. To address this issue in regard to the newborn infant, it is first necessary to identify the interests that come into play. Relevant interests to consider are usually those of the infant, parents and other family members, society, and health care professionals. The newborn infant possesses a value, which, in our society, is thought to entitle it to a right to life and hence to the medical care necessary for its development. Parents, on the other hand, have interests in the well-being of the newborn infant, as well as in their own well-being and that of the newborn's siblings. The enormous financial, emotional, and psychological investment necessary to raise a defective newborn makes it essential to consider the family's interests. Society also has interests in the decisions concerning the care of newborn infants: in allocating scarce resources equitably throughout the population and in areas in which clear benefits are to be gained. Furthermore, medical personnel have interests in using their skills with patients who will most benefit and in acting in accordance with their own personal and professional value systems.

The issue then becomes which of the many relevant interests should be considered rights. For example, does a defective newborn have a "right" not just to life but to a life of a certain minimum quality and hence perhaps a "right" to death if a life of this quality is impossible? This argument has been presented in some cases claiming wrongful life [3]. Do parents have a "right" to or simply an interest in a certain degree of financial security? Does society have a right to impose its standards regarding the care of defective newborn infants? Alternatively, does society have the obligation to support and care for individuals if it has demanded that their lives be maintained? Do medical personnel always have a duty to prolong life? To what extent do the responsibilities of medical personnel extend to the family as well, and do medical personnel have the right to refuse to participate in medical interventions with which they disagree?

The third issue is how conflicts among human rights can be resolved. Although human rights may be absolute in the sense of being universally applicable, they may not be absolute in the sense of never being overrideable. A claim that people have a right to life, for example, should not be taken necessarily to imply that other rights can never override this right. The dilemma about conflicting rights frequently takes the form of a question regarding the locus of decision making. If human rights are the result of human decisions based on value judgments, whose decisions should they be? The newborn, of course, is incapable of deciding for himself. In most situations, parents are the primary decision makers for their infants. By law parents are given the right to "proxy" consent on the assumption that they are the ones to decide in their child's best interest. They are not given this right on the assumption that *parental* interests or wishes, per se, form the legitimate basis for a decision. Society places some restrictions on life and death decisions that parents can make concerning their children. If the state believes a parental decision is not in the best interest of the child, the decision can be overridden.

It might be argued that parents should have the right to make decisions about their infant because they will ultimately be the ones to assume the responsibility for the child's care. In addition, if the child will be raised within the family, decisions that are in the parents' best interest can be seen ultimately as also in the best interest of the child. In accord with this view, parental interests are seen, in most cases, as intimately intertwined with those of the child, rather than as conflicting. It might, however, alternatively be argued that, because of the possibility of a conflict of interest, parental involvement in the process of decision making should be limited. The burden the parents might have to bear might prevent them from being able to consider adequately the independent rights of the infant as a unique individual. Some socially agreed upon standards should be applied when individuals must make such important decisions for others. In this complex balance of rights, it is an oversimplification to assign decision making solely to parents without recognizing the overlapping but distinct interests of parents and child.

Although the interests of newborn infants are usually protected by the benevolence and good intentions of individuals who form the intimate community on which the child is dependent, recognition of the child's interests should not be allowed to rest solely upon the benevolence of others, even that of the parents. The language of rights, therefore, seems especially necessary when dealing with dependent individuals, such as infants, to assure that the interests they cannot assert for themselves are adequately considered.

REFERENCES

1. Farrell, P. M., and Avery, M. E.   Hyaline membrane disease. *Am. Rev. Respir. Dis.* 111:657, 1975.
2. Freeman, J. M.   To Treat or Not to Treat. In J. M. Freeman (ed.), *Practical Management of Meningomyelocele.* Baltimore: University Park Press, 1974.
3. *Gleitman v. Cosgrove.* 49 N.J. 22, 227 A. 2d 689 (1967); *Stewart v. Long Island College Hospital.* 35 A.D. 531, 313 NYS 2d 502 (S. Ct. App. Div. 1970); *Zepeda v. Zepeda.* 41 Ill. App. 2d. 240, 190 N.E. 2d. 849 (1963).
4. Hodgman, J. E.   Chronic Lung Disorders. In G. Avery (ed.), *Neonatology.* Philadelphia: Lippincott, 1975.
5. Macdonald, M.   Natural Rights. In P. Laslett (ed.), *Philosophy, Politics, and Society* (first series). Oxford: Blackwell, 1967.
6. Rawlings, G., Stewart, A., Reynolds, E. O., and Strong, L. B.   Changing prognosis for infants of very low birth weight. *Lancet* 1:516, 1971.
7. Stahlman, M. T.   What Evidence Exists That Intensive Care Has Changed the Incidence of Intact Survival? In J. F. Lucey (ed.), *Problems of Neonatal Intensive Care Units* (report of the Fifty-ninth Ross Conference on Pediatric Research). Columbus, Ohio: Ross Laboratories, 1969.
8. Usher, R. H.   The Special Problems of the Premature Infant. In G. Avery (ed.), *Neonatology.* Philadelphia: Lippincott, 1975.
9. Vapaavuori, E. K., and Raiha, N. C. R.   Intensive care of small premature infants. I. Clinical findings and results of treatment. *Acta Paediatr. Scand.* 59:353, 1970.

# CHAPTER 21

# Pediatric Intervention: Odds Are the Parent Knows Best

*Deborah Michelle Sanders*

*Only a person* who is above the age of majority or who has been legally emancipated is presumed at law to be competent to contract for or consent to medical services. Since a physician who renders services without consent would be liable for the tort of battery, the physician needs to obtain third-party consent for pediatric services. In Anglo-American law, the parent is considered to be the natural guardian of the child's person (although not of the child's property), so that the physician looks to the parent to give legal consent to the pediatric intervention.

The term "proxy consent" is rarely literally applicable in these sorts of situations, although it has a useful figurative or symbolic use whenever parents accept the risk that their child may suffer from the procedure. Privacy within the family is a fundamental legal right, which leads me to advocate an absolute parental authority to give or refuse consent when a physician advises treatment or lawful experimentation, so long as it is not designed to save the child's life. I also believe in limiting enforcement of the parental duty to provide lifesaving treatment to children to situations where the child is above neonatal age. Because this book's format requires brevity, however, I shall not discuss these issues in depth but will introduce you to the problems by describing one position, my own, with special reference to one set of children, neonates.

Among those who conceive of these problems as being political questions (as I do), the position I am here espousing is often called the minimist position. A maximist reformer would attempt to give decision-making authority in pediatric consent to the state, either in the short run — allowing a revision of the law so as to grant children the right to contract and consent for themselves — or in the long run — delegating authority to public guardians to make all decisions concerning pediatric consent.

## PARENTS BEAR THE RISKS

First, to my mind, what is really at stake in the parental decision to give or refuse consent for pediatric intervention is that the parent has *borne the risk* involved in the procedure and has to make the best of the situation should the procedure go wrong. Children cannot accept risks for themselves. If they could, the state would not have the right to remove them from their home when their parents torture them but they want to live at home. Parents have to cope with their children's distress and want to minimize it. Second, parents shape their children's values about what kinds of health care should be obtained. The parent is in a better position to know what his child is growing up to value than is the local trial court judge. A child's ideas about whether or not it is appropriate to volunteer to be an experimental sub-

The original version of this paper was written with the support of National Endowment for the Humanities Grant No. AV-10434-75-76 to the Institute of Society, Ethics, and the Life Sciences through its Post-Doctoral Fellowship Program.

ject derive from his parents' beliefs. Beyond all this, parents want to continue to enjoy their children, totally apart from wanting their children's personal happiness. Parents want the companionship and the sense of a sort of immortality that children can bring, and they seek to ensure that those pleasures are not cut off or attenuated.

Pluralism is the root value implied by our Constitution. The least detrimental means for the state to use in protecting the lives of children without impeding pluralism or familial life-styles would be to restrict overriding of parental choice to situations where the child's life is at stake. That is not the norm today. At contemporary law, courts sometimes override parental decisions to give consent to experimental procedures or to refuse consent to nonlifesaving treatment.

## COURTS DO NOT KNOW ENOUGH TO DECIDE

The paternalistic, or *parens patriae,* power of the state is frequently exercised today in overriding parental decisions about medical procedures to be performed on their children. But a problem that has to do with whether or not a child should get medical treatment is not suitable for judgment by a court. The issue for decision — whether the parents are giving the child the health services desirable for a good life — is not a legal question. What constitutes the "good life" is an important ethical question that can only be answered with reference to the consciousness of the individual. Courts do not enjoy that intimate access. Furthermore, the court generally has a distorted view of the immediate facts about the proposed intervention. The case involves a familial dispute about whether a child should be given certain care. The lawyers for each party and the judge himself are likely to be reminded of times in their own lives when they wished they had got care different from what they had been given. They are likely to try to push such memories out of their minds in order to reduce bias. Occasionally this tactic and other maneuvers will serve the cause of objectivity, but often bending over backward to be fair will only serve to make the resolution of the problem unfair. And never can all the facts about the familial value structure and relationships be brought to the attention of the harried and totally judgmental court.

While parents may be shortsighted or even blind to their child's best interests, the court's blind spots lead to even more trouble. When society relies on the court to make all the decisions, parents are not going to take the common precaution of feeling out where the best path of decision lies, and misery may result. Only if the child's life is in imminent danger should the court overrule a parental refusal of consent to treatment.

And parental consent to pediatric intervention should never be overruled. No exception to this principle should be made on the ground that parents often give their consent to useless or even dangerous procedures. In trying to correct these sorts of errors, the court would institute other errors because it does not know the individual needs of the child. Nor should an exception be made for mentally retarded children — even though they often do not understand the source of their pain and are thus especially vulnerable to meaningless suffering. Familial privacy is not something that can be allowed to vary according to whether a parent has brought bright or slow children into the world.

## COMPENSATION FUND

However, it would be appropriate to organize a compensation fund for those children who are injured through experimental procedures, so as to provide money to divert the children from their problems and rehabilitate them. This fund is needed because parents do not have enough information to be able to truly accept the risks of an experimental procedure. The government might well restrict the entry of children into experiments except in circumstances where only children will do. Researchers would understand the symbolic meaning of the restriction — and would limit their experimental protocols to procedures that respect human dignity, so as to be likely to win individual informed consent. Governmental policy would be a collective decision about all children. However, a court should not make an individual determination about a particular child's service in an experiment, once it has been decided that children are eligible as subjects in it.

## THE ADOPTION ALTERNATIVE

The problem of withholding treatment from newborns has only become pressing within the last decade or so. Although technological advances have not healed birth defects, they have made it possible to save the respiratory and circulatory systems of almost every newborn and to perform sometimes subtle insults to the child's central nervous system as well.

Say that the mother of a baby born with Down's syndrome and duodenal atresia refuses consent to the repair of the baby's digestive tract because she is unwilling to suffer the fate of parenting a retarded child who is likely to die as a young adult of heart disease. When a child neglect petition is filed — as is the traditional event in these cases — and the court appoints a guardian for the child, the predictable outcome will be that the child's intestinal defect will be repaired and that its custody will be returned to her. Now she has been assigned the task of parenting an unwanted child till doom, without recourse to sufficient social service funds to allow the child to experience consistent and devoted education and for the mother to find respite from caretaking tasks at intervals. She might decide to place the child in a custodial institution.

I am convinced that it would be far better to find out at the time when parents refuse consent to the lifesaving treatment whether there is a set of potential adoptive parents available who will be eager to nurture the developing, though handicapped, child. The court's jurisdiction over refusal of consent to lifesaving treatment cases should be restricted to terminating the parental rights of the biological parents and simultaneously authorizing the parental rights of adoptive parents. This policy would require adoption agencies to search for potential adoptive parents for newborns handicapped by birth defects and to communicate regularly with hospital social workers. As long as a child is wanted in the family, focused attention and a sensitive heart enable the parent to guide the child in embracing the world, although to be sure often adequate community services have made the favorable outcome possible.

This change in the law would result in the deaths of many children who would otherwise know only institutional custody. This change in the law would also result in the adoption of some children who would otherwise be isolated behind state hospital walls. Today it is notoriously difficult to find adoptive parents for handicapped foster children. Most such children are beyond the crucial years of babyhood when early infant stimulation could spur their development to the best of their potential. Furthermore, most of these children have suffered the fate of the typical foster child who is moved from home to home, being left at risk for emotional illness. Such disturbances transform handicaps into disabilities. In contrast, adoption of handicapped infants at birth affords continuity that makes acceptance possible and emotional health more likely. Because the prospects are hopeful for these newborns, their adoption can be attained. Since lifesaving treatment is often needed immediately, it would be necessary to develop procedures to expedite the adoption of such children (and to make the adoption final) within 24 hours of the biological parent's refusal to consent. Potential adoptive parents would be thoroughly screened, with especial care taken to determine what sort of handicapping condition they can accept in their potential child, and their names would be placed on a waiting list pending the event of a consensual crisis.

# CHAPTER 22

# Parents, Children, and Medical Decisions
*William Ruddick*

*Parents,* not nurses or physicians, are a child's primary health care providers — and withholders. Often parents act out of ignorance or indifference, and it falls to the school nurse to educate and alarm them about their children's health needs. On occasion, however, parents refuse treatment on metaphysical principles or incorrigible eccentric beliefs, and it falls to a court to consider coercive medical intervention.

Judges typically frame their deliberations in terms of conflicting rights. They weigh, for example, a child's right "so far as possible, to lead a normal life" against parental rights of autonomy and privacy in family matters. Lately children's rights have begun to prevail over parental rights, a shift that reflects increasing legal concern for children and improved surgical techniques. It does not, however, show progress in moral or legal thought about the relationships between children and parents.

For the most part, children are currently viewed as another oppressed group, along with blacks and women, and children's rights are modeled upon civil rights established by the Constitution or recent legislation. But children require separate consideration: Their needs and capacities differ from those of oppressed adult groups. Children are, unlike women or blacks, *necessarily* dependent on other adults for life's necessities. Their rights are primarily "claim-rights" against adults who have correlative obligations to provide them with food, shelter, and the like. By contrast, the civil rights sought by adults are "liberty-rights," or rights of noninterference from others.

Moreover, we must take special care in enforcing children's rights. Since children depend on adults for life's benefits, as well as life's necessities, and since they seem to benefit from continuity of care and of caretakers, we do not want to enforce their claim-rights in ways that alienate parents and thereby deprive children of nonnecessary benefits that even negligent parents can and often provide. Adults, by contrast, rarely if ever are so dependent on a given set of people (not, for example, on employers, voting officials, welfare administrators), and therefore in enforcing adult rights it is not necessary to preserve or foster more congenial relationships between right-bearers and right-violators.

These two differences between adult and children's rights connect with a general fact of human life: A child's body, life, and interests emerge within and diverge from the body, life, and interests of his or her parents only gradually. This slow and often uncertain individuation of child from parents has not, I think, been given proper weight in legal and moral thought. It casts doubt not only on the civil rights approach to child welfare but on the general rights approach itself.

After reviewing some of the rights judges have cited in medical cases, I shall return to this matter of slow individuation, sketch a principle that respects it, and examine some cases in the light of this principle of parenthood. The result, I hope, will provide a better sense of what parents do and do not owe children than does the attempt to define and balance rights.

## RIGHTS OF PARENTS AND CHILDREN

For a need or interest to count as a claim-right, it must meet at least three conditions: It must be within the power of particular persons or agencies to fulfill the need or interest; it must be the obligation of particular persons or agencies to do so; and the failure to do so must be a wrong or injustice to the person with the need or interest.

Clearly, even a healthy child cannot have a right against his or her parents or parent surrogates to "a normal, useful, happy life." Even wealthy, leisured, and devoted parents do not have it within their power to provide such a life, for such a life depends too much on happenstance, both genetic and social, beyond anyone's control. But even were there such a right, it would be irrelevant in many pediatric cases: The best medical care could not improve the chances for such a life when the child in question bears a massive deformity and the psychosocial side effects it has already produced. What is at issue in these cases is a *tolerable* life, not a normal, happy life.

More reasonably, one judge (*Matter of Seiferth* [3]) cited a child's right "so far as possible" to lead a normal life. But even this qualification leaves parents with an onerous burden. If a "normal life" includes a certain material standard of living, parents toward the bottom of the economic scale will bear the double burden of poverty *and* ceaseless toil to raise their children to that standard, not as a labor of love but of duty. Or if the child has a physical handicap or malady, the "normal, so far as possible" principle seems again to burden a parent unreasonably. It requires a devotion to the child that leaves little room for any other interests; such a demand more often produces resentment and negligence than benefit for the child.

We might balance such demands, however, against oft-cited parental rights of autonomy and privacy in child rearing and other family matters. On their face, these rights are puzzling. They look like "liberty-rights" of the sort invoked to permit adults to act alone or in consort in ways that threaten harm to no one but themselves. But a child does not voluntarily join or consent to parental activities.

Paradoxically, what may justify and limit a parental right of autonomy and privacy in child rearing are the child's own interests. By giving parents this freedom from interference, they are more likely to take pride and interest in their children, or at least treat them more consistently than they otherwise would. And there is evidence to suggest that continuity of care and caretakers benefits a child, even when the care is substandard.

Just as we above used parents' interests to search for a plausible child's right, so here we use children's interests to justify a parental right. This mutual accommodation, however, casts doubt on this search for rights altogether. It is actual or likely competitors who need their interests legally defined and protected. A right, as it were, is a ticket to court. But even in our litigious and Freudian culture, we still cling to ideals of family life that celebrate trust, loving care, and common interests.

There must, of course, be legal remedies against abusive or negligent parents. But it is not such parents who are brought to court for refusing permission for medical treatment for their children. Technically, their refusal does count as prima facie negligence, but by all other standards they are responsible parents acting in good, even if (by current social standards) benighted, faith. This is what makes the cases perplexing and the legal model of adversarial parent and child distorting. To free ourselves from the courtroom drama and jargon, let us turn to some facts of parenthood, including that of a child's gradual individuation from the parent(s).

## PARENTHOOD: MOTIVES AND PRINCIPLES

People often become and remain parents in order to reproduce some aspect of themselves or their lives. For example, they want to embody their love for one another, to perpetuate name or business, or to express fully their sexual powers and preferences. But even when there are no such "reproductive aims," a fetus enters a preexisting world of adult desires and interests. It enters gradually, developing within the body and lives of others.

I use "lives" here in the biographer's, not the biologist's, sense. The two meanings are closely related, for a life in the historical sense causally depends upon, and gives value to, life in the biologist's sense. In English, unlike many languages, we depend on context rather than on distinct phonemes to make the distinction. We need to be self-consciously clear here, however, for the biographical notion of *a life* is needed to state a basic problem of parenthood and to formulate a helpful principle.

The basic problem is this: When and how is a parent obliged to enable a child to lead a life of his or her own? Children tend to develop the requisite capacities for forming projects, pursuing them, and reflecting on them with pride or shame, all of which is part of leading a life of one's own. Parents have the power to retard or foster these capacities, as well as the power to adapt them for certain kinds of lives rather than others. What are a parent's obligations and liberties in these matters? To what extent may a parent make a child's life, or a child's range of possible lives, an extension of the parent's own life? How many life possibilities must a parent foster or allow to develop? Must a parent allow a child to develop life possibilities of which the parent would disapprove if realized? (Or, to shift the moral burden for the moment, must a child in making life choices seek to satisfy a parent's expectations or hopes?)

By "a life of one's own," I do not mean the life of a self-absorbed or isolated individual, rugged or leisured. A life is typically, if not essentially, social. It is partially defined by the lives of those we care for, work with, and live with. Nor do I mean to suggest that parents must encourage or force children to live apart. They need never kick them out of the parental nest, so long as they teach them to fly and to build nests of their own if need be. Since parents often do die before their children, they must prepare the children for a life without them, and since death is usually unpredictable, parents cannot procrastinate.

The time of a parent's death comes later and is more predictable than in previous times and cultures. But the technological change that delays death and makes it more predictable also makes the world of the next mature generation less predictable. The more rapid the cultural change, the more difficult the parental problem of preparing a child for it.

Given the uncertainty of death, the system of taxation, and other social factors, it may be difficult to act on the following principle, but it would be neither impossible nor alienating to do so:

A parent should foster life possibilities for a child (1) each of which a child could realize and find acceptable without the parents' aid or existence; (2) each of which, if realized, can be expected to be acceptable to the parents as well as to the child; and (3) all of which jointly cover the range of possible future worlds predicted by people whose knowledge of the world the parents respect.

This principle will not suit those who subscribe to either the traditionalist principle or to the smorgasbord principle. The traditionalist says: What's good enough for me is good enough for my children. (Traditionalists include carriage makers who looked forward to the day when they could hang out a sign, "Smith & Son: Fine Carriage Makers," and trained their sons accordingly.) The smorgasbord advocate says: A parent should set out as many life possibilities as available within his or her means and let the child choose. (Certain child libertarians, often childless, hold this view.)

Nor will the principle suit secular rationalists who wish to hold everyone to a certain set of authorities trained in science, economics, political theory, and futurology. My principle allows parents a free choice of authorities when it comes to predicting the future. It only requires that they give serious thought to the future in the rearing of children.

Although much more could be said to clarify and justify this principle, perhaps we should now turn to the pediatric cases. Indeed, it may seem that even this sketch is superfluous, for the questions of how many life possibilities and which life possibilities do not arise for parents of children with maladies of the kind that invite public and medical intervention. Normal anatomy is of use to pathologists, however, and thus a principle based on normal parent-child conditions can help to make sense of our moral intuitions and legal decisions about abnormal cases, and vice versa.

## SOME PEDIATRIC DECISIONS
Our principle suggests the following fourfold classification of parental refusals, namely, for:

1. Children with *no* independent life possibilities acceptable to both parent and child without treatment, but with some life possibilities (at least one) with standard medical treatment. (Examples: *Seiferth* [3], a child with severe cleft palate and lip, and *Sampson* [2], a child with massive tumor of head and neck.)
2. Children who, if treated, have independent life possibilities but none the parents can accept, even though there are parent substitutes who can accept them and bring the child to accept them. (Examples: infants with Down's syndrome, or mongolism, and a surgically correctable gastrointestinal defect, which would otherwise be fatal.)
3. Children who, even with treatment, have no life possibilities acceptable to the parents or to parent substitutes (other than custodial institutions). (Examples: severe cases of spina bifida.)
4. Children who, even with treatment, have no life possibilities at all. (Examples: cases of massive cerebral cortex defects or damage as in Karen Quinlan's case.)

In which cases have parents violated our principle of parenthood? The parents in both *Seiferth* and *Sampson* seem to be in violation. In *Sampson* the mother wanted the tumor removed, but as a Jehovah's Witness she would not consent to blood transfusion (without which the child would not survive the operation). Without the operation, the child would remain secluded at home, illiterate, and with no life prospects outside the family. Nonetheless, the mother would have refused the operation rather than agree to transfusion — even though she did not seem to think that transfusion would forever condemn her son and ruin his eternal prospects. The court overruled her.

In *Seiferth* the father wanted the cleft palate and lip healed by "natural," or psychic, means rather than by surgery. His desire (in which his bright, adolescent son concurred) was respected, on the ground that the son could (and almost certainly would) choose surgery when older, even though delay would submit him to serious social prejudice and would make surgery less effective. The case is not a clear violation of our principle in that the father and court seemed to allow that there were some acceptable life possibilities even without correction, although they were far fewer. Had the father and court disagreed on this point, the case would have raised an issue with which clause (3) of our principle would have to cope, namely: Are parents allowed to invoke eccentric metaphysical beliefs in determining what life possibilities can be expected? If not, parental autonomy is compromised by current (conventional?) standards of rationality. Courts do not allow parents to risk the death of their children in occult religious rituals. Should the courts allow parents to risk the life possibilities of their children on metaphysical grounds, as in Amish practice and in *Seiferth?*

Parents in the second type of case do *not* seem to violate our principle. For example, in a publicized case at Johns Hopkins, a mother refused to allow surgical repair of a defective intestine in her mongoloid newborn, and the infant was allowed to die. She presumably rejected parent substitutes, and she cited her duties to her other five children to show that she could not give it the care that might eventually give the child some capacity for an acceptable life of its own. The case raises another point for reflection: Are parents to have unlimited discretion as to whether a child growing up with parent substitutes can lead a life acceptable to the parents? The courts allow children who are flourishing with foster parents to be reclaimed by biological parents with far less promising conditions for the child. May the biological parents claim that all life possibilities provided by others, even though superior by objective standards, are unacceptable to them and hence are to be rejected in favor of the life possibilities they can provide? Our principle seems to allow this claim.

Class three is distinct from class two, in concept if not in application. Parents of spina bifida infants, like parents of Down's syndrome infants, may in their initial disappointment, resentment, and guilt too quickly assume that (1) no acceptable individual parent substitute can be found and that (2) institutional life is unacceptable. But if, after suitable counseling and reflection, they refuse surgery for their infant, they do not violate our parent principle, even if death ensues.

The recent Karen Quinlan case is an instance of the fourth type. Physicians and others agreed that, without divine intervention, her brain damage left her with no life possibilities. (The appellate court judge wrote of "no sapient, cognitive life.") Her parents did not hold out for divine intervention. Indeed, they took forced respiration to be secular intervention with God's will for their daughter. In withdrawing her respirator at the high risk of her death, they did not violate our principle. Moreover, they and the higher court rejected the lower court claim that "life was Karen Quinlan's most precious possession and, as such, not even her parents could be permitted to take it away from her." The final judgment made clear that biological life without hope of biographical life was of little, if any, value.

These cases are not neatly classifiable or decidable by our principle. But I have not tried to find a single ground for moral or legal judgment of these cases. Parent-child relationships, let alone parent-child-physician-state relationships, are too complex for any such decisive principle. But the complexity is, I think, only compounded by that of the rights approach. I have sketched here what I hope is a more promising tack.*

REFERENCES
1. Ruddick, W. M.  Parents and Life-Prospects. In O. O'Neill and W. Ruddick (eds.), *Having Children: Philosophical and Legal Reflections on Parenthood.* New York: Oxford University Press, 1978.
2. *In re* Sampson.  317 N.Y.S. 2d. 641 (1970).
3. *Matter of Seiferth.*  309 N.Y. 80, 127 N.E. 2d. 820 (1955).

---

*This approach is followed further in Ruddick [1].

# CHAPTER 23

# Special Children
*Florence Lieberman*

*Children are* an endangered species. Though society expresses a sensible interest in birth control, so that space and resources will not be overtaxed, and though there is increasing concern about parental abuse of children, there is inadequate social attention to the provision of services needed to nourish, cherish, and foster children, including many that enable their growth into happy and productive adults.

The major problems of present-day children derive from negative environmental situations, among which the most severe difficulties evolve from the poverty in which more than 20 percent of the children in the United States live [16]. Poverty leads to malnutrition and retardation, developmental attrition and developmental disorders, and a variety of severe psychopathologies. These problems are exacerbated by discriminatory, restricted access to resources of prevention and treatment. Limited available services tend to be distributed by a variety of professionals who have to determine which services will be provided for or withheld from specific children and their families. Yet too often professionals do not exercise sufficiently their influence or do not have the necessary political power to determine the nature and scope of services that must be provided. As a result the children of poor parents are the most poorly served.

A child is dependent upon the goodwill, intentions, and abilities of others. Children's lack of political power results in limitations of their rights to those rights granted to them by their caretakers and influenced, through expansion or restriction, by the mandates of the larger society. Children are at risk and endangered whenever their social reality cripples their potentialities by not providing for their right to adequate care and the meeting of their needs.

Children whose families cannot care for them become special children with labels such as abused child, foster child, and so on. Then, often instead of being understood as having the right to have their special needs met, they become children with less than normal rights. What may happen is that priority is given to the needs of the new systems in which they are placed, the foster home, the child welfare agency, the residential treatment center; the child's needs too often are subordinated to the needs of others.

## THE RIGHTS OF CHILDREN
In all societies, past and present, childhood has been recognized as a time of training for adulthood, for that period when the fully grown person will be expected to assume the roles, obligations, and privileges that are defined by a particular society. Children have always been understood as being unable to care for themselves and as needing to be cared for, but the definition of the age span of childhood has varied. With the increased complexity of social systems, the duration of childhood and the need for care has increased proportionately, as has the need for societal assistance to families in their task of child rearing. Today the legal definition of who is a child varies within the United States [9], the laws of different states being inconsistent,

anomalous, and having conflicting age levels. Yet modern psychology and child development studies consider that childhood extends through adolescence.

Only recently has a concept of children's rights been advanced, rights similar to those of adults but differentiated by special needs. Historically, children have been treated with heartlessness and cruelty for centuries, child abuse and infanticide being normal aspects of life [2, 3, 5, 7]. Children have been considered extensions and possessions of their parents without any intrinsic and individual rights. In addition, children's natural behaviors often were considered dangerous, so much so that nineteenth-century America reacted similarly to the mentally ill and to children [6]. There was a common belief that both groups needed to be protected against their own dangerous willfullness and had to be taught and enforced to have the manners and obligations of the sane adult world.

Though services developed to meet the needs of dependent, uncared for children, they tended to focus more on physical and moral needs with little attention to or understanding of affective, emotional aspects. Today with the development of deeper knowledge of child development and need, the consequences of neglect and deprivation have been broadened to consider emotional and psychological influences; there is increasing awareness that childhood experiences have a lifelong impact.

The child is now thought of as a person with a full range of rights, including the right to societal mechanisms for personal enhancement [21]. The rights of children include the basic needs of good food and shelter, constancy of care with a secure family, and where necessary, developmentally oriented therapeutic help and facilitation of growth as a developing human being [13].

Needs and rights have different policy implications. A need does not locate the responsibility for meeting of the need, whereas a right suggests that abridgement is an injustice that the institution of the state must correct [17]. The intention of the United Nations "Declaration of the Rights of the Child" [20] was to establish children's rights to special protections, opportunities, and facilities without distinction or discrimination, whether of himself or of his family, and to enable physical, mental, moral, spiritual, and social development. In addition, as a result of the White House Conference of 1970 that focused on children's rights, a National Center for Child Advocacy was established in 1971. Though children have always required advocates, today advocacy has been defined as intervention on behalf of children in relation to those services and institutions that impinge on their lives [14].

Children need advocates for their rights because even today too little is done, too late, and for too few. A major problem relates to the mismatch between services for children and their needs. Too many children are lost between services, and too many are compartmentalized. Too little taxpaying money for basic and immediate needs often results in child caretaking services, which produce chronicity because of temporization about basic and early needs [10]. In addition to the fact that too many children do not have their right to have the basic essentials of life, more than any other group, children, especially if poor and of minority group status, suffer from being seen by people who do not understand them or are not trained sufficiently to be helpful.

Children are the "parapeople"; they are the recipients of paraservices and are most often assigned to paraprofessionals [1]. The needs of individual children and their families are often neglected; the most complicated human problems are left to harassed, untrained staff members. For example, in the child welfare field that deals with neglected, abused children or with children whose families are having difficulty with them, the initial and most serious decisions involving removal from the home are made by untrained, inexperienced, and very young workers. A study [18] of this service found tendencies to underestimate the family capacity and to overestimate the child's strengths. As a result children were too often absorbed into the system rather than helped to stay at home or at least to return home quickly. Though many new services have developed, including community psychiatry, there is a continued lack of workers with the appropriate experience, knowledge, or skill for therapeutic work with children [4].

## LABELING

Too often intervention consists of diagnosing and categorizing the children without supplying remediation. The problem of labels in relation to mental status and intellectual capacity has been of wide concern. Some labeling interferes with access to opportunity by confining a child to a narrow track, too early in life.

Identification, classification, and labeling of children [12] as handicapped, disadvantaged, or delinquent may result in stigma, rejection, exclusion, assignment to inferior programs, deprivation of liberty, or placement in special programs or classes that provide inadequate opportunities for stimulation and learning. Sometimes the labeling processes lead to commitment to institutions that evoke rather than reduce the very behaviors that are being treated.

Another thing about some diagnoses and labels is the odd fact that at different times some are more popular or more common than others. Today "learning disability" seems to afflict an inordinate number of children. The emotional impact upon the child who feels different and not helped is poignant.

## CASE ILLUSTRATION

Manuelo, age 12, was referred to a clinic because of unmanageable behavior at school and at home, including lying and stealing from his parents. His mother said that until the fourth grade, two years earlier, her son had been a straight A student. Then his father was severely injured in an automobile accident, which necessitated many operations. Mother also was away several times that year because of her mother's cancer illness.

Manuelo located the fourth grade as the beginning of his problems, because of a "bad kid" he was with then. Since then his work has been poor. He knew his parents said he was bad. He complained, just a little, that things had been tough since father's illness; he has to sleep in the room with father to be of help during the night. Mother sleeps alone, because father disturbs her too much.

A focus upon Manuelo's behavior would neglect his rights and begin a labeling process. The marital, economic, and sexual difficulties of his parents, the father's despair, the mother's anger, and all the parental unmet needs made it impossible for them to fulfill the child's right to be properly cared for and for *his* needs to be met.

Children rarely complain directly about their own parents. There are many reasons for this, identification, love, fear, need, and a lack of the cognitive sophistication to understand and the verbal skill to explain. It is more common for children to behave and to communicate through misbehavior, withdrawal, and a decline in functioning. These activities are a child's way of telling of despair and crying out for help.

Working with parents to change their circumstances to enable them to provide the support and help their children need is often difficult, because a myriad number of bureaucracies and uncoordinated services need to be confronted and the reality problems of the parents eased before they can have the energy to cope again with the needs of their child.

PSYCHOLOGICAL PARENTING

There are parents who have such severe inadequacies that they are unable to care properly for their children; sometimes parents die. Some parents are so abusive to their children that the children's lives are in jeopardy.

All children have the right to be a member of a family where they feel wanted, where they will have the opportunity on a continuing basis to receive and return affection, but also be free to express anger and, therefore, to learn to manage aggression. Continuity of care and the established need of children for affectionate, stimulating, and expressive relationships with adults is subsumed under the concept of psychological parenting, as advanced by Goldstein, Freud, and Solnit [11]. It may or may not be biological parenting. The concern is with decision making in laws involving selection and manipulation of a child's external environment as a means of improving and nourishing his internal environment. Yet the three writers recognize that the law is incapable of effectively managing, except in a very gross sense, so delicate and complex a relationship as that between parent and child.

The advancement of this concept fails to recognize sufficiently the power of biological ties. Manipulation of a child's external environment is not sufficient to manipulate the internal, unconscious environment, although emotional restructuring for a child cannot occur without manipulation of the real world in which the child lives.

When alternate plans such as foster care are arranged for a child, there is not always sufficient thought of the consequences to the child or to possible alternatives to the high cost of foster care [8], which might include less expensive but intensive work with and support of parents. The traumatic effects of placement [15], the mourning of the loss of the child's own parents, and removal from familiar surroundings [19] have been well documented. Yet sufficient resources, money, time, and professional skill are not devoted to the prevention of separation and placement and for treatment to ameliorate the effects when separation is unavoidable.

Parents who are uneducated, inadequately aware of their rights, without appropriate advocates or advisors, and often respecting or fearing the authority of physicians, teachers, and other professionals may agree to procedures they do not understand, some of which may violate their moral, cultural, and religious beliefs and opinions about child rearing. In some cases their rights as guardians are violated when they are not consulted.

There are many problems that obscure the rights of children and that interfere with these rights being activated. For all children their most primary right is abridged when there is interference with the ability of their families to care for them. It is a first priority that the families in which children need to grow be assisted in their child-rearing work through opportunities enabling access to physical supplies and necessities and to psychological supplies through meaningful, respected work and dignified community status. Where families are debilitated by physical or emotional crises, or even long-standing disturbances, a priority is the allocation of remedial and supportive services to enable children to stay with their own families.

There should be no special children. There should only be children with the right to the meeting of special needs, the needs of all children, and the particular needs of any one child.

## REFERENCES

1. Adams, P. Children and paraservices of the community mental health centers. *J. Am Acad. Child Psychiatry* 14:18, 1975.
2. Aries, P. *Centuries of Childhood: A Social History of Family Life.* New York: Knopf, 1962.
3. Bakan, D. *Slaughter of the Innocents.* Boston: Beacon Press, 1971.
4. Berlin, I. The myth of child treatment in GMHC's. *J. Am. Acad. Child Psychiatry* 14:76, 1975.
5. Bremner, R. H. *Children and Youth in America: A Documentary History.* Cambridge, Mass.: Harvard University Press, 1971.
6. Caplan, R. B. *Psychiatry and the Community in Nineteenth Century America.* New York: Basic Books, 1969.
7. DeMause, L. The Evolution of Childhood. In L. DeMause (ed.), *The History of Childhood.* New York: Psychohistory Press. Pp. 1–74.
8. Fanshel, D., and Shinn, E. B. *Dollars and Sense in the Foster Care of Children: A Look at Cost Factors.* New York: Child Welfare League of America, 1972.
9. Forer, L. The Legal Vacuum. In A. E. Wilkerson (ed.), *The Rights of Children.* Philadelphia: Temple University Press, 1973. Pp. 24–35.
10. Fraiberg, S. Legacies and prophecies. *Smith College School for Social Work* 1:1, 1974.
11. Goldstein, J., Freud, A., and Solnit, A. *Beyond the Best Interests of the Child.* New York: Free Press, 1973.
12. Hobbs, N. *The Futures of Children.* San Francisco: Jossey-Bass, 1975.
13. Joint Commission on Mental Health of Children. *Report of Crisis in Child Mental Health.* New York: Harper & Row, 1969.
14. Kahn, A. J., Kamerman, S. B., and McGowan, B. G. *Child Advocacy.* Washington, D.C.: United States Department of Health, Education and Welfare, Office of Child Development, Children's Bureau, DHEW Publication 120 (OCI), 1973.
15. Littner, N. The challenge to make fuller use of our knowledge about children. *Child Welfare* 53:287, 1974.
16. Richmond, J. The needs of children. *Daedalus* 106:247, 1977.
17. Rodham, H. Children under the law. *Harvard Education Review* 43:487, 1973.

18. Shapiro, D. *Agencies and Foster Children.* New York: Columbia University Press, 1976.
19. Thomas C. B. The resolution of object loss following foster home placement. *Smith College Studies in Social Work* 38:163, 1967.
20. United Nations. General Assembly Resolution 1386 (XIV), November 20, 1959, *Official Record of the General Assembly,* Supplement 16, 1960.
21. Wilkerson, A. E. *The Rights of Children.* Philadelphia: Temple University Press, 1973.

# CHAPTER 24

# The Rights of Parents and Children in a Biomedical and Social Context
*Susan Howard Webb*

*In all the years* of my work with children and youth, I have been constantly aware of the parents. Their rights have to be recognized and their needs and problems met if the children are to be helped. Both groups, parents and children, have a right to the pursuit of happiness promised by the Bill of Rights. It is in the context of the family that I wish to consider the right of parents and children to live as individuals with the help needed to be able to accept their responsibilities.

## CHANGING MORES OF SOCIETY
In any consideration of the rights and responsibilities of parents and children it is important to have as accurate a picture as possible of what both are up against in our modern world. As the mores of our society have changed, the mobility of families has increased and sex patterns relaxed, and thus has the family pattern been transformed.

Fifty years ago by common law children were the exclusive property of the parents. No interference between parents and children was possible. Most young, newly married couples lived near their parents. Grandparents and great-aunts were always available just a few blocks away. Children could go to them when mother was out. Americans have never given to aging relatives quite the reverence that is inherent in other cultures, but still there was a shared responsibility that seems no longer to exist.

The extended family became the nuclear family, smaller in number, although still a family unit, living farther and farther from relatives and dependent on hired baby-sitters or day-care centers to take the place of grandmother down the street.

Even the nuclear family has become less stable. As a result there is a far greater demand on parents to provide the warmth, sense of security, and the very important sense of success that is vital to the well-being of children as they grow up.

## THE CHANGING ROLE OF WOMEN
In the present era are the added problems of the working mother and even more of the single parent home. Over 40 percent of American women hold jobs, not only to meet financial needs but often to meet psychological needs. As women are gaining equal status with men, many feel the need to use their talents, education, and training. They often are highly skilled persons who have had successful careers before marriage. Fewer and fewer households can afford live-in help, even if it were available. Who then fills the role of grandparents, mother's helpers, and the like?

According to a study of the role of the family in early childhood, the family is the primary influence in a young child's development. The kind of care and guidance a child receives in the first three years is critical. Curiosity, social and motor skills, language abilities, sense of security, self-esteem, ability to cope with a variety of situations, moral values — all develop much faster in the early years than later in childhood [5]. The Joint Commission on Mental Health, in a study of child development, states that one-third of handicapping conditions could be prevented or corrected by appropriate care in the preschool years [11].

In the 1960s our values and attitudes were shaken, but what new values and attitudes have replaced the old? Correct behavior is no longer clearly defined. Religious principles have lost their power. Fewer families share a religious experience. Fewer fathers and sons go fishing together.

Rufus Jones, professor at Haverford College, writes about family prayers when every member of the family gathered together after breakfast to listen to the father or mother read a passage from the Bible and share matters of concern with the entire family, such as the serious illness of a child, the loss of crops, or the danger of a flood [12, 13]. Few families today even gather for meals, much less for a family council to discuss each member's rights and responsibilities, or to provide a time when children can speak freely of their disappointments and their expectations as well as to hear those of their parents.

The recent trend of having family vacations, often camping together as a family travels to an interesting part of the country, is an encouraging sign that some of this shared experience is returning. The roles of father and mother are not so clearly defined as they once were. Yet without clear concepts of these roles how can girls and boys grow into strong, able, clear-thinking adults?

First, let us consider the young family. Everyone loves a baby. The picture is one of a young mother and father proudly bringing their first baby home, surrounding that small person with love. But there this idyllic picture ends. Love and happiness may still be there but with it comes frustration and exhaustion from sleepless nights with a demanding job to face the next day. There are piles of laundry. The young parents have never been prepared for all these new experiences. They often are not secure in their own relationships. Changing mores often dictate that baby comes home to a single parent. The baby cries and cries. The young mother or for that matter both young parents get more and more frustrated. The first thing they know the baby has been given a good, hard slap or has been kicked across the floor. Scared, the parent or parents pick up the baby and rush it to a doctor. Remorse sets in. Not only are there serious medical problems to be met, but also psychological ones. Too many young parents, overwhelmed with their new creation, just cannot cope. I would suggest that the help for this sort of situation should begin long before there is a baby. Parents have one of the most responsible jobs in the world and are least prepared to assume it.

## COMPREHENSIVE HEALTH EDUCATION

The American Public Health Association has come out in support of the concept of a national comprehensive sequential program of health education [10] for all students in the nation's schools, kindergarten through twelfth grade. Many health care providers, educators, and organizations are interested. Health education includes studies in personal, family, and community health, covering safety, disease prevention, nutrition, anatomy, physiology, human growth and development, mental health, physical fitness, dental health, and alcohol and drug abuse prevention.

At present, 10 states mandate comprehensive health education programs; 3 others offer optional programs. There are very few courses in sex education and in growth and development. Only 26 states require a course in the use of drugs, alcohol, and tobacco. Nineteen states require school nurse certification. A very few states have certified health educators [16]. Yet there is no question that the courses should be part of the required basic studies along with reading, writing, arithmetic, and history.

Courses in parenting should also be required. School age parents and unmarried pregnant high school girls are high risk groups. Courses in parenting should include education and guidance in health, sexuality, nutrition, childbirth, family life, and preparation for the role of parent.

One independent school in California saw the importance of a course in parenting and child psychology and developed a curriculum with three major facets: theoretical foundations of child development and behavior, direct observation of children, and exploration of the stresses and demands of parenthood. There is a good balance of boys and girls taking the course, with as much discussion of fathering as mothering. This experience points up the need for such education for the higher achievers as well as for the lower achievers, who more frequently become the parents at risk [4].

## DAY-CARE CENTERS

Day-care centers are a whole subject in themselves. Their importance cannot be underestimated with so many working mothers. Good family care centers where children can be left for longer hours are important. Funding is uncertain, and there are too few centers for children who need them. An important step has been taken by some industries, which are providing day-care centers for their workers by helping to support privately run centers or publicly funded centers or setting up their own.

## PROBLEMS OF ALCOHOLISM

Alcohol frequently compounds the child care situation. Listening to counselors of alcoholics no one can miss the impact of heavy drinking on children or parents. The need for preventive education in the use of drugs and alcohol is imperative. One counselor observed that this should start in kindergarten. Children of alcoholics are estimated at 28 to 30 million in the United States [3]. The special needs of these children deserve careful study. Their emotional neglect is as severe as physical neglect. These children often later appear in court as social problems.

The use of alcohol is increasing, and the lonely, single woman, confined with a crying baby, is more apt to turn to alcohol, a sedative drug and the only one with calories to give her a lift [14]. The battered child syndrome is frequently the result of drinking by parents. There is evidence that if these children later appear in court, it is most often on homicide charges.

Support services must include preventive education as well as centers where these parents can go for help. These services must offer foster care for children, so that parents in need can receive necessary attention in alcohol treatment centers.

The problem of alcohol leads directly into that of neglect and abuse. In 1974 the federal government passed the Child Abuse and Prevention and Treatment Act. States are now passing laws in compliance, which require the reporting of cases of abuse and neglect by doctors, nurses, social workers, school administrators and teachers, with immunity for those who report the cases.

Even the most intelligent parents can inflict on their offspring some forms of neglect and abuse. It may be just by not hearing a child when he is trying to express his feelings or by turning him off with the quick comment of "Don't bother me now." Later, when he is a student in high school and wants to talk about a special interest or problem, there may still be nobody with time to listen.

## ADOPTION LAWS

Children still have to be placed in foster homes, even in institutions, to get the care they need. However, children should not be left in foster homes indefinitely. This need requires good adoption laws and well-considered laws covering other aspects of child welfare. Vermont and New York State have just passed laws requiring that when a case of adoption comes before the court, the decision must be made with the best interest of the child [8] as the first criterion.

With any adoption, the right of the mother who is a minor to make her own decision about relinquishing her baby for adoption is important. No longer should her parents have the right to decide for her. It is important that agencies dealing with adoptions be able to guide her choice. A single parent, tied down by an infant, unable to get a job since she has no skills, begins once again the circle of abuse, neglect, frustration, and all the rest. Good laws are essential as well as counseling to meet specific needs.

Legal rights of children were first established by the Gault decision [9] of 1967, which gave procedural rights to children in the juvenile courts, the right to be represented by their own attorneys in all juvenile decisions. In Texas the court decision in *Morales* v. *Truman* closed two training schools, which has led to legislation to establish small community-based facilities. These moves were important toward assuring the legal rights of children. Court action brought by the Children's Defense Fund has resulted in giving a child the right to the least restrictive placement, to rehabilitation in the juvenile court system, to be protected from cruel and unjust punishment, and to equal protection under the law [15]. The Juvenile Justice and Delinquency Prevention Act of 1975 has added more recognition of the rights of children in these areas. It provides funding for much needed programs through the Law Enforcement Assistance Administration.

In the 1976 session of Congress, education for the handicapped has been strengthened by the Education for All Handicapped Children Act. By 1978 all handicapped children in a school district must be found, observed, and provided with a free education patterned to meet their special requirements. Early screening of very young children for developmental disabilities is of growing importance to educators and parents. Well-baby clinics are familiar. This new law provides funds for larger numbers of children, but it first requires the searching out of these children in every community. It will also help the parents of handicapped children to recognize the role they need to play.

## LEGAL RIGHTS OF CHILDREN

What do we believe the rights of children to be? James Whittall, the nineteenth-century Philadelphia Quaker, in a conversation with M. Cary Thomas, President of Bryn Mawr College, put it this way: "Every child should have a happy childhood tucked under his belt" [6]. I would add that parents also should have this.

A happy childhood certainly cannot be achieved without providing every child with caring adults, decent housing, awareness of special needs, and an appropriate education [1]. Required also is a place where the child can achieve his full potential and where health services make it possible to grow into a physically healthy adult. A sense of belonging, of self-esteem, and security are necessary too.

Parents have the same rights. Probably because these needs were not met in their own childhoods, these parents have children who suffered similar deprivations. Parenthood is a major undertaking for the best of parents and an almost impossible task for those parents in need, medically and socially. It is a job for a team of nurses, social workers, doctors, psychologists, therapists, school counselors, and parents working together.

## PARENTS RIGHTS LEAD TO RECOGNITION OF RIGHTS OF CHILDREN

Parents also have a right to all the help they need to resolve their problems and to give their children their rightful support. They need courses to understand parenting, a way to learn what marriage and family encompass, so that they will be ready for that experience. They have the right to learn how to become more positively involved in the lives of their children and to understand the processes of child development. Most important of all, they must be able to find help when necessary, help that will increase the possibility of the family staying together. This emphasis is the new approach of all who work with families and children. The desire is to get children at risk into good home situations as quickly as possible, but if possible to keep them in their own homes.

Both groups, parents and children in need, are becoming more visible. Many advocates, such as supportive family members, medical providers, teachers, administrators, and social workers, are bringing to people's attention these new concerns.

The position of children has come a long way since the days of Dickens and the first child labor laws [2]. Sometimes those of us who deal with legislation or minister to parents and children get discouraged. Changes come slowly. New problems continue to confront us. Nevertheless, now that the social, medical, and legal rights of these two groups of people are recognized as valid, life is improving for children, and we hope that family units are becoming stronger. If I were to choose a subtitle for this paper it would be: "When one hears a battered child crying somewhere, somewhere a parent is crying also" [7].

## REFERENCES

1. *Brown* v. *Board of Education.* 347 U.S. 483 (1954).
2. Children's Bureau. *Child Advocacy – A National Baseline Study.* Washington, D.C.: Department of Health, Education and Welfare, No. 73-18, 1973.
3. Couture, J. Speech by founder of the Other Victims of Alcoholism, Inc., at Conference on Women and Alcoholism, Mendon, Vt., September 21, 1976.
4. Dolder, S. The Athenian School. *The Athenian* 22:3, 8, 1976.
5. Education Commission of the States. *The Role of the Family in Child Development. 1. Public Policy and the Family.* Denver, Col.: Report No. 15, December 1975.
6. Flexner, H. *A Quaker Childhood.* New Haven: Yale University Press, 1940.
7. Fontana, H. *The War Cry.* Salvation Army Publication, September 1976.
8. Freud, A., Goldstein, J., and Solnit, A. J. *In the Best Interests of the Child.* New York: Free Press, 1974. Also, 33 V.S.A. 58 (Supp. 1977), The Best Interests of the Child.
9. *Gault.* 387 U.S. 1–13 (1967).
10. Governing Council. *Education for Health in a School Community.* Position paper by American Public Health Association, New Orleans, 1974.
11. Joint Commission on Mental Health of Children. *Crisis in Child Mental Health, Challenge for the 1970's.* New York: Harper & Row, 1970.
12. Jones, R. *Finding the Trail of Life.* New York: Macmillan, 1926.
13. Jones, R. *Small Town Boy.* New York: Macmillan, 1941.
14. Nye, F. Speech by faculty member of Dartmouth Medical School at Conference on Alcoholism and Women, Mendon, Vt., September 21, 1976.
15. Polin, J. In defense of children. *Child Welfare* 4:80, 1976.
16. Vermont House Committee on Education. Draft No. 2-76-350. Vermont State Legislature, March 1976.

# CHAPTER 25

# The Right to Consent and Confidentiality in Adolescent Health Care: An Evolutionary Dilemma
*Adele D. Hofmann*

*A 15-year-old* sexually active girl is seeking contraception; a 16-year-old runaway has a fever and cough; a third adolescent is concerned about his drug abuse problem; and a pregnant teenager wants an abortion. One thread connects all four; each wishes treatment for a significant health problem, but in fully confidential terms. This treatment is possible, however, only if care can be obtained without their parents' knowledge or consent. Frequently adolescents are unwilling, even unable, to confide in their mothers and fathers about what they perceive to be personal and private affairs. This reluctance derives partly out of concern for possible parental disapproval and distress and partly out of the conviction that earlier childhood protections are no longer justified in light of their emerging maturity. Regardless of motivation, many adolescents elect to remain silent about their need rather than tell their families as a prerequisite to care.

The common and customary practice of axiomatically requiring parental consent poses serious dilemmas for adolescent health care. Pragmatically, the health of a goodly number of young people will be compromised by the resultant delay in treatment. Nor does such a rigid approach developmentally respond to the adolescent's emerging maturational capacity and need to make decisions and judgments on his own. It fails to allow for the kind of flexible, individualized approach essential to optimal management.

Alternatively, even permitting selected minors to consent to health care on their own is no less consequential. It certainly has implications for the legal liability of health care professionals under the still somewhat ambiguous state of the law. It also has a significant impact upon parent-child relationships and the family unit, a legitimate sociological concern.

More fundamentally, this conflict raises significant questions over the nature and extent of parental rights in directing the raising of children and dictating a moral course, which is the basis of the ethical conflict in adolescent health care. Should parental rights be absolute? Should a minor have any rights in determining his personal values and governing his own life and, if so, to what degree? Are parental rights derived from a form of ownership of their offspring, from an obligation as caretakers, or both? What factors should rationally define the termination of parental supervision if the stroke of midnight on the eighteenth or twenty-first year is an insufficient criterion?

## THE LAW AND MINORS' CONSENT
These considerations would have been idle debate just a decade ago. Legal requirements for parental consent were then viewed as binding on virtually all medical

183

services for a minor except in an emergency. Although few states had specific statutes in this regard, common law holding that a minor could not enter into a contract of his own was widely applied. In this interpretation, the doctor-patient relationship is deemed a contractual one, and the treatment of a minor without a valid contract (e.g., consent) constitutes unauthorized touching, or assault [5].

Recent shifts in the law, however, have invaded the absoluteness of this principle. A number of lower court rulings have held that minors indeed can validly consent (contract) for their own care when sufficiently mature to give *informed* consent. In fact, all known cases involving adolescents of 16 years or more have been resolved in this way [10]. However, a recent federal district court decision in Michigan (*Doe* v. *Irwin*), while not invalidating a minor's right to consent to contraceptive services, did affirm an equal right of parents to be notified of this fact.

A spate of recent state statutes [9] also addresses this issue in three general ways. A first group of laws provides that adolescents may consent on their own to such specific, usually sensitive treatment needs as venereal disease, pregnancy, pregnancy prevention, and drug abuse. The second group enables selected categories of minors to consent to all health care by virtue of an emancipating life-style. Emancipation variably may be defined by such factors as being a parent, married, in the armed forces, earning one's own support, or simply living away from home. Age alone can be enabling, and in some states young people of 16 years (or even at age 14 in Alabama) are now statutorily entitled to consent to all types of care. The third direction is evidenced by a small handful of somewhat vague laws simply according the right to consent to any minor judged sufficiently intelligent as to be capable of understanding the nature of the proposed treatment and its relative benefits and risks. This latter point is of particular significance in that it attempts to define maturity in developmental rather than categorical terms.

The scope of these laws that permit minors to consent to their own health care varies from state to state, and no single set of rules applies across the board. Some states have only a single provision, usually drawn from the first category above and relating to venereal disease. Others are more comprehensive. But regardless of the status of the law in a given locale, there is a significant national trend toward a lowered age of consent that pokes vast holes in the former absoluteness of parental control.

A particularly noteworthy moment in the evolution of minors' consent law came in 1976 with the United States Supreme Court decision in *Danforth* v. *Planned Parenthood of Central Missouri* [3], ruling on a Missouri statute governing the performance of abortions in a highly restrictive way. One section mandated parental consent for a first trimester abortion in a female under 18 years of age (second trimester procedures were virtually done away with all together). This provision and nearly all other provisions were struck down as being impermissibly inconsistent with the 1972 decisions in *Roe* v. *Wade* and *Doe* v. *Bolton* [11], which established a female's constitutional right to an abortion prior to fetal viability. *Danforth* affirmed that this right was for minors as well as adults. While the court did not go quite so far as to totally preclude some sort of special protection for minors nor fully resolve the issue of their right to privacy per se, it clearly prohibited states from legislating parental consent requirements for first trimester abortions because this would be tantamount to giving a third party an absolute and possibly arbitrary veto. The decisional right over an abortion now rests with the young patient and her physician alone.

## PARENT-OFFSPRING RELATIONSHIPS: A HISTORICAL PERSPECTIVE

The particular significance of *Danforth* is that for the first time the Supreme Court saw parents and offspring in an adversarial relationship; the wishes of one pitted against those of the other. In all previous rulings on a minor's constitutional rights, primarily in relation to schools and juvenile justice, the child and an institutional structure external to the family unit were adversaries [5]. Parents stood firmly behind their young and even led the way. Although potentially subject to modification consequent to appeals in *Doe* v. *Irwin, Danforth* incepts a new direction that specifically speaks to concepts of a minor's maturity and independent rights, establishing that age alone is no longer a sufficient criterion for an adolescent's subjugation to parental determination.

It is difficult to consider the implications of *Danforth* and minors' consent laws in relation to the relative rights of parent and child in the absence of historical perspective. At the core is the pervasive impact of the American Dream, holding that each generation can and should strive to be something more than the last. The role of the young in this country came to be characterized by a search for new and "better" ways, rather than by the perpetuation of past traditions [4, 8]. Although primarily directed toward economic and social advance, the valuation of change also became a valuation of change's prime mediators, youth. In consequence, the minor young gradually shifted from a position of nearly total submission to parental dominion to one of being relatively separate, independent entities unto themselves.

Three distinct concepts of parent-child relationships emerge in this evolution, respectively emphasizing, first, parental autonomy, second, child welfare, and, last, minors' rights. The era of parental autonomy extended from colonial times to the last quarter of the nineteenth century. In these early times, young people were viewed as firmly subject to parental control under terms of strict obedience until their twenty-first year, almost as possessions. Although children assumed an adult work role in the community at a tender age, this role was as an extension of the family. All profit of a minor's labors belonged to his parents; he was accorded no possessive or determining rights of his own. Parental obligations were limited to preparing their young for a trade, teaching them the three Rs, and raising them in the way of the church [2].

But this stage of relatively unrestricted parental autonomy over the direction of a child's affairs became substantially modified out of a growing social concern for the welfare of the child himself. This change derived both from simple outrage at children's exploitation in labor and from an economic awareness that further advance in the American Dream was increasingly dependent upon those who were technically skilled. In turn, greater educational attention was given to the young to prepare them for this role. The first compulsory school attendance and restrictive child labor laws were enacted in the latter part of the nineteenth century; although it took well into the 1920s for this movement to expand into all states and to include adolescents up to 16 years of age as it does today [7].

Parental rights and responsibilities in this second period shifted to emphasize nurturance and the growing child's developmental needs, rather than primarily preparation for the future in economic and spiritual terms. While mothers and fathers continued to be the prime arbiters of their offspring's fate, compulsory education and child labor statutes imposed significant limits upon their authority. Neglect and child abuse laws added further erosions. The state thus became quite willing to step in and take over when parents were deemed remiss in providing a child with what the state thought best. It should be noted, however, that the young continued to have no voice or representation of their own; all decisions still were made by adults on the minor's behalf and solely from an adult perspective.

The third phase, that of minors' rights, was initiated in 1967 with the United States Supreme Court's decision *In re Gault* [6]. This ruling endowed youths subject to the juvenile justice system with much, albeit not all, of the Bill of Rights that had been formerly denied them in toto. But it was not until the enactment of laws that permitted minors to consent to their own health care and the *Danforth* decision that minors' rights impinged upon intrafamily relationships.

This direction invoked a new concept of the parenting role. In absolute and unmodified terms, fathers and mothers primarily serve as their child's advocate, temporarily acting on his behalf and protecting his interests, but only until such time as he can act for himself. The capacity of a minor to understand the issues involved and to make an informed decision is the enabling factor, rather than age or economic independence. A child is no longer considered part parental possession; instead each child has separate, individual interests and rights. Parental rights are valid only when they are in concert with those of the child and act in affirmation. When a minor achieves the requisite level of intellectual maturity, he becomes free to exercise his independent choice as would an adult. This means, then, that full rights of privacy, confidentiality, and consent rest with those minors capable of understanding the benefits and risks of the treatment involved. Only a minor's competency to give an informed consent is the determining factor; no other aspects enter in.

## A CONTEMPORARY PERSPECTIVE ON MINORS' RIGHTS IN HEALTH CARE

The full expression of this concept of minors' rights does not exist in fact today. Current attitudes and approaches toward the young are an amalgam of both past and present. For example, early parental autonomy concepts may be seen particularly in litigation over adoption and foster care. While law guardians are beginning to be employed to protect the independent interests of children, custody decisions still may be based on an inherent right of biological parents to the possession of their child and the validity of any surrender they may have made. The minor's nurturance needs, much less his wishes, may be wholly ignored [1]. The child welfare perspective is widely used in such instances as seeking parental consent for a youngster's needed health care and obtaining a court order mandating treatment if parents refuse.

A valid question is whether the absolute and full expression of minors' rights concepts should prevail in unmodified form. Singular pragmatic problems are posed by attempting to define maturity and understanding; yet definitive yardsticks are necessary for drafting implementing rules. From another perspective, there is some defense to the argument that a minor's economic dependency gives rights to those who pay. If mom and dad are to pay an adolescent's medical bill, they have a point in wanting to know what it is for. But, more importantly, significant benefit can accrue to many adolescents from parental guidance and emotional support, even in those young people intellectually capable of giving an informed consent.

Nonetheless, there are rational and ethical reasons for a maturationally based minors' rights concept in adolescent health care in which selected minors can indeed exercise their inherent rights to privacy, confidentiality, and consent. This position affirms that no one person, not even parents, can own another and that a minor's dependency upon the decisions of adults is determined by developmental incapacity rather than by arbitrary categorical definitions. A parent's role is to act on behalf of the young out of advocacy and altruism, protecting their rights until maturity has advanced sufficiently to permit the young to assume this responsibility for themselves. This shifting of responsibility will occur in a graduating, differential manner according to the complexity and consequences of the decision involved. While parental consent or involvement is often desirable, it is by no means axiomatic for all adolescents at all times.

The decision as to whether a particular minor is competent to give informed consent and can do so without developmental or psychological disadvantage rests primarily with the treating physician, or an appropriately assembled adolescent health care team. Factors to be assessed and integrated are the patient's particular level of maturity, the quality of the parent-offspring relationship, the implications of requiring parental consent, the nature of the health problem itself, the relative consequences of treating or not, the complexity of arriving at informed consent in terms of comprehending benefits and risks, and the degree of the treatment's reversibility.

Maturity, the most difficult element to measure, is best determined by such behavioral evidence as the following: An adolescent initiates his own health care contact and comes by himself; indicates that his medical affairs are to be confidential and gives good reasons; is a competent historian, asks valid questions about his health affairs, and is motivated to comply with recommendations; conveys that he understands directions and can appropriately weigh benefits and risks; enjoys relative independence in his usual comings and goings from home and, public facilities permitting, takes care of his own transportation; makes the majority of his own decisions in the conduct of his daily affairs and in relation to such personal purchases as clothing, books, records, and the like. In summary, the mature adolescent gives reasonable evidence of being able to stand on his own two feet and of enjoying this role.

Unfortunately, litigational aspects and the omnipresent threat of malpractice tend to block the use of this flexible approach. Without more expansive and definitive enabling statutes than now exist, the health care professions will inevitably favor the conservative and entrenched old common law precedents, which support parental determining rights to the exclusion of rights for the adolescent. But

*Danforth* gives notice that the courts are increasingly ready to support a significant application of minors' rights in health care, even to the point of considerable self-governance.

The law has created a new environment in which adolescent consent in health care can be properly recognized, with the best of "minors' rights" and "child welfare" concepts applied in balance. It places minors' rights to consent, confidentiality, and privacy as paramount, to be invaded only when careful scrutiny concludes that youth will be specifically advantaged thereby. This flexibility is precisely what is required in responding appropriately to the variable and diverse nature of adolescence and in enhancing the ultimate emergence of responsible, autonomous young adults.

## REFERENCES

1. American Civil Liberties Foundation. The conflict of rights among children, foster parents and the state. Part I: Summary of recent legal developments. *Children's Rights Report* 1:4, 1976–1977.
2. Commentary of James Kent. Chief justice of the New York Supreme Court and chancellor of New York State, upon the rights and duties of parents under common law, 1867. In R. Bremmer (ed.), *Children and Youth in America – A Documentary History.* Cambridge, Mass.: Harvard University Press, 1970. Pp. 363, 364.
3. *Danforth* v. *Planned Parenthood of Central Missouri.* 44 U.S. L.W. 5197 (1976).
4. Ginzberg, E. *The Optimistic Tradition and American Youth.* New York: Columbia University Press, 1962.
5. Hofmann, A. D., and Pilpel, H. F. The legal rights of minors. *Ped. Clin. No. Amer.* 20:989, 1973.
6. *In re Gault.* 387 U.S. 1 (1967).
7. Manning, L. Child labor legislation in 1940. *The Child* 5:149, 1940.
8. Mead, M. *Male and Female: A Study of the Sexes in a Changing World.* New York: Mentor Books, 1955.
9. Paul, E. W., Pilpel, H. F., and Wechster, N. F. Pregnancy, teenagers and the law, 1976. *Family Planning Perspectives* 8:16, 1976.
10. Pilpel, H. F., Wechster, N. F., and Paul, E. W. Sex-related health care for minors. *New York Law Journal,* February 27 and 28, 1975.
11. *Roe* v. *Wade* and *Doe* v. *Bolton.* 410 U.S. 113, 179 (1973).

# The Rights of Aging Persons: Issues of Care, Support, and Independence

# CHAPTER 26

# Older People: Issues and Problems from a Medical Viewpoint
*Philip W. Brickner*

*Many older people* in our country are treated with contempt or are ignored. They are deprived of choices about their own lives, and of independence, in the name of medical necessity. This often occurs for no clear reason, at great cost in human dignity and at vast expense to society [2—8, 10, 18].

A program is described in this chapter that provides an alternative to nursing home care. It is included to show that all of our words about issues and problems, and about human rights, are meaningless unless they produce viable, practical results that help people.

## WHAT DO WE MEAN BY "OLD"

The word "old," as applied to human beings, has no regularly understood definition. By "age" do we mean simply the number of years a man or woman has lived? While chronology is a useful rough guideline, we know that one aged individual may be "old" and feeble, while another is "young" and vigorous at the same age.

To understand these differences helps us to recognize that the elderly are not a monolithic group. They are heterogeneous, with diverse needs and capacities — a set of individuals. Some, like Grandma Moses, Casals, Picasso, Churchill, and Casey Stengel, are highly effective at an advanced age. At the opposite pole are children suffering from the unusual disease progeria [12], or accelerated aging, in which they die of arteriosclerosis before adolescence. Are these children old?

The bureaucracies under which we all live have adopted an artificial age — 65 — at which we are officially old. The Social Security system, Medicare rules, and commonplace business practice are obvious examples. The result is an arbitrary end point to productive lives by forced retirement.

Age measured by biological change has more meaning than official definition or simplistic chronology. Alterations in the body caused by age are common knowledge: loss or graying of hair, change in bone structure, with differences in posture and diminution of stature; lessened elasticity of the skin. There also is an increasing age-related incidence of cataracts, glaucoma, and hardening of the lens as well as presbycusis.

In normal aging, endocrine gland secretion changes, with noteworthy alterations in structure and function of the body. The pituitary, thyroid, testes, ovaries, and adrenal glands, which control highly important biological functions, are involved.

The brain also changes. A human being by the age of 20 is already losing brain cells by the hundreds of thousands each year — and these cells are not replaced.

Clearly, any attempt to understand the process of aging requires us to consider each individual, thus denying us the ease of considering aged people as a stereotyped group.

191

## ATTITUDES TOWARD AGING

In our culture it is logical to fear becoming old. The aged people among us — and those who look old — are generally poorer and more isolated than the general population. They are ignored and often abandoned.

But this need not be so. Other societies have developed views about older people that show respect for the dignity of age. The Eskimos, according to Peter Freuchen [11], include older members of the family in daily activities until the biological burden of age becomes too great for the individual to bear. Then the aged person walks off into the snow.

The reverential attitudes of people in the Orient toward their aged are well known [19, 21]. The accumulated wisdom and sage advice of elders are treasured and sought out. An important position is maintained for the elderly in the family structure and in society at large, even though strength and beauty may have faded. These examples contrast in a startling manner with the desperate, often ridiculous, struggle in our culture against looking old.

We value youthfulness. It is unusual to watch an hour of commercial television, for instance, without noting offers of creams, lotions, dyes, and potions designed to rid our bodies of the signs of aging. The purpose is, of course, to help us maintain the appearance of youth. The message: Old age is synonymous with ugliness.

Once we understand this message, it is easy to recognize the reasons that seem to justify our treatment of aged people. There is a motif that pervades our everyday lives: The elderly are inferior. This attitude has allowed us to develop the belief that older people, as a group, are different from the rest of us; next we consider them not equal; and finally we think of them merely as objects to be passed over or stored away.

The origin of this viewpoint stems in part from the great migration from farm to city during this century. In 1910 less than half of our population lived in metropolitan areas; in 1977 almost three-quarters do. Furthermore, the population of the country has doubled since 1910, and the percentage of people over age 65 has risen from about four percent to over 10 percent [23].

Population movement from farm to cities has helped destroy the extended family structure that gave older people a special importance. In the three-generation family, grandparents are valued for their advice and life experience; but this arrangement is dependent on adequate housing, where all members of the family can live together. During the adaptation to more crowded city conditions, the older, less productive family members have lost their purpose. They are often seen as ineffective, superfluous, to be tolerated if necessary, but preferably to be disposed of in some acceptable manner. Older people have come to see themselves as being in the way and unproductive. They feel guilty and are willing to participate in their own disenfranchisement as members of the family and of society. They have been made to accept isolation, abandonment, or incarceration.

A further cause of this problem is the pattern of ethnic change within cities, a result of the flow of immigrants to the United States. Typically, new arrivals settle first in the central city and then move to suburban areas as they become assimilated; but some, often the aged, remain behind. In time they are surrounded by a new wave of immigrants, people who are younger, of different habits, speaking other languages. The aged become strangers in their own neighborhoods. They are alone.

## THE HAZARDS OF INSTITUTIONS

The combination of poor health and social isolation is hazardous for the aged. They cannot go out for help, and medical services rarely go to them. They are truly "stranded by time on the bleak shore of a forgotten or friendless old age" (Lord Shaftesbury) [14]. Yet any attempt to remove them from familiar surroundings where they live to the anonymity of a hospital or nursing home produces depersonalization, a sense of isolation and separation from society, and a higher death rate [1, 6, 13, 15—18].

Homebound aged people form a relatively small proportion of our total population — about 5.2 percent [20, 24, 25] or slightly more [23]. However, they are at a highly significant moment in their own lives and carry with them important connotations for society at large.

An isolated older person may be able to keep up an independent existence alone for a time; to shop, clean, cook, eat, and wash. But any physical change that impedes any of these simple daily functions may be disastrous.

Commonly, an aged person does not know how to reach help, has no contacts, and is not noticed. As a result, physical changes occur, and institutionalization becomes imperative, or death takes place before there is an opportunity to change this outcome.

If comprehensive medical and social services can be provided to homebound, isolated aged people in a timely and appropriate manner, they may be able to maintain independence in their own homes. This desire is often their desperate wish [5]. These services also meet the purpose of society at large, because the costs of institutional care are markedly higher than those of home health services [12, 22].

## HEALTH CARE FOR THE HOMEBOUND AGED: A PROGRAM IN OPERATION

In January 1973 the Chelsea-Village Program was started by Saint Vincent's Hospital in New York, in association with the people and agencies of the Chelsea and Greenwich Village communities surrounding the hospital. It is designed to bring medical, nursing, and social services into the homes of homebound, abandoned aged people.

Our patients are among the medically unreached and are representative of people living in all the urban areas of the United States. Their problems are complex, involving medical and psychological disorders, poor housing, poverty, and social isolation.

### Goals

The stated goals of the program are to help patients remain in their own homes and community, out of institutions, in adequate housing, in the best possible state of health, and at the maximum possible level of independence.

Case finding is among our most difficult tasks, because many of the people who need help are hidden and unable to draw attention to themselves. Individuals and organizations in the neighborhood of Saint Vincent's have been responsible for two-thirds of our referrals and hospital staff for the remainder.

The program entered its fifth year of operation in January 1977. In the first four years, 466 individuals were referred and 3526 home visits made. The average age of these patients is 80; two-thirds live alone; two-thirds are women.

*Financial Savings to Society*

The least expensive nursing home in New York City that accepts Medicaid patients costs about $800 per month in 1977, exclusive of physician's services. On the average, care for our patients costs slightly more than half this amount, including all medical services, rent, and food.

The following case report is illustrative:

M. D. is an 82-year-old disabled man living alone on the fifth floor in a one-room, walk-up apartment. His only resource for food and social contact was a neighbor who had recently died. The medical factors in this case consist of medication for heart disease, reassurance, and regular visits as needed. We have arranged for a Meals-on-Wheels program to deliver a hot meal each day and for homemaker services for regular shopping, cooking, and cleaning. This patient has been sustained at home for the last 16 months, as he wishes, instead of in a nursing home.

His living costs are as follows:

|  | Monthly Costs | Source of Payment |
|---|---|---|
| Rent, utilities | $ 94 | Patient |
| Food (including Meals-on-Wheels) | 70 | Patient |
| Telephone | 6 | Patient |
| Drugs, supplies | 12 | Patient |
| Other expenses | 10 | Patient |
| Chelsea-Village Program costs | 108 | Chelsea-Village Program |
| Homemaker | 160 | New York City |
| TOTAL | $460 | |

Additional savings are produced because to a degree we succeed in keeping our patients out of hospital beds for acute care through early treatment of illness in the home before the patient deteriorates.

People under our care are served without charge or reimbursement. The Chelsea-Village Program has been supported by Saint Vincent's Hospital and grant funds, both public and private.

Hospitals are not permitted to bill Medicare or Medicaid for health services given outside the institution. As a result, few programs similar to ours will develop unless there is a change in legislation designed to support home health services [9].

CONCLUSION

Health care of homebound, isolated aged people should be a national health issue. Their need for medical and social services is almost totally unrecognized, and yet these patients can be found in all communities. Health professionals and government, along with community agencies and individuals, must make a conscious decision in favor of these people; otherwise their needs will continue to be ignored amidst widespread demand for assistance by groups with louder voices.

REFERENCES
1. Bell, W. G.  Community care for the elderly: An alternative to institutionaliza-tion. *Gerontologist* 13:349, 1973.
2. Brickner, P. W. (ed.).  *Care of the Nursing Home Patient.* New York: Macmillan, 1971. P. 1.
3. Brickner, P. W.  Finding the unreached patient. *J.A.M.A.* 225:1645, 1973.
4. Brickner, P. W., Duque, T., Kaufman, A., Sarg, M., Jahre, J., Maturlo, S., and Janeski, J. F.  The homebound aged – A medically unreached group. *Ann. Intern. Med.* 82:1, 1975.
5. Brickner, P. W., Janeski, J. F., Duque, T., Stahl, I., Ruether, B., Kellogg, F. R., Madeira, S., and Stall, J.  Hospital home health care program aids isolated, homebound elderly. *Hospitals* 50:117, 1976.
6. Cantor, M.  Life space and the social support system of the inner city elderly. *Gerontologist* 15:23, 1975.
7. Cantor, M., and Mayer, M.  Health and the inner city elderly. *Gerontologist* 16:17, 1976.
8. Carp, F., and Kataoka, E.  Health care problems of the elderly of San Francisco's Chinatown. *Gerontologist* 16:30, 1976.
9. Chelsea-Village Program.  Four-Year Report. Department of Community Medi-cine, St. Vincent's Hospital, 153 W. 11th St., New York, N.Y., 10011, 1977.
10. Clark, M.  Patterns of aging among the elderly poor of the inner city. *Gerontol-ogist* 16:58, 1971.
11. Freuchen, P.  *Book of the Eskimos.* New York: World Publishing, 1961.
12. Gilkes, J. J. H., Shavill, D. E., and Wells, R. S.  The premature aging syndromes. *Br. J. Dermatol.* 91:246, 1974.
13. Hammerman, J.  Health services: Their success and failure in reaching older adults. *Am. J. Public Health* 69:253, 1974.
14. Hodder, E.  *The Life and Work of the Seventh Earl of Shaftesbury, K. G.* (vol. 1). London: Cassell, 1886. P. 280.
15. Lawton, M. P.  Social Ecology and the Health of Older People. *Am. J. Public Health* 64:257, 1974.
16. Lawton, M. P., Liebowitz, B., and Charon, H.  Physical structure and the beha-vior of senile patients following ward remodeling. *Aging Hum. Dev.* 3:231, 1970.
17. Lieberman, M. A.  Institutionalization of the aged: Effects on behavior. *J. Ger-ontol.* 24:330, 1969.
18. Markson, E. W., Levitz, G. S., and Gognalons-Caillard, M.  The elderly and the community: Identifying unmet needs. *J. Gerontol.* 28:503, 1973.
19. May, W.  Orientation to self. *Ann. Intern. Med.* 82:227, 1975.
20. Population Reference Bureau.  The Elderly in America. *Population Bulletin* 30:3, 1975.
21. Rosow, I.  One moral dilemma of an affluent society. *Gerontologist* 2:182, 1962.
22. Scharer, L. K., and Boehringer, J. R.  *Home Health Care for the Aged: The*

*Program of St. Vincent's Hospital, New York City.* New York: The Florence V. Burden Foundation, 1976.

23. Shanas, E., Townsend, P., Wedderburn, D., Friis, H., Milhoj, P., and Stehouwer, J. *Old People in Three Industrial Societies.* New York: Atherton Press, 1968.

24. United States Department of Commerce, Bureau of the Census. *Demographic Aspects of the Aging and the Older Population in the United States.* Washington, D.C.: U.S. Government Printing Office, May 1976.

25. United States Department of Health, Education and Welfare, Public Health Services. Limitation of Activity and Mobility Due to Chronic Conditions. In *Vital and Health Statistics* (series 10, no. 96). Rockville, Md.: Health Resources Administration, 1972.

# CHAPTER 27

# Is It Right to Allocate Health Care Resources on Grounds of Age?

*Harry R. Moody*

## AGE AS A CRITERION FOR RESOURCE ALLOCATION

Is it right to allocate scarce health care resources to persons on grounds of their age? On the face of it, this question sounds like an invitation to the most ruthless form of "ageism" imaginable [1]. Should the "right to health care" not extend to all members of our society, regardless of race, sex, income level, or age? In fact, should we not be doing more to see that older people get *more* health care resources in view of their greater unmet needs? Suffice it to say that, at bottom, this claim for a "need conception of justice" in health care allocation expresses an ethical ideal: the pricelessness and incomparability of human life, an ideal associated, in the philosophical tradition, with the Kantian doctrine of the infinite worth of the individual and the dignity of man. As against every variety of utilitarianism, thinkers such as Paul Ramsey [4, p. 268] and James Childress [2, p. 202] have upheld a version of this tradition amidst the "hard cases" and thorny issues of contemporary biomedical ethics.

And yet the category of age is different from such suspect categories as race, sex, or income level. At present, the aged, constituting 10 percent of the population, consume 30 percent of the health care resources. As advances in medical technology postpone disease and death until later in life, there is no reason why this proportion of 30 percent could not be extended indeterminately. The needs of older people for health care are obviously greater; at the same time, private resources to pay for this care are smaller. If we focused on a utilitarian criterion for care such as potential future contributions to society, fewer resources would be committed to care of the elderly, for they have less time left in which to contribute. The free marketplace criterion also will not do, because most old people do not have the money to pay for their care. We are left, in the end, with a "need conception of justice" ("to each according to his needs"), perhaps buttressed with a qualification of "similar treatment for similar cases," as Outka argues [3, pp. 81–85].

But in the case of the very old and ill patients, this criterion seems to fall down, for reasons offered by Robert Veatch:

> . . . if everyone has a claim to the amount of health care needed to provide a level of health equal to other persons' health, the system will collapse as soon as the person most in need of health care is in need because he has a condition that cannot be treated . . . the group of the incurably sick who are the most ill must end up with *all* the medical resources. This is certainly inefficient. Furthermore, if they do not benefit from the commitment of resources, it is hard to see why it is just that they get those resources [5, p. 134].

Veatch's critique has special importance for the case of allocating health care resources for the aging. The problems of the incurably ill, as Barbara Yondorf has written [6, p. 465], are shared by many aged patients. But our health care system is largely designed to prevent, cure, or rehabilitate people from illness. When we focus on the aged and incurably ill — the category Yondorf calls "the declining and the wretched" — we may need a very different approach to the rights associated with an involuntary and irreversible state of declining health [7, pp. 2ff.] . One example here may suffice.

## CASE STUDY: FAILURE TO PROVIDE RESOURCES

In New York City there is a large and very well-endowed voluntary nursing home, spectacularly equipped with all the medical equipment and staff support that could be imagined. However, as the former assistant director of the facility observed, this nursing home lacks the basic equipment of an intensive care unit, despite the likelihood of need for immediate use of such equipment, sometimes within a matter of minutes. Failure to provide intensive care facilities was not an architectural omission but rather a decision on where to allocate the resources of the nursing home: namely, to spend money enhancing the quality of life for the residents (mostly in their 80s) rather than providing exotic lifesaving technology to extend life during a medical crisis. The cost of investing heavily in the "life extension" approach for the aged could be exhorbitant, and as Veatch suggests, it is hard to see why vast resources should be expended during the last days of life of these patients.

But, one might reply, in making this argument are we not approaching dangerously near a kind of cost-benefit calculus that would explicitly stipulate age as a criterion for withholding health care resources from people? Unfortunately, such a simpleminded form of utilitarianism expresses very well the prejudice that many health care providers exhibit toward elderly patients. Several points must be noted here. First, the example of the intensive care unit and the nursing home suggests that the option of withholding scarce health resources arises, not in the course of normal or chronic long-term illness, but in states of acute crisis coming after extended decline and in situations where "heroic measures" or exotic technologies are involved. The question "Is it worth it to go on?" is asked not only by health care providers but also by the elderly themselves, who are among the most vociferous in rejecting the idea that they should be kept alive when all hope has been lost. This situation is not unlike that of the middle-aged parent who donates a kidney to a stricken offspring and says, "I've had my chance to live," when considering the risks to the donor. The feeling that age *is* a legitimate category in allocating scarce health resources seems to express a metaphysical claim about the appropriate or "natural" course of the human life cycle. In this sense, aging itself constitutes the inevitable finitude of human existence, and it is properly a consideration in bioethical decisions.

## THE LOCUS OF DECISION MAKING

A second point about the intensive care example concerns the nature of the decision making involved. Failure to build an intensive care unit is not like putting the elderly Eskimos out on the ice floe, nor is it comparable to an explicit decision in favor of triage (as in the famous Holmes case where passengers were chosen to be thrown overboard from the lifeboat). In other words, the nursing home example involves neither active euthanasia nor an explicit decision about who shall live.

The nursing home case is more comparable to "allowing to die," but — and this is a crucial point — the decision for passive euthanasia is *not* made by the health care staff caring for the patient at the time of death. By failing to provide intensive care equipment, we *know* that some of the elderly ill patients will die, but members of the attending medical staff are spared this decision.

In contrast, doctors in many hospitals often write "do not resuscitate" (or signs to that effect) on the patient's chart for such patients who are declining and wretched. But putting this decision at the level of the care provider places attending staff in a difficult professional situation, being forced to make a terrible choice. For exactly the reasons that Ramsey gives, it is wisest *not* to make such choices [4, p. 268]. The withholding of the scarce resource in the nursing home case *is* a choice made on grounds of age (in part), but it is a choice left unacknowledged, left, so to speak, in the margin.


REASONS FOR WITHHOLDING RESOURCES

Finally, let us consider the reasons for making such a choice. Outka argues that a just distribution of scarce health care resources requires the principle of equal access. Yet he acknowledges that the formula "similar treatment for similar cases" contains an "allowance of no positive treatment whatever [which] may justify exclusion of entire classes of cases from a priority list. Yet it forbids doing so for irrelevant or arbitrary reasons" [3, p. 92]. The whole question here, though, is whether or not age itself constitutes an irrelevant or arbitrary reason.

Childress argues [2, p. 202] that decisions for the allocation of scarce medical resources require separate treatment according to two distinct kinds of criteria: (1) medical acceptability; and (2) a decision principle such as equality of access, contribution to society, or some other principle of allocation. Those writers, like Ramsey or Childress, who favor randomized access to scarce resources (such as a kidney machine) do so by explicit rejection of any utilitarian arguments. Still they would allow decisions of exclusion to access based on the first kind of criterion (medical acceptability) even while opposing distinctions of access under the second category that would attempt to weigh the value of alternative human lives. Yet a medical committee in Seattle ruled out accepting for dialysis patients over 45 years of age. As we extend this example to other cases, we recognize that the problem is that "medical acceptability," as Childress himself suggests, can often "serve as a facade behind which other manipulations take place" [2, p. 202]. This is precisely the situation with the ambiguous category of age as a criterion for allocating scarce health care resources.

If we extend this criterion of medical acceptability to the problem of providing health care for the elderly, we recognize that age obviously plays a role in an *actuarial* sense: Certain operations that might be recommended for a 40-year-old patient would not be recommended for an 80-year-old. On these grounds, on an actuarial (probabilistic) version of "medical acceptability," it is entirely reasonable to withhold certain resources from persons because of age. Note here, however, that this *actuarial* criterion, while partially utilitarian, is in no way based on a presumed contribution to society of a patient, which, in my view, would be an unjust basis for allocating life and death resources except in the extreme cases of threat to a whole society (e.g., wartime emergency), which even Ramsey acknowledges.

The actuarial criterion qualifies the notion of medical acceptability to recognize that age itself may be associated with a probability of survival. But the actuarial criterion comes into play in precisely those situations where difficult decisions must be made between competing claims for scarce resources. To impose such an actuarial criterion apart from tangible medical conditions of a determinate disease is to import a dangerous "ageism" of the most ruthless kind. One is reminded of the recent science fiction film, *Logan's Run,* in which citizens of a futuristic "utopia" were executed at the age of 30 to maintain population equilibrium. The danger of any explicit or formalized version of the actuarial criterion is that elderly, or even middle-aged, patients could no longer trust doctors or the health care system. A lingering question in each patient's mind would always be "Will I someday become too old to be treated?" Nevertheless, as an implicit (nonformalized) element, it is proper to take age into account, provided (1) the decision making is as far as possible removed from the interpersonal context of medical treatment, (2) the age factor is treated as part of an issue of medical acceptability, and (3) the withholding of scarce resources comes at a moment of crisis (e.g., the intensive care example) and is not itself a contributing element ("benign neglect") in precipitating the death of elderly patients.

## CONCLUSION

These observations are only the barest beginning of a sustained argument for the circumstances in which age is properly a factor in allocating health resources. Nevertheless, observations of contemporary medical practice help us recognize important qualifications to any open-ended claim of universal access based on "need." Issues of quality of care, not simply life prolongation, and problems of the management of chronic illness are most important in considering the rights of the old to health care [7, pp. 2 ff.]. What I have called the actuarial criterion does not lead to any hard and fast cutoff points or decision rules: hence the difficulty, and the dehumanizing consequence, of *automatically* associating age *itself* with illness or inevitable decline. The boundary shifts with individual conditions of health, longevity, and the will to live, but the factor of age is there and must be considered. In any theoretical discussion of "rights" we ignore age at our peril, because in practice it does play a role and, I would argue, ought to play a role. The problem is to formulate in an adequate way what that role should be and, in so doing, to see the process of aging as an essential factor in the ethics of health care delivery.

## REFERENCES

1. Butler, R. N. *Why Survive?* New York: Harper & Row, 1975.
2. Childress, J. F. Who Shall Live When Not All Can Live? In R. M. Veatch and R. Branson (eds.), *Ethics and Health Policy.* Cambridge, Mass.: Ballinger, 1976.

3. Outka, G.   Social Justice and Equal Access to Health Care. In R. M. Veatch and R. Branson (eds.), *Ethics and Health Policy*. Cambridge, Mass.: Ballinger, 1976.
4. Ramsey, P.   *The Patient as Person*. New Haven: Yale University Press, 1970.
5. Veatch, R. M.   What Is a "Just" Health Care Delivery? In R. M. Veatch and R. Branson (eds.), *Ethics and Health Policy*. Cambridge, Mass.: Ballinger, 1976.
6. Yondorf, B.   The declining and the wretched. *Public Policy* 8:465, 1975.
7. Zeckhauser, R.   The welfare implications of the extension of life. *Gerontologist* 14:2, 1974.

# CHAPTER 28

# Growing Older in America: Can We Restore the Dignity of Age?

*Matthew Ies Spetter*

*We live* in a culture in which growing older is experienced as a calamity, a culture that has contempt for the frailties of age. Uncounted millions consider age a defeat, and both old and younger generations dread the process of aging as a disease. The aged are therefore not granted socioeconomic equality and, but very rarely, respect. Those growing old know that they are in danger of being discarded by their children and by society as such, as if they were essentially worn-out objects. They are lumped together into one anonymous gray category.

Aging panics many people who have seen what happened to their own parents. Thus guilt and fear often make them all the more impatient with the old, angry and not infrequently overtly hostile. Instead of sanely preparing for growing older, we sadly try to banish the emotional reality factors by camouflaging them with code words such as "senior citizen," "golden age," and so forth. By thus attempting to treat aging as an alien rather than a common experience, many have learned to use denial mechanisms, pretending that "old folks" are a nation by themselves, isolated from the stream of social continuity. The individual psychological status is of course deeply influenced by the sociomoral presuppositions of the culture.

Our particular culture supports the notion that functionality is central to human worth. It is a framework characterized by Albert Camus as "the cult of efficiency." When an individual can no longer function "efficiently" in terms of money-making and consuming, he is made to feel worthless, a hindrance, a nuisance [5].

Suddenly a man or woman at age 65 is supposed to exemplify changes of behavior and social aspiration to fit within the stratum of the defeated. If they resist this degradation of their life expectations, they are treated as bothersome at best or as "old fools." The aging person seeking continued vocational, artistic, or sexual gratifications is often treated with patronizing amusement, indifference, or neglect.

By communicating to the aging person that he or she is "over the hill," the social order in fact transmits the message that it is time to leave the stage, to become invisible, to die. Younger men and women often experience profound personality disturbance when society requires essential changes in role expectations. How much more so for the older person who, in addition to factors of physical decline, is now also burdened by the notion that his presence or participation has become superfluous? The typical elder person is portrayed as unhappy, ill, helpless, and "living in the past." This very stereotype permits the creation of social distance, by which retirement or widowhood is felt as a terminal point. Once this trend of thinking is internalized, the stereotype is then paraded as a "reality," and negative projections seem to be justified.

The older person is therefore seen as inferior, unattractive, not "worthwhile" to be with. He becomes a pariah even though he may have a soundly integrated personality and an eager open mind. The sociomoral presuppositions of our youth-crazed culture maintain, however, that the sense of self-respect of the older person is a quaint stubbornness.

Undermined in this way in their individual strengths and spiritual attainments, many parents are slowly but surely convinced by their children that they can no longer, in fact, have mastery over their own lives, that their experience and wisdom are not needed. Hopelessness is thus foisted upon the older person, and with his economic base narrowed, his income reduced, and his old friends dying one by one, he feels cut out from the living, rebuffed into powerlessness by socially sanctioned cruelty, whether overt or hidden. Instead of strengthening the spiritual resources of the aging person as he contemplates his own nearing death, a murderous and brutal indifference grinds him into the dust.

One only needs to visit any nursing home to verify these statements. In my pastoral work I see these separated men and women. I hear their cries in the halls of institutions where they have been relegated to die. When a nation allows contempt for its aged, it is already in the process of accepting the proposition of "necessary evil." Behind this looms the soulless grin of brutes who have come to accept the concept that *some* lives are expendable.

I consider this desacralization of human existence the central ethical issue of this epoch. A nation in the end is not judged by its technology, its affluence, but rather by the way in which it treats its young and its old. Shall we accept the thesis that some members of the human species can be discarded?

By 1980 it is predicted, there will be 25 million citizens in the age group of 65 and over. Aged persons today already account for the largest single sector of the poor who live by themselves. "Poor" means a median income of about $2800 per year! Of all families defined as "poor," one-third are headed by a person over the age of 65. Only dependent children form a larger category than persons aged 65 and over who are dependent on public assistance. In addition, it must be kept in mind that of all first admissions to psychiatric institutions in this country, 45 percent are persons 60 years of age or older. What these figures from a report by the Group for the Advancement of Psychiatry indicate is the crying need for a revision, a radical revision, of the place of the aged in our society.

The psychological reality of aging is very private. Each of us has to deal with its accompanying phenomena alone; there is no help but one's own resources. Because of our socioeconomic arrangements, this private process is denied innumerable older persons. By clustering them together and treating them as if they were worthless or children at best, we degrade their existence. Again and again, older men and women in nursing homes or other institutions are forced to participate in programs so. utterly meaningless in nature that their effect is depressing. Increased anxiety and isolation are frequently the result of not being allowed time or space for the solitary travail of inner preparation for the farewell that lies ahead.

Overmedication with tranquilizing drugs is still another aspect of this trend [1]. Visiting any nursing home and seeing fine men and women nodding as in a coma can be a vision of a particular hell in which those condemned to vegetation have been singled out by the socialization of cruelty and rejection. Old age as it stands today in America is the outcome of the values and structures that this civilization accepts as the "given" of our social organization.

As Simone de Beauvoir put it so well in her book, *The Coming of Age,* "to understand the sadness of old age, we have only to look at the shortcomings of life itself, for old age is life's end product" [3]. Agism, others have pointed out recently, "is the racism of the future" [2, 4, 6].

When a child is born we celebrate the arrival of a new life. All of us believe in ascending, in beginning. Can we not also learn to honor and celebrate the ripening of life, as it descends and yet serves to enrich the gift of a life well lived? Human beings continue. Their highest nobility lies in being able to hand on a certain knowledge about how to live, how to age, and also, how to die.

REFERENCES
1. Cohen, S.  Geriatric Psychopharmacology. Paper given at A.M.A. conference, San Francisco, June 1964.
2. Curtin, S. R.  *Nobody Ever Died of Old Age.* Boston: Little, Brown, 1972.
3. de Beauvoir, S.  *The Coming of Age.* New York: Putnam, 1972.
4. Goldfarb, A. I., et al.  *Psychiatry and Aging* (report no. 59). New York: Group for the Advancement of Psychiatry, 1965.
5. Heschel, A.  To grow in wisdom. *The Christian Ministry* 2:2, 1971.
6. Spetter, M. I.  Growing Older in America. Paper given at the Riverdale-Yonkers Society for Ethical Culture, Bronx, N.Y., March 4, 1973.

# The Rights of the Mentally Ill

# CHAPTER 29

# The Right to Change Behavior: Rights of the Mentally Ill
*Hildegard E. Peplau*

*In this chapter* the simple definition that rights are claims will pertain, recognizing, however, that there are complexities and appreciating that there are many subtleties in both terms already considered in the literature [1, 3, 5, 8]. The subject for discussion is the legal concept of a "right to treatment" for mentally ill persons confined against their will in mental hospitals. The Supreme Court has recently ruled against such involuntary incarceration without treatment. Judge Johnson in Alabama set the initial direction for this important judicial action when he required that state to supplant custodial retention with treatment. More recently in New York, Dr. Morton Birnbaum (who is often credited with originating the phrase "right to treatment") filed suit in the federal district court. His suit raised two issues germane to this discussion: [1] segregation of poorer and sicker patients in inferior state hospitals while wealthier and less ill patients go to better private hospitals; and [2] the contention that New York States' commitment laws are unconstitutional inasmuch as involuntary patients are not given active psychiatric treatment for which they were presumably admitted and which that law does not require. The gist of all of these actions is that all patients have a "right to treatment" and that custodial care is not treatment.

Ordinarily, rights are claims to be exercised on the initiative of a person when his need is evident. The right to health care means that any sick person will have access to the particular health services that his illness requires, irrespective of race, creed, color, age, sex, politics, or socialeconomic class. Ability to pay is not a criterion; the right to life is to be safeguarded if an individual presses his claim to health care. The option to press the claim, except in the case of infants and those with highly communicable diseases that threaten the community, belongs to the individual. In this sense, the right *not* to press a claim, there is an inherent right to refuse treatment. The pursuit of health rights, with respect to organic (medical) diseases is largely a matter of demand, choice, and supply of services. Claim options are choices — to inform oneself of resources, to select and seek access, and to choose to accept services or to reject them.

In the United States, two federal acts have enhanced the availability of treatment resources for the mentally ill and for treatment of milder sociopsychological disturbances of people. The Mental Health Act of 1946 provided funds for research and for preparation of psychiatrists, psychologists, psychiatric nurses, and psychiatric social workers. As a consequence, there is now much more knowledge about the nature of such human disturbances, and there is a larger pool and greater variety of qualified, professional treatment personnel for psychiatric services. In 1963 the Community Mental Health Centers Act was passed. It spurred the more rapid development of a wider range of methods of treatment in community-based centers where services could be sought voluntarily. The act also stimulated a trend toward deinstitutionalization of the "mentally ill." Moreover, it helped to promote a shift from almost exclusive focus on the pathology of "mental illness" to more serious consideration of the term "mental health." The simple assumption, previously held, that curing the "illness" at least in part promoted "mental health" is more open to question now than before mental health centers were so readily available. But in going to such centers voluntarily, individuals and families exercise a claim option — their right to treatment — however they construe their psychiatric problems for which help is sought.

There are still, however, many patients in public mental hospitals. In 1973 there were 444,777 such patients in 337 state and county mental institutions. It is in relation to this group particularly that the "right to treatment" is in need of clarification. Many of these patients do not seek hospitalization but instead have it forced upon them by family informants or social agents — the "social screen" that initially defines the "illness," the hospital admitting personnel serving as confirming definers and recommending admission, which is often involuntary. Under these conditions the claim to a right to treatment (and therefore need of it) is no longer a matter of individual choice, but as in the case of infants and highly communicable disease, it is a decision of social agents in behalf of the individual and largely for protection of the community.

The nature of "mental illness" is a compelling factor when community agents act to hospitalize a person against his will. The views of the ordinary layperson rarely keep pace with professional understanding of the phenomenon called mental illness. Tolerance of deviance from social norms of behavior is easily strained in any social system. The plea for hospitalization of a person who behaves too differently from others in a given system or community is in effect a request: "make him more like us."

Within the intellectual community — among the psychiatric professionals — there are fresh discussions and contradictory views on the nature of "mental illness." So-called mental illness is of a very different order from organic or physical diseases. There is a serious question if "mental illness" ought to be called a "disease." Once there was an advantage to such a designation, when the causes were thought to be "devils" and the "treatment" was incarceration in dungeons or jails among other cruel forms of care. Now with contemporary understandings, there is the question of whether human perplexity is a sickness. The difficulties of the "mentally ill" are conceptualized in different ways by health professionals, and each viewpoint gives rise to a different form of treatment.

The author of this paper includes, among others, these difficulties of persons labeled "mentally ill": problems of self-identity, of relationships with people, of usage of language-thought processes to derive and designate meaning of experience, and of functioning as a continually evolving person in social systems. These problems and their resolution do not fit neatly into the cause-effect, disease-oriented medical model. Clearly a rubric more distinctive of the phenomena in question ought to be substituted for the phrase "mental illness." Perhaps human dilemmas would be a more apt term.

New theories always open up possibilities of fresh explanations of phenomena. General systems theory — in particular its constructs of closed systems — suggests that "Psychopathology clearly shows mental dysfunction as a system disturbance rather than as a loss of single function" [6, p. 1101; 9]. The work of Satir and others, in family therapy, is demonstrating how "closed family systems" produce and maintain the dysfunctions traditionally called "mental illness"; nonmedical techniques, used to assist those families to become "open systems," thus enable all family members, including the "identified patient," to get on with growing, evolving, and developing as persons.

The conception of mental health becomes an issue if the aim of treatment of the "mentally ill" is to change the behavior of the "ill" person into that direction. But as Smith points out, the construct of "mental health" is far from clear. He refers to the term as ". . . an evaluative concept and . . . science has not yet learned to deal sure-footedly with values" [10, p. 166]. He suggests that the value issue remained implicit so long as the focus was on "mental illness" but that with the shift to consideration of the meaning of "mental health . . . even pathology becomes problematic as value differences become explicit." Smith points out that the only value-free definition of normal is statistical and that averageness is not optimal functioning. Similarly, autonomy, creativity, and self-actualization are value-laden concepts but are more relevant to human functioning than "adjustment." In fact, adjustment generally refers to accommodation to the demands of a situation in which a person participates, which is what an individual diagnosed as "schizophrenic" has done in terms of his closed family system.

"Mental health" is more often than not defined in terms of prevailing values – individual and social; in usage by some persons the defining dimensions may go beyond current values. Presently, however, all social and cultural values are in flux from "absolute values" to more relative ones – relative to a culture, a context, or a period in time. As Smith indicates, there is no consensus, either in this society or worldwide; anyone can formulate or choose, persuade others, or even dispute values [10, p. 184].

The foregoing comments on general systems theory and on values are only two of the many competing efforts at clarification of the terms "mental illness" and "mental health." The discussion is germane to the question of "right to treatment," for there is as yet no certain method to determine or to treat "mental illness." Personal testimony from family and social informants and vague behavioral criteria are hardly reliable. As Chessler has shown, the label "depression" is all too quickly applied by male psychiatrists to women who complain about their children, housewifely role, or sex-stereotyped expectations of them [4]. The precision in diagnosis in other instances is similarly open to question.

Once diagnosed and institutionalized, however, the question of "right to treatment" arises as an alternative to custody. Unlike diagnosis in medical diseases, the psychiatric diagnosis does not provide guidance for the treatment. By definition, a "sick" person is not supposed to be in a position to diagnose and treat himself or even to be able to judge the necessity for or quality of the treatment. This is even more the case in relation to "mental illness." The "mentally ill" have long been declared "incompetent" in this and other respects. Moreover, patients who do not pay for services are supposed to take and be grateful for what they get. Clearly, institutionalized "mentally ill" persons are in a most vulnerable position with respect to a "right to treatment."

Treatment presently available covers a wide *range*, but most of it, following the "medical model," is construed as doing something "to" the patient or "for" him, rather than "with" him. It is "other determined" and applied presumably to eradicate causes or ameliorate symptoms. Even today, treatment in this vein includes electroshock, insulin coma, lobotomy, use of mood-changing drugs, and more recently the use of "token economy" to effect behaviors in the patient that the treater has decided upon. None of these treatments eradicates causes, for the causes are not known. If the word "cause" applies at all, the causes are multiple and interacting. The effects of such treatments — immediate or long range or both — are to change behavior in the direction of quiet, obedience, compliance, conformity, and generally toward lower integrative levels.

In contrast, there are forms of treatment, the psychotherapies, based upon minimally definitive understanding of the human intrapersonal, interpersonal, and systems processes, related to the difficulties of the "mentally ill." As Frank has shown, there is a question of whether these various forms are equally effective or whether some forms are better for certain individuals [6]. Since the Frank publication, many additional forms of psychotherapies, including various kinds of "sensitivity" and meditational patterns of "treatment," have appeared [7]. All of these psychotherapies have the merit of doing something with the patient as a participant, but differences in the expected extent and quality of such participation vary widely.

Since the Supreme Court action, institutionalized psychiatric patients have a "right to treatment." What does and what should such treatment include? What is presently available depends upon the numbers, qualifications, and values of the professional staff in a given hospital. It is a well-known *fact* that public mental hospitals have very few, if any, board certified psychiatrists and very many, often unlicensed, foreign-born and foreign-educated physicians. Some hospitals have recently employed more and better qualified professionals other than physicians: psychiatric nurses, psychologists, and social workers. In some hospitals their expertise is respected and utilized; they are given autonomy and latitude in their direct work with patients. Services in some hospitals are moving toward excellence; others are still custodial warehouses for unfortunate people. The level of care generally reflects what any given state is willing to give financially in support of some quality of assistance to its troubled citizens. Given the present state of affairs — the lack of theoretical understanding and consensus on the nature of mental illness — the quality of assistance must reside in the educational qualifications of the employed professional personnel, a fact that Judge Johnson in Alabama took into account when spelling out criteria for treatment.

Does an involuntarily admitted patient by the mere act of entry make a claim on the "right to treatment"? The law holds that otherwise he cannot merely be "held" in the institution. However, the prevailing custom in many public mental hospitals is to agree to discharge of a patient only if the patient's family will accept his return. There are, of course, foster care and other kinds of community placement of patients, but placement generally applies only after treatment. Under these circumstances, the "right to treatment" is in effect a mandate — that the institution must treat the patients. So far no literature has been found that modifies the patient's vulnerability, at least in the respect that the patient has some say — a choice — among various treatment methods: a "right to refuse," for example, electroshock or excessive medication and a choice among other available treatment options. Without such a condition, the patient does not have the options that are open to both private and community mental health center patients. In fact, the treatment right is imposed.

Characteristically, anyone in the "sick role" is at the mercy of treatment methods and philosophical beliefs that underlie them, which are held by particular treating professionals. The "mentally ill" patient, however, is often accidentally assigned to a particular ward and staff; whatever values and methods operate in that situation may be said to constitute treatment. How to overcome vulnerability to arbitrarily applied methods of treatment — that is the question. In recognizing this problem, Whittington puts it this way:

. . . a solo practitioner's evaluation of a patient he may treat may not always be objective, particularly when his caseload is low. Further, a major part of clinical efficacy is the practitioner's belief in his method; this belief renders him less than objective in deciding on a treatment approach best suited to the patient. Consequently, the institute sought to safeguard the patient by having a highly competent evaluation staff not committed to any particular treatment modality. During the initial assessment, specific behavioral goals and expected outcomes were defined in cooperation with the patient. A prescribed course of treatment specifying type, duration, cost, and probable outcome — was recommended to the client. If the client accepted the recommendations, the evaluation staff then placed him with the proper treatment program or clinician to carry out the plan [11, p. 25].

Whittington suggests: "Perhaps it will be a good experience for all of us, also, to admit that we frequently do not know what is best for people and that it is really more honest as well as more appropriate clinically to give people a fuller range of choices in trying to correct the life dilemmas that bring them to us" [11, p. 27].

What most often collides with the foregoing propositions is the self-concept of professionals that is antithetical to the idea of being judged by other professional colleagues; the idea of "mentally ill" patients judging a professional's capabilities and refusing his or her treatment is well-nigh abhorrent. Patients who refuse certain forms of treatment may be coerced, drugged, or transferred to "back wards." Staff indifference to patients who do not respond to treatment provided is well known — perhaps most clearly illustrated by dying patients. The "right to refuse" treatment, when exercised by mentally ill patients, is a slap at professionals who are likely to view it as one more instance of the patient's incompetence associated with his "disease" — mental illness.

The right to refuse treatment poses a social problem as well. The difficulties associated with "mental illness" will tend to evolve in the same direction, that is, get worse, for most patients who do not have (and eventually use) constructive, growth-provoking help. As with anyone's behavior, the *tendency* is to maintain present patterns, varying in the same direction except under extreme anxiety or panic, which forces pattern changes. To the extent that the "mentally ill" are permitted to exercise a right to refuse treatment, with the exception of so-called "spontaneous cures," it can be predicted that such patients would become "chronic." In that event, they would continue to be public charges – an economic burden to society – as indeed years of lack of treatment has already demonstrated in public mental hospitals. To release such patients untreated into the community is less a matter of their "danger" to society than a question of their being a burden, modeling or communicating their difficulties to others as well as their possible exploitation by others, which their release imposes. On the other hand, colonies set up to shelter such patients might be preferable to a lifetime in institutions.

The "right to treatment" quite possibly should be modified to the right to choose from an array of available treatments. The right to refuse treatment might follow conditional participation in sessions designed to investigate conceptions of treatment offered and reasons for refusing. The right to treatment, however, pinpoints the urgency to assess the present state of knowledge about mental illness and mental health; to pull together in a substantive way the emerging consensus and to pinpoint the philosophical and ethical issues concerning the nature of disturbed human processes. Sharpening the antithetical claims and the viable knowledge on which they rest would clarify rationales for treatment.

## REFERENCES

1. Bandman, B.   Rights and claims. *Journal of Value Inquiry* 7:204, 1973.
2. Bertalanffy, L. Von.   General Systems Theory and Psychiatry. In S. Arieti (ed.), *American Handbook of Psychiatry* (2nd ed., vol. 1). New York: Basic Books, 1974. P. 1101.
3. Benn, S. I.   Rights. In P. Edwards (ed.), *Encyclopedia of Philosophy*. New York: Macmillan, 1966. Pp. 195–196.
4. Chessler, P.   *Women and Madness.* New York: Doubleday, 1972.
5. Feinberg, J.   The nature and value of rights. *Journal of Value Inquiry* 4:243, 1970.
6. Frank, J.   *Persuasion and Healing.* Baltimore: Johns Hopkins University Press, 1961.
7. Harper, A.   *The New Psychotherapies.* Englewood Cliffs, N.J.: Prentice-Hall, 1975.
8. Peplau, H. E.   Health care rights. *Image* 7:4, 1974.
9. Satir, V.   A Family of Angels. In J. Haley and L. Hoffman (eds.), *Techniques of Family Therapy.* New York: Basic Books, 1967. P. 719.
10. Smith, M. B.   *Social Psychology and Human Values.* Chicago: Aldine, 1966.
11. Whittington, H. G.   A case for private enterprise in mental health. *Administration in Mental Health* 25:23, 1975.

# CHAPTER 30

# Legal Rights of the Mentally Disabled
*June Resnick German*

*For far too long,* the mentally disabled were treated as a class apart from the general population with respect to basic human rights, taken for granted by all other members of society. They were locked away in large institutions, generally isolated from populated areas, and virtually forgotten. Recently, however, due to the advent of psychotropic medication, which enables hospitalized mentally ill patients to return to their communities within a short period of time following admission [1], and due to the increasing concern for the rights of the disadvantaged in society, attention has been drawn to the plight of the mentally disabled.

Within the last decade, the recognition by some that mental disability provides no basis for deprivation of fundamental rights has spurred a flood of litigation to restore rights to patients who should never have been denied them in the first place and to recognize new rights that flow from the severe deprivation of liberty that is often imposed on these patients. State legislatures, too, are beginning to recognize the rights of the mentally disabled by enactment of laws to ensure that such rights are protected.

In the American legal system, in addition to those rights accorded by common law, legal rights take two forms. They may be derived either from statutes that are enacted by the legislative branch of government, which makes the rights subject to repeal, or from the Constitution, which makes the rights immutable. Most litigation in mental health involves the application of the Fourteenth Amendment constitutional guarantees of equal protection of the laws and due process of law to the mentally disabled.

## THE RIGHT TO EQUAL PROTECTION OF THE LAWS

At the outset, it must be emphasized that the mere presence of mental disability alone is an insufficient basis to deprive a person of any rights to which other members of society are entitled. Unless it can be clearly demonstrated that the individual is unable to exercise his rights, even if he is hospitalized for his disability, he should not be deprived of rights he would otherwise have as a citizen.

Equal protection generally requires that all persons be dealt with equally unless there is a rational basis for a legislative determination that a class of people be treated differently [6]. However, if a statute deprives persons of a "fundamental interest" [7], such as liberty in the case of the involuntarily hospitalized mentally ill, or employs a "suspect classification" [6], which might occur in the case of the class of mentally ill [10], then the statute would be subjected to strict judicial scrutiny. It would be the obligation of the state to prove both a compelling state interest and that the classification is precisely suited to serve a permissible state objective. Therefore, it can be argued that the mentally disabled as a class should not be denied rights afforded others unless a compelling state interest can be proven.

Historically, though, it has been the general practice to take rights away from all persons hospitalized for mental disability. Such patients were isolated, not permitted to vote, and often had the control of their property taken from them. While these deprivations were not legally justifiable, they were the accepted practice in many states. Presently, however, the practice is changing. Many states now recognize that rights may not be denied merely as a consequence of hospitalization for mental disability. New York law, for example, expressly requires that no person shall be deprived of any civil right, including the right to vote, hold office, have civil service status, and licenses, merely because of treatment for mental illness [11, Section 15.01]. Furthermore, the fact that a person is hospitalized is not an adjudication of his competency [11, Section 29.03], and therefore he is entitled to control his property unless in a further proceeding, he is shown to be unable adequately to manage his property.

Thus there appears to be increasing recognition that equal protection of the laws requires that persons receiving treatment for mental disability be afforded full rights afforded other members of society unless compelling reasons can be demonstrated for their deprivation.


## THE RIGHT TO DUE PROCESS OF LAW
### Procedural Guarantees

Much of the recent litigation in the mental health field centers around the Fourteenth Amendment requirement that no state shall deprive any person of life, liberty, or property without due process of law. Since involuntary hospitalization entails a severe deprivation of liberty, any person subject to involuntary hospitalization must be afforded the basic due process rights of a court hearing, adequate notice of the proceedings, the privilege against self-incrimination, and the right to be present at the proceedings and to cross-examine witnesses. The right to counsel must also be afforded, including counsel provided by the state if counsel is not otherwise available. Many states provide these rights by statute, but if such were not the case, persons subject to the hospitalization process would be entitled to them anyway [5].

In addition to the above rights, the issues of the criteria to be used at a court hearing to determine need for involuntary care and treatment is gaining increasing attention. Most states now allow involuntary hospitalization if a patient is shown to be either a danger to himself or others or in need of involuntary hospitalization but, as a result of mental illness, is unable to recognize the need for such treatment. However, such vague criteria are coming under increasing scrutiny. Several court decisions have set aside statutory standards that are not strictly drawn and have imposed higher standards that must be met before the severe deprivation of liberty that involuntary hospitalization entails is permitted to occur.

Not only have courts acted to ensure that the criteria for involuntary hospitalization be made more specific and narrowly drawn, but courts have also acted to require that the state meet a high standard of proof that the criteria are present. Therefore courts have held that a state must prove that the patient meets the criteria for involuntary hospitalization by clear and convincing evidence [9] or beyond a reasonable doubt [2].

Once a patient is hospitalized, his confinement does not end his right of access to the courts to determine whether the need for involuntary hospitalization continues.

Thus patients are entitled to a writ of habeas corpus to test the legality of continued confinement.

In addition, patients should have the right to periodic judicial review to make sure their continued hospitalization is scrutinized, regardless of whether or not they actively seek release. Some patients do not know they have alternatives to hospital-ization, although they actually need not be subject to continued involuntary treat-ment. Indeed, long-term hospitalization may have more detrimental than beneficial effects. Patients may become so dependent on a hospital regimen that they become institutionalized to the extent that they may never want to leave the sheltered con-fines of the hospital. At the time of the judicial review, the continuing need for hospitalization can be examined, and alternatives to hospitalization can be explored.

### The Least Restrictive Alternative
Another basic constitutional right is the right not to be restricted in one's liberty to an extent greater than necessary in furtherance of a legitimate governmental pur-pose [14]. This means that if involuntary hospitalization is seen as a legitimate exercise of governmental power to protect the health and welfare of its citizens, in-voluntary hospitalization and treatment must then be in the least restrictive facility. Outpatient care, therefore, must be considered before inpatient care is involuntarily imposed. If inpatient treatment is the least restrictive option, such treatment must be in a setting least restrictive of liberty, including the treatment ward or facility that impinges least upon freedom.

In New York the constitutional principle of least restrictive alternative is incor-porated into the statutory scheme by requiring any doctor who signs a certificate that serves as a basis for involuntary hospitalization to swear that he has explored all less restrictive alternatives and has found none adequate to care for the prospec-tive patient's needs [11, Section 31.27 (d)]. In addition, the law requires that in-voluntary hospitalization is permissible only if, among other requirements, it is essen-tial to a person's welfare [11, Section 31.01].

The United States Supreme Court has endorsed the least restrictive alternative principle in one of the few, and thus one of the most important, civil cases in the mental health field decided by the Court thus far. In *O'Connor* v. *Donaldson* [12], the Court stated: "... a State cannot constitutionally confine without more a non-dangerous individual who is capable of surviving safely in freedom by himself or with the help of willing and responsible family members or friends" [12, p. 576]. Therefore, if the severe deprivation of liberty that involuntary hospitalization en-tails is not necessary, it is not constitutionally permissible.

### The Right to Treatment
Another highly important right of mentally disabled persons who are involuntarily hospitalized as a result of their disability is the right to treatment. This right also derives from the due process clause of the Fourteenth Amendment.

The recognition of a right to treatment received nationwide attention in the two Alabama cases brought to federal court on behalf of the mentally ill [16] and men-tally retarded [17]. In a landmark decision, Judge Johnson determined that Ala-bama patients "have a constitutional right to such individual treatment as will give each of them a realistic opportunity to be cured or to improve his or her mental condition" [16, p. 784].

Mentally retarded patients were determined to have a constitutional right to "such individual habilitation as will give each of them a realistic opportunity to lead a more useful and meaningful life and to return to society" [17, p. 390].

In accompanying orders, Judge Johnson set forth specifically what constitutes minimum adequate treatment dealing with such matters as physical plant of the facility and staffing patterns. He further required treatment plans suited to each patient's individual needs and required that patients receive adequate medical treatment for their physical needs.

An additional aspect of the right to treatment that emerged in *Wyatt* v. *Stickney* is the right to compensation for institutional labor. The court found that "nontherapeutic, uncompensated work assignments . . . constituted dehumanizing factors contributing to the degeneration of the patients' self-esteem" [18, p. 375]. In its order [18, p. 381], the court required that if patients agreed to engage in therapeutic labor for which the hospital would otherwise have to pay an employee, such labor would have to be part of the patient's individual treatment plan, be approved by a qualified professional, be supervised by a staff member to see that the activity had therapeutic value, and be compensated in accordance with minimum wage laws.

To date the issue of the right to treatment has not specifically been recognized by the United States Supreme Court, although *O'Connor* v. *Donaldson* [12] presented an opportunity for the Court to consider it. However, the Court expressly chose not to decide the issue of the right to treatment on the facts presented in that case, neither accepting nor rejecting the doctrine [12, p. 571]. The Court did state, however, that where treatment is the announced purpose of involuntary hospitalization, the courts are not powerless to inquire into the adequacy of treatment being provided [12, p. 574, n. 10].

While no other right to treatment case has reached the Court in the area of mental health, in a related context, that of narcotics addiction, the Supreme Court assumed that the mentally ill cannot be punished for their illness just as narcotics addicts could not be punished for addiction [13]. Since without treatment, a hospital is nothing more than a prison for those involuntarily retained therein and an involuntary patient who receives no treatment is actually being punished for illness, this kind of confinement would seem violative of the Supreme Court dictate. Thus if the right to treatment issue were squarely presented to the Supreme Court, it appears likely that the Court would endorse it.

## The Right to Determine Course of Treatment

People generally are entitled to seek or refuse medical treatment for their ailments: mentally ill persons should also have this opportunity. The mere presence of mental illness without further demonstration that the person is unable to make his own decisions should not strip the person of his right to determine what is to be done to his body or his mind.

Any mentally disabled person, just as any other patient undergoing treatment for his disability, should be accorded respect for his human dignity. To as full an extent as possible, a patient should be advised of recommended treatment, including an explanation of what can be expected from the treatment and what both its beneficial and adverse consequences are. Alternative treatments should be similarly explained. The patient should be permitted to participate in devising a treatment plan suited to his individual needs, and his objections to a particular course of treatment should be considered and suitable alternatives employed when possible.

Except in rare emergency situations, no person hospitalized for mental disability should be forced to undergo treatment against his will before he has had an opportunity to challenge the need for involuntary treatment at a court hearing [4]. It is quite possible that the patient may be able to show at the hearing that he is not in need of involuntary treatment and should be discharged from custody.

With respect to such extreme forms of treatment as surgery, shock therapy, certain forms of behavior modification, use of experimental drugs, or psychosurgery, additional considerations exist. Since these treatment forms are extreme in nature and may have severe adverse consequences, the express consent of the patient should be required. If the patient's condition appears to be such that he is unable to competently give or withhold consent, this issue should be determined in a court proceeding, and if he is found not competent for this purpose, the court should order appropriate measures to protect the patient's interest.

At least one court [8] has held that certain forms of therapy can never be imposed on involuntary patients. In a case in which his doctors recommended psychosurgery to enable a violent patient to return to the community and in which the patient consented, the court found that the fact that the patient was in a coercive environment impeded his ability to freely consent, and thus the court refused to allow him to consent to the procedure. The court was also concerned with the lack of evidence that such procedure could result in substantial benefit to the patient and the uncertainty of the risks that would be posed.

Religious beliefs must also be evaluated to determine if a patient may be given coercive therapy. The governing principles provide that religious freedom is not to be impaired unless it can be shown that a "clear interest, either on the part of society as a whole or at least in relation to a third party . . . would be substantially affected by permitting the individual to assert what he claimed to be 'free exercise' rights" [15]. Since treatment for psychiatric illness primarily affects only a specific patient's welfare, it would seem that in most cases involuntary treatment should not be administered in the face of a valid religious belief.

## FIRST AMENDMENT FREEDOMS

The First Amendment provides fundamental rights of freedom of religion, speech, press, and right to assembly. These rights are afforded to all people, irrespective of mental condition. Patients are entitled to practice their religion and cannot be forced to take medication or undergo medical treatment that contravenes their religious beliefs unless additional and compelling interests are present.

The rights of patients to correspond freely and to receive visits, while generally incorporated into statutes, are actually derived from the First Amendment freedoms of speech and assembly. Despite statutory and constitutional requirements, hospital doctors often regard these rights as privileges that a patient must earn. This practice is a blatant denial of patients' rights that should never be tolerated.

Although the rights of the mentally disabled have improved significantly in recent years, much more remains to be accomplished. Of particular concern are the rights of various subclasses of the class of mentally ill, including juveniles, women, and those involved in the criminal justice system. Litigation [3] involving the latter group has in some respect improved their status, but efforts must be continued to erase the deprivations and inequalities that remain.

## REFERENCES

 1. Arieti, S. *American Handbook of Psychiatry* (2nd ed., vol. 5), p. 444, 1975; Kolb, L., *Modern Clinical Psychiatry* (8th ed.), pp. 623–624, 1973.
 2. *In re Ballay*, 482 F. 2d 648, 662–669 (D.C. Cir. 1973); *Lessard* v. *Schmidt*, 349 F. Supp. 1078, 1095 (E.D. Wis. 1972).
 3. See *Baxstrom* v. *Herold*, 383 U.S. 107 (1966); *Jackson* v. *Indiana*, 406 U.S. 715 (1972); *United States ex rel Schuster* v. *Herold*, 410 F. 2d 1071 (2d Cir.) cert denied 396 U.S. 847 (1969).
 4. *Bell* v. *Wayne County General Hospital at Eloise*, 384 F. Supp. 1085 (E.D. Mich. 1974); *Winters* v. *Miller*, 446 F. 2d 65 (2d Cir.), cert denied, 404 U.S. 985 (1971).
 5. *In re Gault*, 387 U.S. 1 (1967); *Heryford* v. *Parker*, 396 F. 2d 393 (10th Cir. 1968); *Lessard* v. *Schmidt*, 349 F. Supp. 1078 (E.D. Wis. 1972), vacated and remanded on other grounds, 414 U.S. 473 (1974), judgment modified and reinstated, 379 F. Supp. 1376 (E.D. Wis. 1974), vacated and remanded on other grounds, 421 U.S. 957 (1975), reinstated, 413 F. Supp. 1318 (E.D. Wis. 1976); *People ex rel Woodall* v. *Bigelow*, 20 N.Y. 2d 852, 231 N.E. 2d 777, 285 N.Y.S. 2d 85 (1967).
 6. *Graham* v. *Richardson*, 403 U.S. 365, 371 (1971).
 7. *Harper* v. *Virginia Bd. of Elections*, 383 U.S. 663, 670 (1966).
 8. *Kaimowitz* v. *Michigan Department of Mental Health*, Unreported, Cir. Ct. Wayne Co., Mich. (1973).
 9. *Lynch* v. *Baxley*, 386 F. Supp. 378, 393 (M.D. Ala. 1974).
10. *Mental Illness: A Suspect Classification?*, 83 Yale L.J. 1237 (1974); *contra* Note, *Developments in the Law – Civil Commitment of the Mentally Ill*, 87 Harv. L. Rev. 1190 at 1229–1230 (1974).
11. N.Y. Mental Hygiene Law, (McKinney 1976).
12. *O'Connor* v. *Donaldson*, 422 U.S. 563 (1975).
13. *Robinson* v. *California*, 370 U.S. 660 (1962).
14. *Shelton* v. *Tucker*, 364 U.S. 479 (1960).
15. *Winters* v. *Miller*, 446 F. 2d 65, 70 (2d Cir. 1971).
16. 325 F. Supp. 781 (M.D. Ala. 1971), enforced 344 F. Supp. 373 (M.D. Ala. 1972), aff'd sub nom *Wyatt* v. *Aderholt*. 503 F. 2d 1305 (5th Cir. 1974).
17. 344 F. Supp. 387 (M.D. Ala. 1972), aff'd sub nom *Wyatt* v. *Aderholt*, 503 F. 2d 1305 (5th Cir. 1974).
18. 344 F. Supp. 373 (M.D. Ala. 1972).

# CHAPTER 31

# Social Environment as the Basis for Discerning Mental Disability: The Effects of a New Paradigm on Patients' Rights

*Angela Barron McBride and William Leon McBride*

*In recent years,* evidence has been accumulating rapidly concerning the extent to which environmental factors, particularly aspects of the *social* environment, appear to affect the incidence of mental illness. The studies of M. Harvey Brenner [4] have been widely cited; they show a relationship that is, to all intents and purposes, indisputable between mental illness and socioeconomic changes: During economic downturns, for example, rates of suicide and of mental hospital admissions increase. It is in the burgeoning area of women's studies, however, that the move toward attributing mental illness to environmental, not dispositional, factors is most dramatic, for there is considerable evidence that women are categorically expected to evidence a "normal pathology" as part of their social roles and that these roles in turn make women sick.

A woman's personal inclinations toward either genius or schizophrenia have to take a back seat to the fact that women typically are seen as abnormal because they are not men. Freud, for example, made the phallic synonymous with the human and excluded women from *full* personhood: "For women the level of what is ethically normal is different from what it is in men. Their super-ego is never so inexorable, so impersonal, so independent of emotional origins as we require it to be in men" because "castration has already had its effect" [9]. In what has now become a classic study, Inge Broverman and her associates found that clinicians have a double standard of mental health [5]. They see a man and an adult as one and the same, but they are likely to assume a healthy mature woman is more submissive, less independent, less adventurous, more easily influenced, less aggressive, less competitive, more excitable in minor crises, hurt more easily, more emotional, more conceited about her appearance, and less objective than either a healthy mature man or a healthy mature adult.

The feminine ideal is devalued, as is work labeled as done by women over the same work labeled as normally done by men [15]. Pheterson, Kiesler, and Goldberg found that a male painter tends to be rated more favorably than a female painter when the presented artistic evidence is identical [16]. Deaux and Taynor found a similar tendency to rate the male applicant for a program of study abroad more favorably than the female applicant when their performance is equivalent and high [8]. When presented with scholarly articles on architecture, dietetics, law, and city planning, both sexes judged those written by John McKay more favorably than the same articles written by Joan McKay [10].

The diminished sense of personal potency supposedly characteristic of mental illness is what is normally expected of females from an early age. After an extensive review of stories in elementary school textbooks, Jacklin and Mischel found that when good things happen to a male character in a story, they are presented as resulting from his own actions. Considerably fewer good things happen to female characters, and when they do they are typically at the initiative of others or they simply grow out of the situation in which the girls find themselves [11]. There is no way you can treat women in our society for lack of self-confidence without first comprehending the female's Catch-22: When a woman is ready to achieve the feelings of self-esteem that come from successful femininity, she is simultaneously losing esteem because neither she nor the society really values her traditional roles. Yet traditional roles are considered essential to normality [2].

Not only are "well-adjusted" women normally expected to share the flaws of the mentally ill, but also the traditional roles of wife and mother themselves lead to further psychological impairment. Sociologist Jessie Bernard, for example, came to the startling conclusion that marriage is good for men physically, socially, and psychologically, but it literally makes thousands of women sick [3]. Having a support system cushions him; being a support system drains her. Elizabeth Janeway contends that motherhood demands of women an emotional tour de force of which few are capable:

First, they are asked to regard the bearing and raising of children as at least a very large and significant concern of their lives and, perhaps, as the crown and center of their existence, although, in the nature of things, this undertaking will demand their full efforts for something less than two decades out of a life that will run to seventy years. Second, they must fit their children for a society whose needs and aims are at best uncertain, and which may in fact seem to the mothers as well as the children morally unjustified and emotionally unsatisfying. At the same time, the most admired goals of society are pretty well closed to these women themselves. Third, they are expected to do all this only by means of an emotional relationship, instead of (as in the past) with the help of economic activities and social processes that relate to the larger world . . . . Fourth, having called forth this relationship, mothers are aware that they should maintain it in such a delicate balance that the child can grow out of it without harm to his own psychic strength. This program they are supposed to carry out with little training and little support from society itself, in the belief that any failure will justly be laid at their door [12].

Susan Darley concurs with this indictment of the role of mother: "Not only are the possible rewards available to them for good performance of their salient role less striking, but also the possible punishments are greater than they are for men" [6]. Given a job description that reads, "to keep everybody happy" [1], who would not feel chronically ineffectual?

In considering the move toward attributing mental illness to environmental, not dispositional, factors, it is important to note that the underprivileged have characteristically had their success minimized by attributing it to temporary reasons, while their failure, on the other hand, has regularly been underscored by linking it with dispositional factors [7]. So we find that the success of females is less likely than that of males to be attributed to basic ability; females are more likely to be perceived as just "lucking out." But negatively valued groups are beginning to question such stereotypes. Symptoms of psychological distress need not be evidence of inadequacy, but they may be appropriate responses to external pressures. When a woman repeatedly underestimates her ability to solve math problems is she to be blamed for having an inherent neurotic lack of self-confidence? Or are these establishment words of psychoanalyst Theodor Reik to be blamed for contributing to the demise of self-confidence in 51 percent of the world's population: "Fortunately, no woman believes, deep within her, that a man is much impressed by her brilliant intellect. She attributes no more value to it than as an accessory to a beautiful dress. I refer, of course, to a woman in her right mind; but when is a woman in her right mind?" [17]. When, indeed, is a woman in her right mind?! Given the sort of overwhelming evidence that we have cited concerning women, it seems to follow that the same question needs to be posed concerning many other groups and individuals.

Meanwhile, convergent conclusions, all pointing to a radically new conception of the very meaning of mental illness (and perhaps even, although this lies beyond the scope of the present paper, of the nature of disease in general), have been drawn by theorists with varied political and professional backgrounds. The philosopher Herbert Marcuse developed an ingenious internal critique of Freud's thought in order to show the possibility of there existing future civilized societies that, contrary to Freud's pioneering depiction of our own civilization, could be free from the widespread "sublimation" that is said to be the basis of our mental discontents. Marcuse saw the generation of greater and more widely distributed abundance of material goods as a necessary precondition for such a development [14]. Thomas Szasz has written a much quoted work whose principal thesis is contained in its title: *The Myth of Mental Illness.* Myths, of course, must always have some basis in fact if they are to be widely believed, and Szasz shows how facts of which everyone is aware have been misinterpreted by those responsible for mental health care in the past in order to fit a preconceived framework [18]. The various works of R. D. Laing have made provocative sallies into the political aspects of mental disorders and especially into the issue of how power relationships within our society's smallest political unit, the nuclear family, influence mental attitudes [13]. We do not mean to endorse all the theoretical conclusions, often exaggerated, of these thinkers or of any others writing today in a similar vein, but we do regard the convergence of some of their analyses with the previously mentioned empirical data as being highly significant.

The upshot of such new perspectives on the nature and meaning of mental health and disease should certainly not be to deny the influence of physical or chemical factors on certain types of behavior generally regarded as aberrant, nor should it be to tie the hands of those health professionals or law officers or both who are charged with attempting to minimize the amount of social harm caused by some such kinds of behavior. Finally, we are not necessarily forced now completely to discard the concept of "responsibility" as applied either to those who are denominated as mentally ill or to the rest of the population. But there is much about past ways of thinking about the so-called "mentally ill" that needs to be revised.

In the past and indeed right up to the present time, the dominant way of thinking about the mentally ill has been that they constitute a category of human beings radically (qualitatively) different in kind from all others, a category that had no claim to participate in most ordinary civil affairs. Theorists of democracy, for instance, have typically regarded "the insane" as troublesome exceptions to their ideals of universal participation in political life, sharing this status of exceptionality with children, not very long ago with women, and only a little longer ago (in the United States, at least) with slaves. Women's rights are now a topic of widespread public discussion, and a few hesitant incursions have begun to be made in the area of children's rights in our legislatures and law courts, but the legally recognized and enforced rights of those certified to be "insane" are often, to all intents and purposes, nonexistent.

Consider the case of an accused criminal who decides (or, often more accurately, is persuaded) to plead not guilty to a serious charge on the grounds of insanity. If the judge orders him or her to be incarcerated in a mental institution, the practical effect is often an indefinitely long sentence; whereas our penal system, with all its horrors and failings, is at least characterized by elaborate provisions for maximum-term sentences and for parole. In some other countries, it appears to be a fairly common practice to sentence those accused of certain crimes, of which political dissidence seems to be especially popular with the authorities, to mental institutions rather than prisons. From some accounts of personal experiences that one reads, one may infer that the latter form of sentence is often preferable to the former in terms of quality of life!

While many aberrations of this sort may simply result from cynical manipulation of both mental health and criminal justice systems for the sake of private or political gain, the fact is that they derive whatever apparent justification they may claim from the traditional but outmoded conception of mental illness to which we have referred. It would be impossible to pretend that someone with unconventional opinions merited removal from ordinary society and confinement to a special institution if the "them and us" paradigm concerning "the insane" and "the sane" did not still exert great influence on the thinking of most people (often, of course, even of those in mental institutions). In our own country, the elimination of this paradigm would result in abolishing the whole traditional practice of "certifying" to a patient's insanity, with its connotations of total depravity and hopelessness and its side effect of concentrating God-like powers in the official certifiers, those at the top of the psychiatric professional hierarchy. What we now call mental illness would be reconceived primarily as response to stress — the stress of one or another factor in the social environment.

The emphasis in mental health care would shift, as in fact it is slowly shifting, to preventive measures. But this means, if one takes seriously the findings noted at the beginning of this paper, that this branch of medicine will have to involve itself deeply with matters that have traditionally been consigned to politicians and social reformers. For it is now clear that the incidence of mental illness and the very concepts of psychological inferiority and superiority are in part dependent on such factors as the availability of social support systems, such as counseling services, day-care centers, adequate housing, utilities, and transportation, and ultimately even expendable income.

Thus, while the mental health professions stand to lose some of the unjustified autonomy that they, or at least their elites, have enjoyed with respect to the "certification" and control of a portion of our population, they also have an opportunity and indeed an obligation to expand the scope of their activities (though certainly not in the direction of some new, exclusive *controls*) within society at large. The expansion of the notion of mental patients' legal rights that these developments imply goes hand in hand with a recognition that any abstract conception of "rights" as such is useless or even meaningless outside of a society whose members acknowledge their mutual interdependence and, through all differences of age, sex, physical and mental health, and so on, their common humanity. A state of affairs in which such acknowledgments are universally made is, after all, the definition of a healthy society.

## REFERENCES

1. Ames, L. B. *Child Care and Development.* Philadelphia: Lippincott, 1970. P. 273.
2. Bardwick, J. M. *Psychology of Women: A Study of Biocultural Conflicts.* New York: Harper & Row, 1971. P. 190.
3. Bernard, J. *The Future of Marriage.* New York: World Publishing, 1972.
4. Brenner, M. H. *Mental Illness and the Economy.* Cambridge, Mass.: Harvard University Press, 1973.
5. Broverman, I., Broverman, D. M., Clarkson, F. E., Rosenkrantz, P. S., and Vogel, S. R. Sex-role stereotypes and clinical judgments of mental health. *J. Consul. Clin. Psychol.* 66:157, 1974.
6. Darley, S. A. Big-time careers for the little woman: A dual-role dilemma. *Journal of Social Issues* 32:85, 1976.
7. Deaux, K. Sex: A Perspective on the Attribution Process. In J. H. Harvey, W. J. Ickes, and R. F. Kidd (eds.), *New Directions in Attribution Research* (vol. 1). Hillsdale, N.J.: Lawrence Erlbaum, 1976. P. 341.
8. Deaux, K., and Taynor, J. Evaluation of male and female ability: Bias works two ways. *Psychol. Rep.* 32:261, 1973.
9. Freud, S. Some Psychological Consequences of the Anatomical Distinction Between the Sexes (1925). In J. Strachey (ed.), *Collected Papers* (vol. 5). New York: Basic Books, 1959. P. 196.
10. Goldberg, P. Are women prejudiced against women? *Trans-Action* 5:28, 1968.
11. Jacklin, C. N., and Mischel, H. N. As the twig is bent — sex role stereotyping in early readers. *School Psychology Digest* 2:30, 1973.
12. Janeway, E. *Man's World, Woman's Place.* New York: William Morrow, 1971. P. 57.
13. Laing, R. D. *The Politics of the Family and Other Essays.* New York: Pantheon Books, 1971.
14. Marcuse, H. *Eros and Civilization.* Boston: Beacon Press, 1955.
15. Mischel, H. Sex bias in the evaluation of professional achievements. *J. Educ. Psychol.* 66:157, 1974.
16. Pheterson, G. I., Kiesler, S. B., and Goldberg, P. A. Evaluation of the performance of women as a function of their sex, achievement, and personal history. *J. Pers. Soc. Psychol.* 19:114, 1971.
17. Reik, T. *The Need to Be Loved.* New York: Farrar, Straus, 1963. P. 250.
18. Szasz, T. *The Myth of Mental Illness.* New York: Harper & Row, 1974.

# CHAPTER 32

# The Right of the Mentally Ill to Refuse Treatment
*Elsie L. Bandman*

*Among those provisions* of rights for patients drawn up by such bodies as the American Hospital Association [1, 2], and the New York State Department of Mental Hygiene [10], no right appears more questionable or more difficult of application than the statement that gives the patient the right to refuse treatment.

A number of questions arise. First — is it ever appropriate for a mentally ill person to refuse treatment? After all is not the very organ of judgment and of consent affected by the illness? Is not the very capacity of the person to distinguish reality that which is under question? Therefore what right does the patient have except to comply unconditionally with therapeutic directives?

Second — is it the right of the nurse to advocate that the patient refuse treatment? If so, under what conditions and with what limitation? The question now may be asked, "How do nurses or how does anyone get their just rights?" B. Bandman states that we get our rights in two stages. The first stage he sees as acts of crying out our needs, interests, and desires expressed as claims to rights. These claims to a right "marks out what we do to get others to recognize our right, but is not necessarily our right" [5]. In Bandman's second stage, a legal system judges among competing claims to rights those rights that deserve being designated as rights and conferred within a community. Rules are then needed to give teeth to rights along with enforcement, which enables rights to be granted and protected. Just rights can only be recognized and supported by a moral community. As B. Bandman points out, "Rights are not privileges nor powers; rights are conferred rather on the powerless among us and are designed to place checks on those who have privileges . . . . Rights provide a just basis for claiming one's due" [4].

## AMENDMENTS TO THE MENTAL HYGIENE LAW
On June 12, 1975, the New York State Department of Mental Hygiene amended the mental hygiene law to read that in general, "patients may object to any form of care and treatment and may appeal decisions with which they disagree" [10]. In an emergency, treatment may be given "despite objection in a case where the treatment appears necessary to avoid serious harm to life or limb" [10]. A significant change in the law is that portion which states that "patients who are in a voluntary or informal status may not be given treatment over their objection" [10]. If the patient objects to all recommended forms of treatment, the patient, following notification, may be discharged to either an outpatient status or to involuntary status. Even for the patient in involuntary status, there are provisions for objection in the revised law. These objections may be based on an assertion of treatment conflicting with religious beliefs. In cases of involuntary status, informed consent based on full disclosure of potential benefits and harm is necessary for electroconvulsive therapy, surgery, major medical treatment, and the use of experimental drugs or procedures.

If the patient is 18 years or older with "sufficient mental capacity to give consent, the procedures may be initiated only with the patient's consent" [10]. The client has the right upon request to have a self-selected person present when consent is sought. Limiting conditions to patient consent is for an age less than 18 and for emergency conditions necessary to protect life or limb. Safeguards to the rights of consent are provided in instances of doubt of sufficient mental capacity. An independent opinion is obtained from an outside qualified consultant, and the director makes the final decisions with procedures for objection and appeal. Despite the psychiatric facility director's retention of a great deal of control of the final decision by way of questioning the sufficiency of mental capacity, nevertheless, entry into the whole issue of the rights of the mentally ill has been gained through the limitations imposed and the established procedures for appeal.

## CONTROVERSIES SURROUNDING THE RIGHT TO REFUSE TREATMENT

The right to refuse treatment is possibly the most controversial of patient's rights, since among all rights it is the only one that locates the power and the autonomy of decision within the patient. Other rights present the patient as a petitioner for a decent and respectful quality of care and treatment. This right was recently challenged by a hospital and upheld by the courts in the case of a patient who refused the recommended amputation of a gangrenous leg. This decision follows the classic precedent set by Judge Cardoza in 1914 involving New York Hospital in which he stated, "Every human being of adult years and sound mind has a right to determine what should be done with his own body" [17]. The decision declared the individual's right to one's own body as taking priority over the demands of the physician to perform a medical procedure. Clearly, if a person suffers from a physical illness as life threatening as advanced heart or renal disease or cancer, no one can legally force that adult to secure treatment. The institutionalized practice of "signing one's self" out of the hospital against medical advice is a recognition of that right.

This is simply not the case with mentally ill patients who are hospitalized. Client refusal of prescribed therapies has commonly resulted in consequences ranging from the use of force to secure compliance, to enforced seclusion, to verbal abuse, to loss of privileges, food, visitors, and other retaliatory means, up to and including immediate discharge.

## ISSUES OF PATIENT REFUSAL

Why should the mentally ill person who refuses the prescribed antipsychotic drug or electroconvulsive therapy be treated any differently from the patient who refuses an analgesic drug for his pain or insulin injections for his diabetes or radical surgery for his cancer or dialysis for his failing kidney? The response most likely to be offered is the supposed irrationality of the person with a mental disability.

Several stances in relation to the patient's refusal are possible. The first position assumes that by virtue of mental illness and subsequent hospitalization, the patient is incompetent to make any decision regarding his treatment. The argument holds that the executive function of the person is impaired and precisely that which is in need of treatment. One supporter of this position regards the act of consent from a truly disturbed person as a travesty:

The prerequisites for truly informed consent would seem to be the capacity accurately to assess reality, process information, and based on the processing, consider several options and choose among them. These capabilities are seriously impaired in a severely disturbed person . . . because his cognitive faculties are influenced in his disturbed emotional state [19, p. 661].

There are several questions to be raised regarding this position. One is to question the assumption that these capacities of decision making are in a fixed state of either function or nonfunction [9]. Even if the patient is in a disturbed emotional state are his capacities for choice seriously impaired all of the time? Opposition to this position is based on a dynamic conception of the state of the cognitive faculties and of the extent of the disturbance in emotion.

A common clinical example is that of the hallucinating client who turns away from conversations with unseen others to respond lucidly to the therapist's query, "Are you hearing voices the group doesn't hear?" and resumes constructive participation in group therapy. Persons in emergency rooms in panic pursued by phantoms are often amenable to explanations and acceptance of the helpful effects of medication offered. Disturbed patients may either request or accept the offer of seclusion in a quiet room when anxiety mounts and threatens control. Alternatively, the patient may refuse medication under conditions of stress and elect instead to engage in verbal interaction with a staff member or fellow patient. It is conceivable that even highly disturbed patients may prefer to talk about their sources of discomfort than to be "zonked out" by medication, with the consequent sealing over of the problem. These are only a few examples of the fluctuations of the personality and its openness to appeals of reason and self-interest even when seemingly deeply disturbed.

## INDICATIONS FOR AND OBJECTIONS TO DRUG THERAPY

Although drugs are widely accepted as the most single effective therapeutic agent, nevertheless there are reasons for questioning the sometimes wholesale and undiscriminating use of medications. For example, Davis and Cole report the wide variations in use of antipsychotic drug therapy, both in manner and level, from one psychiatric setting to another. Crisis-oriented facilities and emergency rooms use intramuscular medications freely with the rapid increase of drugs to high levels in two or three days, reduced only when the patients look quiet and sleepy. This is in contrast to the "better staffed, more selective private facilities [where] a more thoughtful *drug free evaluation period of days or even weeks may precede a rather gradual initiation of drug therapy*" [6]. The authors report that in all of the drug studies they evaluated, treatment with phenothiazines did prevent some relapses, which varied with the sickness of the patient population studied. Within a given time period, the sicker patients had a greater number of relapses than the less seriously ill. "Thus about fifty percent of a moderately ill population of chronic, hospitalized schizophrenics will relapse *within six months after discontinuance of drug therapy. . . . The corollary is equally true, namely, fifty percent do not relapse.* Thus, half of this patient population *are taking drugs they do not need*" [6, p. 455].

## THE KNOWLEDGE OF THE PHYSICIAN AS THE FINAL SOURCE OF AUTHORITY

A second objection to the mentally disabled patient's refusal of treatment is in the contrast between the extensive base of knowledge of the physician and the supposed limited acquaintance of the patient with the therapy in question. Sedhev [19] un-equivocally sees the choice of treatment methods as properly residing in the province of professional judgment alone. Exception to Sedhev's position can be made on the basis of its extreme expression of paternalism, however benevolent, which places all responsibility for treatment choice within the province of a physician.

This position denies the rights of the patient to consent or to dissent from the prescribed therapy and, therefore, denies the right of self-determination. It denies the client the benefits of prior experience with the drug and knowledge about intolerable aspects of it. [For example, the description of stelazine, a very commonly used drug, by the 1973 *Physicians' Desk Reference* [13] reports that "Neuromuscular (Extrapyramidal) reactions . . . [as] symptoms . . . seen in a significant number of hospitalized patients" [13, p. 1332], such as motor restlessness, dystonias, pseudoparkinsonism, and even persistent or irreversible tardive dyskinesias.]

Freedman states that "Reports of skin and eye changes and persistent dyskinesias in patients on chronic medication are sufficiently frequent to suggest that every patient's therapy ought to be interrupted semi-annually or annually and reinstituted only if warranted by the symptoms" [8, p. 539]. The question arises why wait six months to interrupt the medication if the client is symptom free? Or what if the client has full knowledge of the side effects of such drugs and refuses on exactly this basis? Furthermore, in the absence of such knowledge does not the Patient's Bill of Rights [1] provide that the patient should be informed of the possibility of "extrapyramidal syndrome of parkinsonism, dystonia and akathisia" due [8, p. 540] to high doses and prolonged use of the antipsychotic drugs?

Freedman [8] documents the careful systematic administration of antipsychotic drugs based on the indications for their use. He repeatedly emphasizes the need for judiciousness and caution both in administration and withdrawal. He recommends a drug-free observation period after six to eight months for the chronically ill followed, if indicated, by a different drug of another class. After the initiation of therapy if syndromes occur, he recommends a reduction in dose or the use of an antiparkinsonian medication. He concludes with a sobering statement:

About 25 percent of patients treated with long term neuroleptic drug therapy develop extrapyramidal dyskinesias and hyperkinesia that persist after drugs are discontinued. A recent comparative study of autopsy material demonstrated anatomical brain changes in patients displaying such a syndrome. The treatment of this delayed syndrome is unsatisfactory: antiparkinsonism agents are ineffective and the symptoms are often worse after drug discontinuation [8, p. 546].

## PRINCIPLES BASIC TO THE RIGHT TO LIBERTY OR THE RIGHT TO TREATMENT

Loren Roth [15] poses the issue of treatment and its refusal in this way. What is really at stake? Is it the principle of due process, which in a free society requires that when liberty is in jeopardy every precaution is required? Or ought the stake to be that the right to treatment is more fundamental than that of unrestricted liberty, as Rachlin insists [14]?

In this paper, both views are regarded in the effort to clarify underlying assumptions as the basis for decision and advocacy positions in nursing. Cases can be cited to support either position dependent on the choice of principle. Does one permit the mentally disabled person to refuse breast surgery for a possible cancer and to be discharged? If not, do we allow a young Park Avenue matron who prizes her physical beauty to reject a radical mastectomy in favor of a lumpectomy for a cancer of the breast?

Do we allow nonsymptomatic patients to refuse medications and to take the 50 percent chance of a relapse and the 50 percent chance of no relapse? Or do we require all patients to stay on drugs and run the 25 percent chance of irreversible dyskinesias?

The basis for the discussion of human rights rests on John Stuart Mill's famous statement on liberty:

One very simple principle as entitled to govern absolutely the dealings of society with the individual in the way of compulsion and control, whether the means used be physical force in the form of legal penalties, or the moral coercion of public opinion. That principle is, that the sole ends for which mankind are warranted, individually or collectively, in interfering with the liberty or action of any of their number, is self-protection. That the only purpose for which power can be rightfully exercised over any member of a civilised community, against his will, is to prevent harm to others. His own good,, either physical or moral, is not sufficient warrant ... over himself, over his own body and mind, the individual is sovereign ... [12].

If, however, as Roth [15] points out, the rights of the mentally ill client were absolutely free, as was the recent case of the adult 25-year-old primipara Jehovah's Witness in New Jersey who refused a blood transfusion and died as a consequence or the New York City patient who refused a leg amputation and whose decision was upheld, there would be no involuntary commitment. All mental patients would be voluntary admissions. Obviously, these admissions are not all voluntary on the grounds that hospital care and protection are greater than the treatment available to that person without hospitalization. Thus, involuntary hospitalization and treatment are required.

The current commitment procedures in most states now require that a person be dangerous to himself and to others on the basis of demonstration, such as a prior attempt at suicide. Furthermore, in some states, the prospective client must be protected by law and lawyers. This procedure is an adversarial approach rather than the medical model of commitment that is charged with vagueness and subjectivity as in the 1976 case *Suznki* v. *Quisenberry* [20]. Thus human rights are given added legal protection at the expense, perhaps, of the trust, or fiduciary relationship, with the physician.

Another consideration stems from the Supreme Court decision regarding *O'Connor* v. *Donaldson* [15], in which the latter was released from a Florida state mental institution after 15 years of custody on the basis of no danger to himself and to others and the ability to survive outside the hospital. Since the court backed away from the issue of treatment, Roth [15] infers that dangerous patients may be given custodial care and allowed to refuse treatment. This method of care is compatible with the management of dangerous patients but incompatible with commitment procedures. These procedures rely on the patient's incompetency to judge his need for treatment and, therefore, strengthen the right of the physicians to carry out the therapy despite refusal.

Another major change in the right to refuse treatment occurred in Alabama because of Standard 9 of the *Wyatt* v. *Stickney* decision [15]. Even before electroconvulsive treatment can be given to a consenting patient, the case must be reviewed by at least four psychiatrists or equivalent medical or mental health professionals, one internist or one neurologist, and there must be an attorney present at the procedure. As a consequence, Bryce Hospital in Alabama has not given any electroconvulsive treatments since then. Through the mechanism of full disclosure and informed consent, the ground has been laid for the client's right to refuse treatment.

Recent court rulings have been favorable to the prisoner's refusal to accept such behavioral control measures as succinylcholine, a breath-stopping drug, apomorphine, an emetic, thorazine, which had been placed on standing orders for juvenile delinquents, and tranquilizers, which had been given to a practicing Christian Scientist [15]. In the case of *Welsch* v. *Likins* [15] in Minnesota, which involved a mentally retarded person, the court ruled that excessive use of tranquilizers to control behavior may constitute cruel and unusual punishment. I wonder how many of you have seen the abuse of tranquilizers in excessively sedated elderly patients and mentally ill patients lying about asleep on floors and chairs most of the day?

Perplexing questions remain, such as the limits of treatment for the declared incompetent and the full meaning of involuntary commitment in terms of refusing conventional treatment. Roth [15] cites a case at the Boston State Hospital in which the court has prohibited involuntary medication in the absence of a serious threat of violence to self or others. The physicians there are not administering medication to any patient who refuses it, and supposedly dangerous patients are transferred out of the hospital. This ruling is bound to be a landmark decision in the care and treatment of the mentally ill and hopefully another measure of success for the recognition of their rights.

## TWO MODELS OF INVOLUNTARY COMMITMENT

Roth [15] offers the argument that the courts representing society must choose between two models of involuntary commitment, either the treatment model or the segregation model. The purpose of the treatment model of involuntary commitment is to give conventional medical treatments indicated for that patient's condition. Under the segregation model, the client is removed from society to prevent harm to self or others because of inability to care for self. Treatment is offered to this group without necessary acceptance.

The strengths of the segregation model are, in Roth's [15] words, that "it emphasizes patient self-determination, protects bodily integrity and forces the patient's physician, if he is able, to establish a working relationship, therefore by over the long run facilitating the patient's voluntary treatment."

The inadequacies of the segregation model, in Roth's [15] view, are that patients are in custody without treatment and in danger of being labeled "helpless and hopeless" or given R. D. Laing's description of "mad or bad." There is always the question of staff willingness to work under these conditions as well as whether or not third parties will pay for the lack of treatment.

For Roth [15], this problem leaves the issue of competency of the person at question, namely, who is to determine competency, how is it to be determined, and who will give consent for the patient. Roth [15] states that much has been accomplished in pursuing patients' rights; they can now vote, marry, and carry on business, but they are not afforded the ultimate in protection, that is, from the physician himself. Therefore Roth [15] believes that a determination of competency, with consent to accept or to refuse treatment, must be made at the time of commitment, thereby securing authorization to treat the patient conventionally.

Alternatives are to seek consent on the basis of one case or one treatment at a time during hospitalization. This procedure can be time-consuming and may require court appearances, but to me it reinforces the principle of patient self-determination. I also see here a golden opportunity for nurse advocacy. Another alternative Roth [15] proposes is for the court to appoint a special attorney, as mental health review officer, to determine the best interests and the competency of the client. Proxy consent is now commonly used, but it places the interest of the patient with the relatives and not with the patient.

Review boards of physicians and hospital administrators have been used in cases of patient refusal, but in Roth's [15] opinion, these review boards serve mainly as a rubber stamp for the physicians' initial decisions. Roth [15] suggests interdisciplinary human rights committees to adjudicate differences between the doctor and patient, make judgments of competency, and give proxy consent. Here again, I see the possibilities of a significant role for nurses on these committees.

Patients make an important point when they refuse treatment. They are insisting on the right of self-determination, the right to liberty, the right to control their own bodies, and the right to control their own behavior. Refusal can precipitate the need for a working relationship between the client and therapist for agreement or at least understanding concerning outcomes of therapy.

Obviously, there are many winds of change in the treatment of all patients, whether medical, surgical, or psychiatric. The movement is clearly away from the extreme paternalistic attitude of the physician conveyed in the classic statement, "Trust me, I know and want what's best for you," toward a contractual relation in which the client and health consumer occupy more of an equal relation. Clearly Mill's dictum that the only reason for interfering with a person's liberty is to protect others is having its effect in the movement of the underserved and the undervalued mentally ill to protect themselves from the excesses of the health care delivery system.

## NURSING ADVOCACY

Where does nursing stand in all this? There is historical evidence that nurses have been clear and courageous about their position regarding the treatment of the mentally ill. Fagin cites Peplau as introducing graduate students to the notion of their right to refuse to participate in electric shock [7].

Should the nurses' position of advocacy of the patient's right to consent or to reject treatment be construed as a simple act of anti-intellectualism or antipaternalism; I hasten to cite limiting conditions. First, knowledge is far from simple, and the range of the concepts of knowing includes familiarity with the subjects, competence in performance, and having truths on matters of fact such as the scientific data concerning the drugs.

A second concept of knowing, according to Scheffler, is to embrace both the "skill and care pertaining to technological control of the environment, and those intellectual arts and experiences whose value is intrinsic to themselves" [16, p. 2]. Finally, Scheffler states, knowledge includes standards, ideals, and taste, including conceptions of truth and evidence and preferences among possible strategies of investigation.

Obviously, knowing is a serious matter that needs to be shared among the participants engaged in the process of making decisions. Only by a clear and open declaration of position is the nursing practitioner's integrity manifest as an integral part of a therapeutic process based on trust.

As Ashley [3] points out, nurses can no longer simply function under physician supervision as physician extenders:

Nurses traditionally and currently assume responsibility for most of the care given to most of the people in almost every health care facility in the country . . . . The continuing professional vigilance of nursing is indispensable to insure the recovery of patients who have received medical diagnoses and are taking prescribed treatment [3, p. 124].

In reality, Ashley points out, medical supervision is a myth, and "Nurses do not practice in the presence of physicians and are not constantly supervised by them. Absentee supervision by physicians is a reality in practice that ought to be recognized by law" [3, p. 129].

Schlotfeldt states the goals of nursing as "to help people attain, retain and regain health (through) concern with man's health seeking and coping behaviors as he strives to attain health" [18]. Ashley points to the potential contribution of nurses in improving health care, "It may be the best cure for what ails the American health care system" [3, p. 133]. The profession of nursing by virtue of its hundreds of thousands of members located in normally every nook and cranny of this country is able to exert a leadership position in supporting the rights of the consumer in demanding access to adequate health care. This movement is a shift from a totally uncritical alignment with the physician and the hospital dominated by the physician. This position supports both the rights of the nurse and of the health consumer to self-determination and control over conditions of vital concern to health [11, 21].

As B. Bandman points out, "Rights single out the most important values, needs and desires of a society; they hold others accountable for carrying out the duties thus implied" [4] . Clearly, human rights belong to everyone, and nursing can realize its potentials best if it seeks to provide an alliance with the health consumer in which there is provision for both assent and dissent to therapeutic goals and means.

REFERENCES

1. American Hospital Association. *A Patient's Bill of Rights*. Chicago: American Hospital Association, 1975.
2. American Hospital Association. *The Right of the Patient to Refuse Treatment*. Chicago: American Hospital Association, 1975.
3. Ashley, J. A. *Hospitals, Paternalism and the Role of the Nurse*. New York: Teachers College Press, 1976.
4. Bandman, B. Do nurses have rights? No. *Am. J. Nurs.* 78:84, 1978.
5. Bandman, B. Rights and claims. *Journal of Value Inquiry* 7:204, 1973.
6. Davis, J. M., and Cole, J. O. Antipsychotic Drugs. In S. Arietti (ed.), *American Handbook of Psychiatry* (2nd ed., vol. 5). New York: Basic Books, 1975.
7. Fagin, C. M. Nurses' rights. *Am. J. Nurs.* 75:82, 1975.
8. Freedman, A. M., et al. *Modern Synopsis of Comprehensive Textbook of Psychiatry*. Baltimore: Williams & Wilkins, 1972.
9. Goffman, E. *Stigma*. Englewood Cliffs, N.J.: Prentice-Hall, 1963.
10. Kolb, L. C. Quality of Care and Treatment, Title 14, Part 27, NYCRR. Albany, N.Y.: Department of Mental Hygiene, June 12, 1975.
11. Mercurio, K. From the desk of the president. *The Calendar* 37:2, 1977.
12. Mill, J. S. *Utilitarianism, Liberty and Representative Government*. London: J. M. Dent, 1910.
13. *Physician's Desk Reference* (27th ed.). Oradell, N.J.: Medical Economics, 1973.
14. Rachlin, S. One right too many. *Bull. Am. Acad. Psychiatry Law* 3:99, 1975.
15. Roth, L. H. Involuntary civil commitment; the right to treatment and the right to refuse treatment. *Psychiatric Annals* 7:50, 1977.
16. Scheffler, I. *Conditions of Knowledge*. Chicago: Scott, Foresman, 1965.
17. Schoendroff, V. Society of New York Hospital, 211 N.Y. 125, 129, 105, NE 92, 93 (1914) (Cardoza, J.).
18. Schlotfeldt, R. Nursing is health care. *Nurs. Outlook* 20:245, 1972.
19. Sedhev, H. S. Patients' rights or patients' neglect: The impact of the patients' rights movement in delivery systems. *Am. J. Orthopsychiatry* 46:660, 1976.
20. *Suzuki* v. *Quisenberry*. 73-3854 (D Hawaii, Feb. 24, 1976).
21. Wilson, S. K. Limiting intrusion-social control of outsiders in a healing community. *Nurs. Res.* 26:103, 1977.

# CHAPTER 33

# The Myth of Mental Science
*Raziel Abelson*

*The trial of Patty Hearst* for bank robbery dramatized the unresolved conflict between determinism and libertarianism. A special feature of that trial was that the defense claimed innocence on the grounds of a newly baptized mental illness called "coercive persuasion" by the experts and "brainwashing" by the layman. The philosophical question involved was whether there really is such an illness and, in general, whether the concept of a mental illness makes good sense. Some critics of psychiatry, led by Dr. Thomas Szasz, say no. So apparently did the Hearst jury.

Sixteen years have passed since Thomas Szasz's heretical manifesto, *The Myth of Mental Illness* [7], burst among the worshippers of psychiatric medicine, denouncing their idol as a fraud. Yet today psychiatry and psychotherapy are more widespread and lucrative than ever. I think the reason for the failure of Szasz's campaign is that it was insufficiently radical. He attempted a palace coup where a full-scale revolution was needed. Other psychiatric heretics, such as R. D. Laing [3] and E. F. Torrey [8] have done the same and have failed equally.

Szasz's critique of medical psychiatry consisted of four main theses that he regards as mutually supporting, although in reality they are inconsistent: (1) The myth — mental illness is a myth, like witchcraft and demonic possession; (2) antipaternalism — those who are forcibly treated for mental illness are cruelly and unjustly deprived of their rights; (3) full responsibility — people who deviate radically from conventional standards of conduct should not be relieved or deprived of responsibility for their actions; (4) psychotherapy as social science — since there is no such thing as mental illness but only deviant behavior, psychotherapy should not be considered a branch of medicine but should be recognized as applied social science [6, 7, 8].

I shall argue that thesis (1), while true, is given an exaggerated importance, that (2) is both false and inconsistent with (1), and that (3) and (4) are half-truths whose partial falsity is more misleading than their partial truth is illuminating.


THE MYTH

Both Szasz and Torrey have argued that bizarre behavior, such as expression of delusory beliefs, sexual perversion, hysteria, and compulsive behavior, is wrongly taken as symptomatic of mental illness, on the grounds that our identification of abnormal conduct is dependent on variable social standards, while bodily abnormalities are objectively identifiable independently of cultural norms. Szasz admits that the concepts of health and illness are quasi-normative concepts, but he holds that the criteria of bodily health, unlike the criteria of mental health, are fairly universal and invariant. He concludes:

> Psychiatry . . . is very much more intimately tied to problems of ethics than is medicine . . . . What people now call mental illnesses are for the most part communications expressing unacceptable ideas [7].

233

There is considerable justice in this complaint. The alarming variation in standards of behavioral normality has produced abominations such as commitment to asylums and even lobotomizing of critics of the Soviet government and of Russian Jews who apply for emigration to Israel, whereas in the West, homosexuality and masturbation have been stigmatized as "disturbances" and "disorders." Szasz, Laing, Torrey, and others have done a great service in bringing attention to these abuses of psychiatric power and prestige and in warning us against excessive trust in psychiatric expertise. It must be granted them that standards of behavior vary more widely than standards of bodily condition. Nevertheless, the difference is one of degree rather than of kind and insufficient to preclude reasonably objective judgments of serious abnormality, such as delusion or self-destructive behavior. In my opinion, the correct ground for rejecting the concept of mental illness is not here but in the fact that human conduct is not completely determined by causal processes.

Evidence of bodily disease consists of abnormal states called "symptoms" that are theoretically explainable as due to a causal process that, if left unchecked, threatens to incapacitate the agent and shorten his life. So-called mental and emotional disorders could properly be called symptoms in this sense only if we had well-confirmed theories about the unobserved processes that produce them. But we have no well-defined concept of a mental process analogous to biochemical processes such as bacterial infection and hormonal imbalance, nor have we any clear idea of causal connections between the actions of a person and any mental or bodily processes occurring inside that person. Thus the idea of mental illness is a metaphor, which it would be foolish to take literally. Nevertheless, it is a useful metaphor, and its critics would be well advised to worry more about the myth of causal laws of behavior and mental processes than about the analogy between psychiatry and medicine.

## ANTIPATERNALISM

Is it an unjust invasion of the rights of the insane or neurotic to compel them to undergo treatment? If it is indeed unjust, that injustice cannot, I think, be blamed on the myth of mental illness, since the analogy to medicine in no way supports compulsory therapy. Medical patients are not compelled to accept treatment, nor are they relieved of responsibility for their actions and decisions.

Szasz, Laing, and Torrey claim that psychoses and neuroses are merely deviant modes of behavior motivated by unconventional beliefs and values. They overlook the fact that such behavior is deviant in a very special way in that, unlike rationally deviant conduct, it is not explainable by the agent in terms of his own beliefs, values, and rules. The disturbed person usually has the same values and accepts the same rules of conduct as everyone else, but he suffers from an inability to act in accordance with them. He is not deliberately deviant, for his own good reasons, but deviant in spite of himself.

Does it follow that the neurotic and psychotic are not responsible for what they do and may be forcibly committed for therapy? The answer depends on many factors and yields no single policy. Do the subjects in question indicate in more tranquil moments that they want assistance? Is their behavior a serious threat to the well-being of others? In the first case, confinement and therapy are not really compulsory, while in the latter case they may be necessary means of protecting others. The practice of automatically committing people on the authority of two psychiatrists is already under legal attack and is in process of replacement by more reliable judicial processes, for which we can thank the efforts of critics like Szasz. But it should be kept in mind that it is not the mere deviance of behavior; rather it is the fact that compulsion, addiction, and delusion are contrary to the agent's own interests and threatening to others that justifies some degree of paternalistic intervention and restriction of freedom. Indeed, if Szasz and Torrey are right in explaining neurosis as infantile malingering rather than illness, then surely society has the right to treat the neurotic like a small child.

## FULL RESPONSIBILITY

Szasz and Torrey maintain that the concept of mental illness works to deprive eccentric people of responsibility for their actions. They ignore the fact that it was not on *medical* grounds that insanity was accepted as an excusing condition in the law. The M'Naghton rules specified mental derangement as defined in terms of ignorance of right and wrong or of the nature of one's actions [2]. The efforts of psychiatrists, as reflected in the 1954 Durham case [9], to have all sorts of mental or emotional disorders accepted as excuses are quite another matter, and Szasz and Torrey are on firmer ground in opposing this tendency. The logic of excusing conditions is indeed muddled in the minds of psychiatrists like Karl Menninger, who claim that any kind of deviant behavior exculpates a person [4]. Szasz, however, compounds this confusion by blaming it on the analogy to medicine instead of pointing out that *illness does not rule out responsibility*. Illness is a disability that can only explain and excuse the *failure* to perform a required action, such as getting to work on time; it cannot explain why a successful action was performed, nor excuse it on the ground that the agent could not have refrained from performing it. Illness as disability can only prevent action; it cannot itself *act* — only persons can act.

## PSYCHIATRY AND SOCIAL SCIENCE

Szasz and Torrey maintain that psychiatry is a branch of social science rather than a branch of biology or medicine [6, 7, 8]. Now the concept of behavioral or social science has been assailed by many philosophers influenced by Ludwig Wittgenstein, who have argued that genuine science requires well-confirmed causal laws and that voluntary human action cannot, on pain of incoherence, be governed by causal laws or explained by causal theories [1]. Whether or not this radical libertarian view is correct, the fact remains that there exist no causal generalizations about human conduct worthy of being called "laws," nor well-confirmed theories about causal processes connecting bodily or mental events with subsequent actions that enable us to predict what people will do.

Thus I think Szasz is wrong to stress the harmfulness of the illness metaphor rather than the far greater harm of the belief he seems to share in causal laws of human behavior. His proposal to shift from a medical model of psychiatry to a "game-playing model" seems to me a leap from the frying pan into the fire. What Szasz means by "game playing" is not the field of mathematical logic developed by von Neumann, but rather common sense strategies for getting along with others and organizing one's affairs [7]. Why then the pretentious title "game-playing model" if not to embellish mundane practical wisdom with the glamour of exact science?

As if afflicted with pangs of conscience, both Szasz and Torrey apologize for their claim that their discipline is social science. In *The Ethics of Psychoanalysis,* Szasz writes, rather defensively:

. . . it is only to be expected that psychoanalysts claim to be scientists . . . . The modern professional is compelled to make this claim for if his work were labelled nonscientific (or unscientific) he would be saddled with a value-negative identity [6].

At this point in his discussion of ethics, Szasz seems to have confused ethics with public relations. Similarly Torrey, in *The Death of Psychiatry,* confesses with remarkable candor:

To call something a science today is as much evaluative as it is an indication of methodology. Science means goodness, purity, efficiency, reliability, fidelity . . . . Everything claims to be based on science if it hopes to be accepted. This in itself is sufficient reason to call it "behavioral science" [8].

The gnat of pseudomedicine is strained at while the camel of pseudoscience is swallowed. Why not strain at both? Psychiatry is a practical art at best, involving experience, training, insight, but it is not a science applying causal theories and laws. The myth we need to free ourselves from is not that of mental illness, but that of causal determinism, the myth of an emperor whose transparent gown is woven of nonexistent laws of human behavior.

REFERENCES
1. Abelson, R. *Persons.* London: Macmillan, 1977.
2. Abrahamson, D. *The Psychology of Crime.* New York: Columbia University Press, 1960.
3. Laing, R. D. *The Politics of Experience.* New York: Ballantine, 1968.
4. Menninger, K. *The Crime of Punishment.* New York: Viking Press, 1968.
5. Szasz, T. *The Manufacture of Madness.* New York: Harper & Row, 1970.
6. Szasz, T. *The Ethics of Psychoanalysis.* New York: Basic Books, 1965.
7. Szasz, T. *The Myth of Mental Illness.* New York: Harper, 1961.
8. Torrey, E. F. *The Death of Psychiatry.* Radnor, Pa.: Chilton, 1974.
9. Wiggins, J. W. Criminology and the Sick Society. In H. Schoeck and J. Wiggins (eds.), *Psychiatry and Responsibility.* Princeton, N.J.: Van Nostrand, 1962.

# CHAPTER 34

# Comments on the Rights of the Mentally Ill
*Emmett Wilson, Jr.*

## THE RIGHT TO TREATMENT

No one would quarrel about providing psychiatric help to all individuals who need it and want it, especially to those who have had the misfortune of ending up in a hospital for their difficulties. Least of all would psychiatrists and members of the ancillary professions skilled in administering psychological treatment object to providing this help. But the hitch is that therapy is a skill, a skill that takes time and training to learn, and there are not all that many therapists around to conduct treatment. If funding were available to provide such therapists, everyone could get treatment appropriate to his illness. But, practically speaking, funding just is not available. How do we cope with the problem then? I suggest that other criteria must apply, or do apply in point of fact. What is done in actuality is the use of some sort of triage system.

Skilled psychotherapeutic ability is one of the scarcest of services. Furthermore, many of the hospitalized patients are quite difficult to treat. Their treatment requires special skills that are not readily available within the psychiatric profession. Their illnesses are sometimes quite severe, and no one is quite sure what to do about those particular forms of illness. Some individuals also come for treatment with minimal motivation for internal psychological change, perhaps with a rather poor capacity to utilize verbalization and insight, minimal psychological mindedness, and so on, and would thus be difficult to reach therapeutically, if reachable at all. In an objective and hard-nosed discussion of selection among those who should get treatment, one could certainly argue that it is best to apply the psychotherapeutic time available to those individuals for whom it would produce the most advantage in the shortest time. Given the scarcity of treatment personnel and time as well as the difficulty of treating severely ill hospitalized patients, we might wonder why money should be spent to help the hospitalized psychotic, rather than providing psychotherapeutic help to some other rather large and obvious groups still out in society who are underserviced psychiatrically; for example, children and adolescents.

None of us likes a triage system or likes to acknowledge its existence. It has enormous capacity for misuse. But budgetary allocations are themselves finite in amount. It might be best to plan budgets around these practicalities. It would be desirable if we could or would afford to make treatment available for all who need it. But it is doubtful that we could ever afford it. What is to be done then? Funding might be directed toward development of programs of brief psychotherapy, family therapy, group therapy, halfway houses, and group homes, measures that might increase the availability of at least some kind of treatment to the hospitalized and severely ill patients. Of course, we would begin to worry whether such programs would degrade and erode the general overall quality of care delivered. If the principle guiding delivery is to give everybody a little, perhaps no one is going to get very much of anything. But there must be a balance we can find that would not completely compromise the quality of the therapeutic programs provided.

## PATIENTS' RIGHTS

Treatment, if it is to be successful, must at times interfere with or ignore the rights of patients. Certainly treatment is going to come into conflict with those rights. Sometimes in a hospital we may have to subdue a disturbed and panicky patient by brute force to sedate him. Again, the hierarchy of privileges (the privilege of going to the cafeteria, the privilege of going to the cafeteria unescorted, etc.) is certainly degrading, but the negotiation of these privileges constitutes one of the means by which milieu therapy works in the hospital. Supposedly, the patient regains his self-respect by being able to handle more and more activities. Visitation rights and the right to receive letters are questionable therapeutically, when separation from a destructive family pattern might enable a patient to gain sufficient ego strength to deal better with the pattern. Treatment may very well not be possible without some of this system of rewards and punishments, restrictions and limitations.

Similar observations may be made about permitting the choice of treatment. There are many modalities of treatment, many different and debatable approaches that have been tried and are being tried. When the range of therapy extends over such techniques as electroconvulsive therapy, insulin shock therapy, psychosurgery, as well as psychotherapy, there is surely reason to expect informed consent from the patient, especially when any of the biological approaches are being employed. But it is not so easy to allow the patient his choice of medication and the regulation of its dosage. With psychotherapy, it may still be detrimental and certainly quite difficult to allow the patient a choice. Sometimes a therapist quite skilled in the use of intensive uncovering psychotherapy, even preferring to use that approach, may nonetheless decide to support a patient and shore up his fragile defensive structure rather than analyze it. The therapist may feel that to do anything else at the moment may run the risk of overwhelming the patient with unconscious material and provoking a psychotic regression. The patient, however, may have heard that only this or that particular approach is the right or the best or the most thorough treatment and may object to anything he considers a lesser alternative. Again, it is in the nature of therapy to be frustrating, for out of this frustration comes psychological growth. If one allows a choice of treatment, with the patient opting for modalities he finds more gratifying, there is an open invitation to therapeutic failure.

## THE NATURE OF MENTAL ILLNESS

Finally, I want to offer some remarks on the nature of mental illness and the right to be different. Some claim that if mental illness is an illness like physical illness, then man can *undergo* treatment, passively, and he has the option to accept therapy or reject it as he wishes. This "medical model" of mental illness is sometimes contrasted with a sort of Szaszian concept of mental illness as a problem in living. In the latter, there is no special role for the physician and no need to invoke a medical model. This is a very difficult issue, of course, and I can only make some suggestions here. Abelson seems to believe that therapists are the possessors of some certain technical expertise, but he feels that the theory by which we support that technical expertise leaves much to be desired. Abelson believes that there is no well-defined concept of a mental process analogous to well-defined biochemical processes or neurophysiological processes. He holds that there is no clear notion of a causal relationship between mental or bodily processes going on within a person, and his actions. Here I shall have to enter a protest. We are much more scientific than Professor Abelson would acknowledge, and there is reason for the medical model to be applied. Yet I do have some reservations. We can explain developmental deviations and, I think, we can say a great deal about the necessary conditions for normal development.

But this does not mean that we can explain what life is about or how to live one's life. I am quite unhappy with any attempt to extend the medical model to all human phenomena. We are entitled to call some deviations "illness," but I am deeply concerned that our science not be construed in such a way that we end up eroding the "right to be different." For if we should move in this direction, the psychiatrist has ceased to be therapist and has become involved in a sort of policing of reality, with implications for the way that life should be lived. I believe this problem is not faced only by the hospitalized underprivileged patient, but by rich and poor alike, by the hospitalized patient or the patient in private therapy. The right to be different must be respected. I believe that there is a scientific veracity in the psychological theory that underlies our therapy, but I believe also, along with some eminent theoreticians in our field, that there is an area of human endeavor that is beyond our science, and which will not be explored, much less explained, by our science. There is an area of personal freedom, of personal action, that lies outside our range of explanation. It is all too easy for the psychiatrist to attempt to prescribe what sorts of ways individuals should live their lives "to be healthy." But the most skillful therapist refrains from exercising this sort of influence on his patient and avoids creating a myth, derived from the therapist, of what the patient should be like and how he should live. The patient is not to end up as a processed individual who has not made his own choices and who has been remade in the image of his therapist. Psychotherapy as a treatment method aims at the removal of developmental conflicts. The treatment goal is the resolution of conflict, so that the individual can freely choose his own life goals. His life goals, outside the scope of therapy, are the goals that a patient would seek to attain if he could put his potentialities to use. Life goals are the patient's concern, not the therapist's. As therapists, we know what is phase appropriate along the way to adulthood, and we know the tasks that each phase of development has to resolve. We are not, however, able to specify what life should consist in. We do not know, for example, what is "phase appropriate" to being a human being. We might at most be able to say something about what is phase appropriate to being an adult, but even the hallmarks of adulthood, that is, successful object and career choice, parenthood, and so on, have not been achieved by some of the most creative and important contributors to our society and to our culture. The richness of life cannot be captured in a metapsychological profile. The theory explains the necessary conditions of normal development eventuating in adulthood, and the theory explains and provides a rationale for technical interventions that might influence that developmental process. But neither our theory nor our therapy can possibly prescribe what life is to be like. That, fortunately, is the patient's concern, and his alone.

# The Functions of Prisons and the Rights of Prisoners

# CHAPTER 35

# The Functions of Prisons and the Rights of Prisoners
*R. Joseph Novogrod*

*American prisons* continue to be in trouble, despite hard denials by reputable experts with good intentions. Perhaps a new look at old facts is in order. In fact, for those with the spirit and hope to move toward change, several directions remain open. One way would be to revive and restore the substance of rehabilitation, first by calling it by its rightful name — habilitation — and then by giving it a full, fair try. Another way might be to maintain the lid on potential disturbances while also preparing to implement certain improvements gradually. And then there are hopeful administrators who merely work harder to make things as they are a little better. It is entirely correct to claim that what we do not know about the *potential* of prisons for rehabilitation is more important than what we do know.

There are sincere critics who say that the entire prison system remains bankrupt of new, vigorous ideas and that no practical, imaginative visions are beyond the horizon. Others assert that the time has arrived for a reexamination of current methods and refuse to accept present conditions as the best that can be done. Meanwhile, the burden on each taxpayer continues to climb. Judges have become increasingly conscious that space in many prisons for future offenders is scarce and many inmates may have to be released to make room. Even the most innocent citizen has been told that prisons house lawbreakers, punish the wrongdoers, claim (though less than before) to repair the habits of inmates, and attempt to set examples for those tempted to enter a career of crime. Conflicts are constant between advocates of old-fashioned custody, which requires security for and storage of inmates as warehoused tenants, and therapists who believe that they are patients in need of treatment.

Reform of prisons has usually been a mild, synthetic attempt, since it merely took into account a single aspect of institutional requirements. As a prominent former warden said in his latest publication:

The past two centuries of prison experience have certainly demonstrated that the usual approaches to rehabilitation do *not* work. As a result, many people, both inside and outside the walls of penology, have become disenchanted with the prison and its various "treatment" attempts and have begun to despair over the absence of any real renovation of the system. For these people, the slogan "tear down the walls" is becoming a rallying cry. Although it is possible that the prison, as now constituted, may indeed one day be abolished, a pragmatic view makes clear that our society may not yet be sufficiently advanced to cope with such an innovation [1].

In the face of escalating crime rates on all fronts throughout the nation, one might believe the most reasonable response is to become stricter with all kinds of punishment and by no means attempt to humanize prison conditions. Of course, there is a severe need for punishment, and many violent inmates do belong inside. But let the average person answer the question: Do you want this type of inmate, or others, to return as hateful, dangerous, and violent as the day he was sentenced? And most inmates return to the streets, since few remain to die in prison.

There have always been severe philosophical swings in respect to changes for the rights of inmates. This inconsistency in attitude is truly at the core of the entire prison problem. As the prisoners' rights are diluted, so are the reasonable prospects for moving ahead in new and healthful directions. In periods when punitive measures were dominant, inmate rights had a low priority. And at times these rights were endangered. During periods when rehabilitation was in vogue — whether authentic or rhetorical — enlightened officials extended the rights of inmates in respect to visitors, recreation, and readings, as examples, but were always careful that punishment was considered as a sound deterrent and that rights could be loaned back quite readily. The historic balance sheet of rights has largely been negative, including those rights considered innate for the inmates. The prisoners were largely at the mercy of those persons in command, whether at the top or in middle levels of administration. What seems to be essential is a charter of rights, in the spirit of the United Nations' statements, which would perform several purposes: (1) permit and encourage the average inmate to share in and make positive contributions to his own spiritual, mental, emotional, and physical repair; (2) provide new directions and dimensions to all those seeking to change, or resisting such an opportunity; and (3) set significant examples to those who have abandoned the impulse to shift desires, wants, and demands. Perhaps the following suggestions can break new ice.

First — can the gulf that exists between the watchers and the watched be reduced? Despite the comparative closeness and the common denominator of imprisonment, the relationship between the inmates and the administrators can best be characterized as adversary. The gains to each group might be uneven, but at least the correctional staff could share in the process of bringing an inmate closer toward the management of his own affairs. It is often said that regardless of institutional aims, the inmate population usually can soften or reduce their impact. Inmate councils have been established, with varying amounts of success, to encourage inmate representatives to express their grievances and settle disputes. However, these arrangements are sometimes in the hands of persons who know what power means and may not always serve the concerns of all members in the population. In many police departments, despite elaborate community relations programs, basically each policeman performs this vital function. Why, therefore, cannot each prison officer maximize his title and make correctional services come alive?

Second — why not strengthen the incentives to excite and reward the energies and interests of the inmates? Quite often, prisoners attracted to so-called programs are merely anxious to appease the authorities and become model prisoners as a basis for early release, and their genuine feelings may be against these programs. But meaningful programs might reverse the trend and actually benefit the participants. I am thinking of sessions in human relations, decision making, and coping with many practical matters that continue long after a person's prison service. These sessions could supplement training courses to achieve high school diplomas or advanced degrees. It must be granted that the only habilitation that lasts is prompted by the person most intimately involved. These programs might excite a prisoner to move ahead once his sentence ends. If the world of self-help finds increasing application among the average person on the outside are there not some elements of self-help to assist those behind the walls? There are many who argue that such a trend would cause a rift between the prison staff and the prison population and that the negative climate of most prisons would reduce such efforts. On balance, these risks have to be taken unless inmates really prefer to sit in their cells, wait for the end of their sentences, and allow one 24-hour period to duplicate the next.

Third — why not encourage the maximum openness of institutions to as wide an assortment of outside influences as possible; including citizen groups, members of the media, research scholars, student groups, and others whose legitimate concern with the prison scene can be tested? The potential benefits from this openness tend to favor gains rather than losses. The isolation of most prisons causes family hardships, removal from real events, and a sense of being inoculated in a colony of the forgotten. Despite this invitation to the general public, it is doubtful that even if the prison gates were completely open, a large audience would respond. But today's policies can be measured by the "we" versus "they" equation, which now prevails on America's streets. If the public wants to be secure, it defies common sense to graduate inmates in worse condition than before admission. The choice seems clear: either seriously attempt change or continue to deceive the public that storaged inmates means safer streets. In short, the public might change its views once it saw for itself what the high price paid for the average inmate actually means on its tax-dollar investment.

Fourth — why not connect colleges and universities, which often provide criminal justice and law enforcement training, to prison centers at state and federal levels throughout the country? It is possible for student interns to volunteer for academic credits and work on certain projects essential to prison administrators. This arrangement usually introduces fresh ideas that can be measured against present programs for their merit. Although many volunteers are considered irritants in a closed system, to refuse such potential service would be shortsighted. On the opposite side, prison managers could be invited more often to campuses for discussions and exposure to new developments for future consideration. In fact, why not reverse roles and have academic faculty members serve during a semester on a prison campus and encourage the wardens to teach what they practice?

Fifth — inmate rights move in two directions: extension or reduction. This fact applies to mail censorship, round-the-clock medical services, visiting hours, recreation, religious services, special diets, and other aspects of an inmate's entire prison experience. Reduced rights reflect attitudes and practices, so that the same rights are used as a prize to be earned or lost in the power struggle. If a prisoner's rights are so frail, this fragility can reduce an inmate's self-respect and shrink his or her appetite to improve. The distinction between rights and privileges is sometimes disregarded, and both are often removed as a deterrent to future misconduct. To bargain for basic rights discredits even the weakest imitation of a democratic organization, in or out of prison systems. How can a prisoner respect the rights of others to life, liberty, and happiness if his own rights are systematically disregarded, disparaged, or crushed by the system that society chooses to prevail? Shouting matches might keep rights as a weapon to ensure good conduct, but this effort to politicize rights is indeed a risky business. Perhaps both the staff and inmates need to test the enduring value of decent respect. It can work both ways.

In sum, it strongly appears that the prison system must either remain passive or meet the risks of change. Actual costs continue to burden average taxpayers, who often put prisons at the bottom of their list of priorities. Progress is spotty, and there are few institutions that truly reflect all the elements enumerated in this paper. Some prisons operate, so it seems, to benefit the staff rather than change the behavior patterns of the inmates. If major breakthroughs exist, they remain locked in the minds of many forward-looking administrators, while they are hostage to the system that they so often inherit. Some staff members resign to promote their cause from the sidelines.

Most prisoners return to the "real" world, whether it is worse or better. And in time, their same behavior is repeated, unless their prison service was an antidote to improved performance in mainstream society. Anything short of this goal is costly and wasteful. Madness can never make sense.

REFERENCE
1. Murton, T. O. *The Dilemma of Prison Reform.* New York: Holt, Rinehart and Winston, 1976. P. 189.

# CHAPTER 36

# The Rights of Prisoners and Moral Reformation
*Juan Cobarrubias and Maria Cobarrubias*

*An editorial* in *The Prison Journal* [7] states that "society has invested its prisons with a multiplicity of functions — retribution, punishment, deterrence, safekeeping, rehabilitation and reintegration —," while recognizing that "there is little consensus as to the priorities of these various aims." The priorities, in fact, spring from assumptions that underlay the penal system.

Only recently have prisoners been regarded as retaining a minimum set of rights, even though other rights have to be withdrawn. For example, a court in Virginia in 1871 regarded prisoners as "slaves of the state" who had "forfeited" not only their liberty but "all personal rights except those which the law in its humanity accords to them" (*Ruffin* v. *Commonwealth*, B). But by 1961, a federal district court judge in Arkansas (*Talley* v. *Stephens*, 61) noted that prisoners "lose many rights and privileges of law abiding citizens, . . . [yet] they do not lose all of their rights." The due process and equal protection provided by the Fourteenth Amendment "follow them into prison and protect them there from unconstitutional administrative action on the part of prison authorities." This shift from *rights lost* to *rights retained* calls for a clarification of the rights persons retain while imprisoned.

The nature of rights has been a matter of debate and controversy in the literature. Various authors have attempted to clarify the nature of rights by discussing their relationship with claims. Some have equated rights with claims, others have defined rights as justifiable claims, while still others [2] have defined rights as "valid claims."

Within the context of a prison, *permissions* may be considered to be more primitive than rights, insofar as they are the kind of prescriptions that generate rights. There are, on the one hand, claims that are justified within a system of permissions and, on the other hand, claims that are not.

To understand the rights prisoners do retain while in prison, it is important to note that they are a *captive society*. All members of a captive society lose some of their individual freedom and rights, willingly in some cases, and unwillingly in others. However, if we were to compare the mentally ill and the mentally retarded confined in mental institutions, or a leper in quarantine, with a prisoner, we could say that none of these individuals "could have acted otherwise," except the prisoner. This ascription of responsibility is a characteristic feature of prisons.

It seems characteristic of most captive societies, with the possible exception of the monastic orders, for society to interfere with personal freedom. However, the justification for this interference is different. In the case of persons in quarantine and persons confined to mental institutions, the justification seems to be based upon safekeeping and care, whereas, in the case of prisoners, the justification for interfering with their freedom is based upon the fact that they have broken a law, have done something wrong, and are responsible for their own actions. The limitation of freedom for prisoners amounts to punishment, as opposed to the limitation of freedom of people in mental institutions.

The rules of prisons are prescriptions that express permissions or prohibitions from a certain authority. These permissions are precisely what generate some of the rights and claims of the prisoners. We should understand that "authority" here means the agent who issues the prescription with the purpose of controlling human behavior. Peters [6] distinguished two senses of the term "authority," namely, de facto and de jure. By de facto authority, he meant the person who exercises authority is regarded as having a right to be obeyed. But Peters also points out that de facto authority is parasitic on de jure authority. The office of the judge stands as the de jure authority, which makes a valid claim in sentencing a prisoner, whereas prison authorities stand as the de facto authority.

Even though the rights of prisoners are affected by both types of authority, the most common problems they face do not come from de jure authorities, but rather from de facto authorities. It is the lack of moral awareness, the misuse of authority, and the arbitrary way in which the staff grants privileges that account for many of the problems.

The role of authority regarding permissions needs still further clarification. By permitting an act, the authority may only be declaring that the act is tolerated. Thus tolerations are special permissions in which the authority decides not to interfere with the subject's behavior but neither does it protect the subject from interferences of other subjects regarding his or her behavior.

Not all permissions are mere tolerations. If permissions are combined or conjoined with the prohibition to prevent the subject of the permission from doing what has been permitted, we can say that the subject of the permission has a *right,* relative to the subjects of the prohibition. In these cases the authority undertakes to protect the subject from interferences on the part of other subjects.

This conception accounts for the different kinds of rights of prisoners. Thus, for instance, if the subjects who are not allowed to interfere with a permission are the other inmates and if the authority of the permission rests upon the prison authorities, we can say that this permission is an *internal* right of the prisoner as a member of a captive society. Most internal rights are conceived of as privileges granted by prison authorities. Examples of these rights may be the right to use prison facilities, participate in furlough, work-release and educational training release programs, and association with other prisoners in order to participate in such different activities as performing groups, sport teams, and the like.

If the subjects who are not allowed to interfere with a permission are the prison authorities and staff and if the authority of the permission rests upon the de jure authority, we can say that this is a right that prisoners *retain.* Examples of these rights are the right to communicate with outsiders through correspondence and visitation, not to be the subject of unreasonable searches, to observe religious rituals, not to be treated in a cruel and unusual way (although "what constitutes cruel and unusual punishment of the imprisoned is seen differently by inmates, courts and prison authorities") [3] , and the right to appeal, which might be regarded as one of the most important rights that prisoners retain.

Permissions may be combined or conjoined with commands to enable the subject of the permission to do the thing that he or she is permitted to do. In this case, we can say that the subject of the permission has a *claim* relative to the subjects of the command. The permission granted to the "Bird Man of Alcatraz" to continue his research and to write a book while in prison, conjoined with the command to enable him to do it, gave him a claim relative to his keeper. Claims that are understood in this way also constitute a right, or a "valid claim" [2]. It is clearly seen that the converse will not hold true, that is, not every right constitutes a claim. Quite often, prisoners have many other claims [1] that are not always granted.

The two types of rights, *internal* and *retained,* account for two well-known difficulties that prisoners face concerning their rights. On the one hand, the rights they retain may be considered similar to some of the rights of law-abiding citizens. But the fact that prisoners are members of a captive society prevents them from exercising their rights. On the other hand, the distinction also accounts for the conflict that commonly arises from the two different levels of authority. For instance, regarding the right to appeal, prison administrators and keepers tend to view inmates primarily as *imprisoned persons,* whereas courts see them primarily as *appellants* whose claims are made more difficult by the fact that they are members of a captive society. Neither the conflict that exists between the de facto and the de jure authorities nor the adjustment that permits the exercise of retained rights in the context of the captive society may justify the complete annulment of the rights retained by prisoners.

Some strong retributivists maintain that society has not only the right but also the obligation to deprive prisoners of all their rights. It seems to us that society has the right to protect itself but that this right does not justify the strong retributive claim, for this claim presupposes that the punishment must fit the crime, which, in turn, presupposes that society can reliably and accurately weigh the amount of punishment that should be applied to a wrongdoer. We know, however, that although the retributive premise that punishment must fit the crime is theoretically sound, there is no way of measuring the reliability or the amount of punishment that should be applied on every occasion. On the other hand, the history of our penal system contains numerous cases of appeals that have proven unfair punishment.

It seems reasonable to say that the prisoner must retain, at least, the right to appeal, but this right is contrary to a strong retributive view that maintains that prisoners do not have any rights. A peculiar version of retributivism holds that wrongdoers acquire with their offenses the right to be punished, and thus it may be said, this right is perhaps the only right that prisoners have. This view goes as far back as Hegel. However, many contemporary scholars, for example, Honderick [4] and Quinton [7], have indicated that this is an odd sort of right.

The utilitarian theory of punishment justifies punishment by the value or utility of its consequences. It also says that although punishment is prima facie an evil in itself, it is justified inasmuch as it can prevent a greater evil. The utilitarian view of punishment considers that one of the most important effects of punishment is deterrence. This view has led to some criticism because adherence to this position could eventually justify the punishing of innocents. However, it is unclear in the literature what deterrence refers to, that is to say, whether it is conceived of as individual deterrence, general deterrence, or both.

Our view is that treatment in prison as moral reformation is not necessarily inconsistent with some forms of utilitarianism insofar as punishment, deterrence, and obedience may constitute part but not certainly the whole function of prisons. In our view prisoners' internal and retained rights explain that prisoners are bound by rules insofar as they are members of a captive society. The fact that some rights are withdrawn amounts to punishment, but the penal system should allow the exercise of prisoners' retained rights. And, finally, treatment as moral reformation is more consistent with prisoners' retained rights.

In a penal system whose functions are basically understood as punishment, deterrence, and obedience, "the inmate is given no basis for developing an understanding of the moral basis for law and society," and "this obviously has implications for the inmate's developing of moral reasoning" [5].

The notion of reformation is certainly not new in the literature of prisons. Reformation was conceived as something that could be achieved through persuasion, obedience, self-control, and respect by sermonizing and rewarding certain forms of behavior. This treatment amounts to a form of indoctrination, whereas reformation as moral reform amounts to a change of the moral point of view of inmates and staff rather than indoctrination. In this approach, inmates are provided with opportunities to arrive at their own decisions, for which they may feel responsible. The major force of behavioral change is displaced from the de facto authority to the moral pressure of the entire group, which creates an internal locus of control and develops motivation. This focus also provides opportunity for interpersonal relations amongst peers, which, in turn, creates a life-style quite different from that now prevalent in prisons. Finally, it also facilitates the offender's reintegration into society.

Moral reform calls for a specially trained staff, a new type of professional who will meet the needs of prisons; well-trained personnel may be one of the best ways to improve the quality of life in our prisons. Our suggestion calls, finally, for a sound aftercare program as follow-up for reinforcement.

## REFERENCES

1. Bandman, B.  Rights and claims. *Journal of Value Inquiry* 7:206, 1973.
2. Feinberg, J.  The nature and value of rights. *Journal of Value Inquiry* 4:252, 1970.
3. Flint, D. P.  Justice through litigation: 1929–1970. *Prison Journal* 2:23, 1971.
4. Honderick, T.  *Punishment.* London: Hutchison, 1969. P. 37.
5. Kohlberg, L., Scharf, P., and Hickey, J.  The justice of the prison and intervention. *Prison Journal* 2:7, 1971.
6. Peters, R. S.  Authority. In A. Quinton (ed.), *Political Philosophy.* London: Oxford University Press, 1967.
7. *Prison Journal.*  Editorial. 2:1, 1971.
8. Quinton, A.  On Punishment. In P. Laslett (ed.), *Philosophy, Politics and Society* (first series). New York: Macmillan, 1956. P. 84.

# CHAPTER 37

# The Right to Punish
*Gertrude Ezorsky*

*Should criminals* be punished just because they deserve it? Strict retributivists (R philosophers) believe we have both a moral right and a moral duty to inflict such deserved punishment, even if no useful purpose, such as deterrence of future criminals, is served. However, according to some milder retributivists, that is, teleological retributivists (TR philosophers), we have a moral right, but not a moral duty, to punish deservedly, unless society benefits as a consequence. I shall argue for TR and against R.* Consider a hypothetical case, devised by an R philosopher, John Kleinig:

A Nazi war criminal finds his way to an uninhabited island where he eventually manages to carve out an idyllic existence for himself. He is discovered thirty years later. He has no desire to leave or to cause further trouble, though he is not repentant. And there is no doubt whether punishing him at this point in time will do any good — it is so long since the event (which shows no sign of recurring). The use of retrospective legislation to convict him (as in the Nuremberg trials) might in fact be socially disruptive [1].

According to the R philosopher, deserved although useless suffering ought to be imposed on the Nazi. Kleinig offers three claims to support his R position.

First, the Nazi (hereafter N) would not be "justified in complaining about any suffering which was imposed on him for his misdeeds."

I agree. Punishing N would certainly not violate N's moral rights. However, this suffices only to show that we have a *right* to punish him, that is, that TR is true. But it does not follow that we have a duty to punish him, that is, that R is true. Hence Kleinig's first claim suffices only to show that TR, but not R, is true.

Second, Kleinig argues that N's suffering for his misdeeds "would be just." I construe this R claim to be such that, if N's suffering were just, there would be a moral duty to impose it. But if such suffering were unjust, there would be a moral duty to prevent it. However, if such suffering were neither just nor unjust, then as a TR philosopher would claim, there would be, not a duty, but a right, to either impose or prevent such suffering.

Imagine that N is in mortal agony and S, a woman whom N had tortured, has a pain-killing capsule that, unless taken by N, would be wasted.

Let us assume the following: If there were a duty to impose suffering on N, S has a duty to refrain from giving N the capsule. But if there were a duty to prevent N's suffering, S has a duty to give N the capsule.

Suppose that S decides: Let N suffer as he has made me suffer. She refuses to give N the capsule. Surely S could not be faulted for having failed in her moral duty. Hence there is no duty to prevent N's suffering.

But suppose, instead, that S is a saintly person, a woman who in the past has shown great moral strength. She is moved to sympathy by N's agony and gives him

---

*A right is construed as a weak right or a liberty. Agents have a weak right or liberty to act in a certain way if and only if they are under no obligation to refrain from so acting.

the capsule. Must we conclude that in this situation S is too weak to carry out a painful moral duty? Not at all. She may have successfully overcome a desire for retaliation. Should S be faulted for having violated her moral duty? On the contrary, I believe that if S spares N, she is even more radiantly good. Hence there is no duty to impose suffering on N.

Since there is no duty to either prevent or impose suffering on N for his misdeeds, such suffering turns out to be neither just nor unjust. Thus there is not a duty but only a right to either impose or prevent such suffering.

Finally, Kleinig construes the denial that N's suffering is just as "tantamount to believing that it is all right," that is, morally permissible, to commit crimes that cannot be usefully punished.

Not true. TR philosophers believe there is a right to punish such crimes. Hence they are not committed to the belief that these crimes are morally permissible. Moreover, they might argue that expenditures required for useless punishment (e.g., prison maintenance) of persons like N be used instead for preventing similar crimes (e.g., to provide rescue for individuals tortured under repressive governments). Surely, to urge that acts of crime be *prevented,* rather than uselessly punished, is not "tantamount" to believing that such crimes are morally permissible.

The R philosopher's claim that we have a moral duty to inflict useless, but deserved, suffering on criminals is, I suggest, wrong. As the TR philosopher claims, we have a moral right, but not a duty, to punish such criminals, just because they deserve it.

REFERENCE
1. Kleinig, J. *Punishment and Desert.* The Hague: Martinus Nijhoff, 1975.

TOPIC IV

# Rights in and to Health Care

## RIGHTS IN HEALTH CARE

The hospital is to health care what the school is to education and what the court is to law. Issues of life and death and the conflicts of patients' rights and institutional duties take place in the hospital. Within its walls all the tension, hopes, passions, fears, and rivalries of sex, race, and authority are played out. The hospital is the central arena of health. It is a research center, the training center for health professionals, the provider of treatment in emergencies, and the last resort of society to alleviate, treat, and cure illness.

The hospital is a different world. Although pain and fear are one's companions, expectations for recovery are high. Even if the hospital is seen as a place for the dying, miraculous cures are anticipated for one's self. The staff members are regarded as miracle workers, and unconditional trust is often placed in them.

Privacy, a provision of the Patient's Bill of Rights, is violated immediately on admission by a number of necessary intruders, each one of whom performs only one or a few of the tasks essential to care and diagnosis. For the patient, "the fear and trembling unto death" consists of not knowing when the physician will come and exactly what will be done. Will it hurt? What tests will be ordered? Will the patient be fully informed and given a choice, including a right to refuse?

The patient may be fearful of the succession of strangers who enter the room: different faces and different kinds of nurses on three shifts, students in the various health professions, interns, residents, and technicians who take blood, x-rays, and do other kinds of tests. This fear mounts as the curtain is drawn and the investigation begins.

The patient's identity is at stake. He or she may wonder, What do the physicians and the nurses think of me? Will they treat me well? Can they do anything to help me? The important questions are: What is my trouble, how big is it, is surgery necessary? Is it cancer or heart trouble or something that will kill or cripple me? There is existential anguish, fear of the unknown. Sickness is an impediment to living with joy. Normal functions and customary roles of ordinary life are suspended. Pain and sickness cause career dislocations, home and family disruptions, and tensions. Illness and pain place a severe strain on personhood and rights. Most of all, they cause the person to feel helpless, infirm, and not in charge of her or his own rights any longer.

The sick, comatose, and dying require other people to represent them. Others, however well meaning, do so only imperfectly and with their own motives at work. At any rate, the patient in fear and dread awaits the tests, dutifully does what the physician orders, and the staff directs and hopes or prays for the best.

If the tests are negative, the patient's relief is enormous and compelling. What did it matter that his or her patient rights were waived here and there? The outcome is good. But a good verdict in health is negative, not like the good verdict of a discovery or a promotion, which is positive. And yet the person suddenly freed of ill health feels temporarily wonderful, and rights seem not to matter.

Rights, however, worth having are based on respect. To be regarded as a person means that a patient "possesses an inviolability founded on justice that even the welfare of a society as a whole cannot override" [16, p. 3]. Rights based on justice provide a baseline for decent care and treatment. Application of this principle implies staff accountability. An element of just treatment for a patient is respect for individual autonomy and the ability to decide for himself or herself. This implies complete and undeviating truthfulness and full disclosure of information relevant to that patient. Autonomy and truthfulness accorded to patients are not utopian ideals; they are the requirements of rights based on justice.

"Therefore," according to Rawls, "rights secured by justice are not subject to political bargaining or to the calculus of social interests . . . . Being first virtues of human activities, truth and justice are uncompromising" [16, p. 4]. A person cannot live as a person without rights based on freedom, truth, and justice. The institutionalization of rights thus becomes the responsibility of those who operate hospitals in the same way that storekeepers, club owners, and municipal and state governments have responsibilities toward their clients. Moreover, if patients' rights to truth and to treatment are to be implemented, the decision-making processes should emanate from an environment of justice.

## THE MEDICAL STRUCTURE AND THE PROBLEM OF ALIENATION

Between the time a patient enters and leaves a hospital, many issues arise over who makes what decisions about the various functions entailed in the process of treatment. Plato [15] was one of the first thinkers to articulate the doctrine of functionalism. According to this doctrine, a thing is what it does. While Plato did not use these exact words, he insisted on giving precise definitions to various social functions, such as those of the physician. Functionalism, as Plato defined it, implies that one functions without error, by which he meant that one *knows* what one is doing. To know what one is doing means to do no harm, because, for Plato, to know is to know the good. To knowingly treat a person consists in applying knowledge, skill, and concern, not to earning wages. According to Plato, justice can never consist in doing harm [15, p. 9], since to harm a person consists in making him or her worse.

Justice is identified by Plato with "the excellence of the soul," and to Plato, the just soul as with the "just man" lives well [15, p. 39]. To Plato "knowledge is virtue" and ignorance is bad [15, pp. 30–37]. This means that nothing is ever done well without knowledge. There are, however, limits to what advanced technological medicine can do in even the most prestigious medical centers.

The medical structure has been characterized as an empire with millions flocking daily to its citadels seeking help. Some seek the pursuit of health and of its priesthood, the physicians, as a religion.

We Americans are health worshippers. We have invested, and continue to invest, billions in our temples of health: our hospitals, medical schools and nursing homes. As no other society, we indulge physicians and surgeons, not only monetarily but technologically [22].

How is this gargantuan structure operated? Several features predominate in medical decision making. One is a tight authority structure with physicians at the helm. They give the orders. One writer characterizes this structure as "MDeity" [20].

Others see this structure as hierarchical with men at the top of the pyramid and women almost exclusively in subordinate positions. Health professionals below the management level have no voice in policy decisions that affect their employment and patient care [12].

Women suffer economic discrimination in the medical hierarchy through lack of sufficient opportunity for upward mobility even when they demonstrate excellence in caring for and giving service to patients. Sexism reportedly enters into factors determining the relatively small percentage of female physicians and women hospital administrators, male social workers, and male nurses.

Feminists contend that many working wives and mothers carry the burden of homemaking and child rearing while also working, for the most part out of necessity. And, for the majority of them, their modest salaries are insufficient to employ household help.

According to the critics of the medical "empire" or "hierarchy," blacks, Hispanic people, and Asians, especially, are said to be underrepresented in medical schools and also in the upper echelons of the medical decision-making structure. Racism most probably contributes to this underrepresentation at all levels of the medical structure, but it is most conspicuous at the top of the pyramid.

Kant's dictum, "Don't treat people as means only," applies to patients and to health professionals. Murphy reports studies in philosophy and nursing that clearly demonstrate that the moral reasoning of persons is affected by the moral atmosphere [12]. She reports, for example, a study by Bueker in which both the role performance and value orientation of the nurse are affected by the social organization of the hospital. Vast differences were found in the quality of patient care therapy given or withheld and in the structure of relations and communication between patient, nurse, and doctor. Shared responsibilities and group decision making occurred in the democratic unit where patients and staff alike were treated with respect and the personal growth of all fostered. It was absent in the authoritarian, power-centered patient unit.

Overall, there are no insurmountable obstacles to a health agency's use of Kant's dictum for treating everyone as both means and ends. A hospital could easily provide an environment of justice in which the claims and rights of each patient or health professional are considered the primary principle in a just social order.

A recent essay in the *New York Times* illustrates one of the major problems that now besets the giant sprawling industry euphemistically known as the American health care system:

Several weeks ago, my father survived a heart attack . . . . While he has survived the heart attack, he may not survive the bills . . . . And he may still need an operation! . . . Although he has never made a great deal of money in his barbershop, he did at least make enough money to . . . "make ends meet" . . . . My father and mother have been good citizens, honest, hard-working, and have in their own small way contributed to the economic good of our country. However, he is now likely to be ruined financially by medical bills . . . . My father — like a good number of other self-employed, working class Americans . . . can neither afford the escalating costs of the monthly private health insurance premiums, nor meet the required means test of Medicaid. Financially, my father and others like him find themselves in the unenviable situation of being in a "no man's land" when it comes to affording adequate health care for their families [1, p. 23].

This essay serves to personalize the most basic issue of the right to health care: To what extent shall it be freely available to everyone as a right?

In 1976, 139 billion dollars was spent on health care. This was "8.9 percent of the gross national product . . . . Government at all levels contributed 42 percent of this money. In New York State, [the] government's share was 49 percent and per capita spending for health exceeded $800 million last year" [22].

Despite these massive expenditures and continually escalating costs, with twice as many physicians and surgeons per person as most Western countries, the United States health care system is beset with problems. Large numbers of people in inner-city ghettos and in rural areas, as well as the poor, have difficulty in obtaining access to the kind of care they need at the time its needed [6, p. 14]. The health levels in this country also are not as high as in other developed countries. Fuchs cites the United States death rate for males ages 45 to 54 as almost double the Swedish rate [6, p. 15]. Infant mortality "is one-third higher than in the Scandinavian countries and the Netherlands" [6, p. 16]. Fuchs also reports the great variation in health levels among different groups. "Black infant mortality is almost double the white rate, and black females ages 40 to 44 have two-and-one-half times the death rate of their white counterparts" [6, p. 16].

These problems of cost and inaccessibility of medical care and unsatisfactory and greatly varying levels of health point to the need for drastic changes. Geiger characterized the American medical care system as a great, impersonal, Hydra-headed technological monster clanking across the landscape, gobbling up the Gross National Product and excreting computerized bills, neither curing nor caring, growing two new hospital wings and a cobalt radiation unit for every general practitioner that is lopped off, engineered by greedy villains [7, p. 1].

The problem can be seen in its entirety as the perennial problems in economics of priorities and the allocation of limited resources among competing needs. Or the problems can be seen, as here, from the moral perspective of whether health care is a right, fully available to all, a purchasable commodity, or a privilege.

## THE RIGHT TO HEALTH CARE

No issue in health care stirs more controversy than the right to health care. Such questions as unlimited right to an artificial heart or kidney, cosmetic surgery, abortion, and the treatment of illnesses due to a destructive life-style are posed as a cost or feasibility factor. If there is a right to health care, what then are its limits?

The issue is sometimes expressed as a choice between "socialized medicine" and "privately controlled medicine," or between publicly supported and taxed health care versus voluntary, third-party, insurance-based programs or care simply based on fee for service. There are advantages and drawbacks to each extreme position and perhaps added strengths to the more moderate, intermediate positions.

An extreme position, which may be called the entrepreneurial model, is that of Sade. According to Sade, health care is not a right [19] ; it is a purchasable commodity on the open market [18]. His argument begins and centers on a person's right to earn a livelihood. He compares the physician's right to conduct a practice and to charge fees in relation to his or her skills with the baker's and candlestick maker's right to do likewise. Sade restricts rights to "freedom of action" [18]. Everyone has a right to make a living, to preserve himself or herself, and to charge whatever a "free" or private economy will bear.

To Sade, "health care cannot morally be granted to anyone. It is a service that must be treated like any other service: It must be purchased by those who wish to buy it, or given as a gift to the sick by the only human beings who are competent to give that gift, the health professionals themselves" [18, p. 13]. Sade's defense of the right to make a living is his interpretation of the right to live — as what has previously been referred to — as an *option right* [3, 8, 9, 19].

Sade identifies rights as option rights, but this view of rights is regarded by others as too narrow. Some argue that there are other rights as well, namely, welfare rights, which include the right not only to make a living, but the right to "a full human life" [10] and to the means necessary for achieving such a life, including adequate food, shelter, clothing, education, and health care.

Opposed both to Sade and to those who uphold a right to health care is the opposite position, which was long ago taken by Plato. Based on the notion of justice in Greek thought, Plato [15] distinguished between the social responsibility of providing health care and the physician's role. According to Plato, the function of medicine, in the strict sense, is to treat the sick:

Now tell me about the physician in that strict sense. Is it his business to earn money or to treat his patients? Remember, I mean your physician who is worthy of the name [15, p. 22].

Socrates answers the question:

So the physician as such studies only the patient's interest, not his own. For . . . the business of the physician in the strict sense is not to make money for himself but to exercise his power over the patient's body [15, pp. 23–24].

The position taken long ago by Plato, which argues for the identification of justice with the health and harmony of the soul, is supported by those who currently defend welfare rights. For Plato identified justice, along with the health of the soul with well-being. As Sade may be identified with the entrepreneurial model, so Plato may be identified with the welfare model. Thus the controversy in health care may be seen as one between option rights and welfare rights [2, 8, 9, 21].

Several intermediate positions come between these two parameters, with hard questions for either extreme. To Charles Fried, for example, the objection is not to health care as an option right or as a welfare right, but to drawing the line between the right and the implied obligations on others to provide for such a right [5]. Fried holds that a right to health care means that someone has to pay and to provide for it. Given the way American society is presently constituted, care that is available to some cannot be available to all, such as an artificial heart for everyone who needs one.

Second, the right to health care as the right to equal access implies "intolerable governmental controls of medical practice [and] to unreasonable burden of expenses . . . " [4, pp. 28–30].

Third, Fried then points to a sense of the right to health care that is defensible, a right that provides a floor for "a decent standard of care for all . . . but" one that is "not . . . equated with the best available" [4, pp. 28–29].

Fourth, Fried is critical of "inflated claims, inefficiencies and guild-like monopolistic practices of the health professions" [4, p. 28].

Fried has posed the question of the right to health care as problems of implementation and cost to society. This stand moves the question of the right to health care away from an "all" or "none" position. Arguments now rest on decisions about the taxation necessary to support a national health program versus the (option right to) alternative of strengthening the present system of voluntary contribution in conjunction with tax-supported financing of health care.

The range of health care needs and the amount needed by the poor in terms of what society can practically and willingly provide is the subject of works by Blackstone [3], Michelman [11], and Nickel [13]. They argue that one can extend the beachhead initiated by Fried beyond the minimum. Blackstone maintains:

. . . the due process clause could arguably protect citizens against fundamental threats to life, including hazards or severities due to economic conditions or natural inequalities [3, p. 403].

Since the "common welfare" clause of the Constitution implies the protection of "basic needs," including health, there is a right to the protection of health. This right is provided for by the "due process" clause of the Fourteenth Amendment [3, p. 403].

A health delivery system that provides for providers but *functions* inadequately to distribute health care is, on Plato's criteria, not fulfilling the requirements of justice. A Rawlsian argument, not dissimilar to Plato's, would be against unjust distribution of "primary goods," including health care [16, p. 62; 17].

A recent argument by Nickel [13] shows that a political option right, such as "the right not to be tortured," is analogous to "the right not to starve," or not to suffer from medical inattention. This right is social and economic and calls for others to provide appropriate help. It is a right of assistance to those in need and thus a welfare right.

Hildegard Peplau [14] argues along with Blackstone and Nickel for the right to health care. She cites an earlier version of the *Code for Nurses,* which points to society's need and mounting demand for "comprehensive health services [which] can be met only through a broad and intensive effort on the part of both the community and the health professions" [14, p. 4]. Peplau also cites the report of the President's Commission on Human Rights held in 1969, which shows how health care gradually became recognized as a right first in local public assistance programs, then for the indigent aged, and then for blind, dependent children followed by aid to crippled children. Peplau maintains that health care as a right has been achievable for special groups and is and ought to be achievable for all.

Peplau cites the World Health Organization's Constitution as stating that "the enjoyment of the highest attainable standard of health is one of the fundamental rights of every human being" [23].

Peplau contends:

If the USA can spend billions on a senseless war, it can surely afford a kidney dialysis machine for everyone who needs one; it can surely provide the most expensive treatment for every hemophiliac; it can . . . provide schools for its mentally retarded, health care for its aging citizens, and treatment for its deviants [14, p. 8].

Peplau concludes that "our country has the wealth; . . . to set its priorities" toward the achievement of health care rights [14, p. 8].

Those in favor of resetting priorities may be tagged for identifying rights with utopian aspirations; those who regard the present as binding may be tagged with the "is-ought" fallacy of assuming that what is ought to be.

One cannot live a full human life when wracked by illness. The greatest of trage-dies is to suffer from a preventable disease simply because of poverty and lack of adequate health care. This view, at any rate, is the position of reformists like Black-stone, Nickel, and Peplau, who would provide more than a floor of bare subsistence in health care that Fried seems to suggest.

## REFERENCES

1. Allegrante, J. P. Well, Who Needs Life Savings? *The New York Times,* April 27, 1977.
2. Bandman, B. Option Rights and Subsistence Rights. This volume, Chap. 5.
3. Blackstone, W. On health care as a legal right: An exploration of legal and moral grounds. *Georgia Law Review* 10:391, 1976.
4. Fried, C. Equality and rights in medical care. *Hastings Report* 6:28, 1976.
5. Fried, C. Rights in health care — beyond equity and efficiency. *N. Engl. J. Med.* 293:241, 1975.
6. Fuchs, V. R. *Who Shall Live?* New York: Basic Books, 1974.
7. Geiger, H. J. Who Shall Live? *The New York Times Book Review,* March 2, 1975.
8. Golding, M. P. The Concept of Rights: A Historical Sketch. This volume, Chap. 4.
9. Golding, M. P. Towards a theory of human rights. *Monist* 52:4, 1968. 1974.
11. Michelman, F. The Supreme Court, 1968 term-forward: On protecting the poor through the Fourteenth Amendment. *Harvard Law Review* 83:7, 1969.
12. Murphy, C. P. The Moral Situation in Nursing. This volume, Chap. 46.
13. Nickel, J. Are Social and Economic Rights Real Human Rights? Paper read at Society for Philosophy and Public Affairs, City University of New York, Gradu-ate Center, New York City, March, 1977.

14. Peplau, H.  Is health care a right? Affirmative response. *Image* 7:4, 1974.
15. Plato.  *The Republic* (edited by F. Cornford). New York: Oxford University Press, 1945.
16. Rawls, J.  *A Theory of Justice.* Cambridge, Mass.: Harvard University Press, 1971.
17. Rawls, J.  Justice as fairness. *Phil. Rev.* 62:164, 1958.
18. Sade, R.  Is health care a right? Negative response. *Image* 7:11, 1974.
19. Sade, R. M.  Medical care as a right: A refutation. *N. Engl. J. Med.* 285:1288, 1971.
20. Schwartz, D. H.  The Patient's Bill of Rights and the Hospital Administrator. This volume, Chap. 40.
21. Sidel, V. W.  The Right to Health Care: An International Perspective. This volume, Chap. 49.
22. Whalen, R. P.  Health Care Begins with the I's. *The New York Times,* April 17, 1977.
23. World Health Organization.  Preamble to the Constitution of the World Health Organization. *World Health Organization: Basic Documents* (26th ed.). Geneva: World Health Organization, 1976.

# The Rights of Patients and the Problems of Hospital Care

# CHAPTER 38

# The Patient's Bill of Rights
*Willard Gaylin*

*On January 8 [1973] the American Hospital Association published a "Patient's Bill of Rights," covering what the association calls the most commonly questioned situations that patients encounter in a hospital. In the following guest editorial, Dr. Willard Gaylin, cofounder and president of the Institute of Society, Ethics and the Life Sciences in Hastings-on-Hudson, New York, argues that the prerogative to define patients' rights lies elsewhere.*

A stay in a hospital exposes an individual to a condition of passivity and impotence unparalleled in adult life, this side of prison. You are dressed in an uncomfortable garment, leaving you exposed and ludicrous; told when you must sleep and when you must rise; informed of what you may eat and when you have to eat it; notified as to when you can have visitors, who they shall be, and how long they can stay. You are discussed in the third person in your presence as though you were some idiot child or inanimate object. If you are unfortunate enough to have an interesting case, you will be presented to a group of strangers who may take the invasion of your privacy as their privilege. Your chart, at the foot of the bed, will contain all the vital information that you would seem to be entitled to have; yet, should you attempt to examine it, you will be treated like a prepubescent caught with a copy of *Portnoy's Complaint.*

Some of this may be necessary for health and some for convenience, but most of it is simply the inevitable result of an authoritative person dealing with people who unquestionably accept his authority.

Hospital regulations are endured by a patient conditioned to seeing his physician as a benevolent father in whose reassuring presence he is prepared to play the role of the child. Beyond this, however, more serious rights are violated under the numbing atmosphere of the same paternalism.

Modern scientific medicine, as exemplified in complex teaching hospitals, has advanced technical skill at the cost of personal warmth. Often there is no one physician rendering care, rather a battery of specialists, and while "treatment" may be superior, "care" is absent. This depersonalization of medicine is having a predictable effect on the patient, causing him to abandon his tendency to romanticize the physician, and, by extension, the medical community. For this and other reasons the patient is now pressing for a reevaluation of the medical contract.

In response to this, the American Hospital Association recently presented, with considerable fanfare, a "Patient's Bill of Rights." It is a document worth examining, for nothing indicates the low estate of current hospital care (as distinguished from treatment) more graphically than the form of the proffered cure.

The substance of the document is amazingly innocent of controversy. It affirms that "the patient has the right to considerate and respectful care" and, beyond that, the right to "reasonable continuity of care." He is told that he may expect a modi-

From Willard Gaylin, Editorial, The Patient's Bill of Rights, *The Saturday Review of the Sciences,* March 1973, p. 22, by permission of the author and the publisher.

cum of personal privacy; that the usual medical concern for confidentiality should be respected; that he has a right to expect "a reasonable response" to his request for service; and, as in any other commercial transaction, that he has a right to receive an explanation of his bill.

In addition, he will be relieved to hear that, as a patient in a hospital, knowledge of the "rules and regulations" that apply to him is manifestly his due — just as it would be if he were a participant in a poker game. Similarly, the right to obtain information "concerning his diagnosis, treatment and prognosis" seems perfectly straightforward — no more than the minimum required of any standard commercial transaction. On the other hand, the patient's right to "obtain information as to any relationship of his hospital to other health care and educational institutions in so far as his care is concerned" is disquieting, for it anxiously suggests that while his exclusive reason for being in the hospital is his personal health, the hospital may have multiple, unstated other reasons influencing its treatment of him.

Finally, when the bill affirms the patient's right to "give informed consent prior to the start of any procedure," his "right to refuse treatment to the extent permitted by law," and his right to be advised "if the hospital proposes to engage in or perform human experimentation" on him, it seems to be merely belaboring the obvious. It says no more than that the hospital is subject to the same laws concerning assault and battery as any other institution or member of society.

The objection to this well-intended, though timid, document is that it perpetuates the very paternalism that precipitated the abuses. By presenting its considerations as a "Patient's Bill of Rights," it creates the impression that the hospital is "granting" these rights to the patient. The hospital has no power to grant these rights. They were vested in the patient to begin with. If the rights have been violated, they have been violated by the hospital and its hirelings. The title a "Patient's Bill of Rights" therefore seems not only pretentious but deceptive. In effect, all that the document does is return to the patient, with an air of largess, some of the rights hospitals have previously stolen from him. It is the thief lecturing his victim on self-protection — i.e., the hospital instructs the patient to make sure that the hospital treats him according to the rules of decency and law to which he is entitled. It would be more appropriate if the association addressed its 7,000 member hospitals, cautioning them that for years they have violated patient rights, some of which have the mandate of law, and warning them they must no longer presume on the innocence of their customers or the indifference of judicial authorities.

Since this is a patently decent document, the fact that the American Hospital Association takes the circuitous route of speaking to the patient of his rights, rather than to the hospital of its duties, reveals the essential weakness of such professional organizations. The AHA, like the American Medical Association and similar groups, is designed to be the servant of its constituent members — and not of the general public. A servant does not lay down the law to his master. In this regard, the AHA can only state that it "presents these rights in the expectation that they will be supported" by the member hospitals. The fact that it feels the need to alert the patient indicates how insecure that "expectation" is.

A reevaluation of patient rights — one that goes beyond the old rights reaffirmed in this bill — is greatly needed. The public should not look to the professional association for leadership here. It is not for the hospital community to outline the rights it will offer, but rather for the patient consumer to delineate and then demand those rights to which he feels entitled, by utilizing all the instruments of society designed for that purpose — including the legislature and the courts.

# CHAPTER 39

# There Is Nothing Automatic About Rights
*Elsie L. Bandman and Bertram Bandman*

*The "Statement on a Patient's Bill of Rights,"* issued by the American Hospital Association in 1973, was regarded by some as the advent of a new and elevated status for patients (1). Others believed it simply recognized that basic human rights would now be protected in hospitals.

Gaylin, an eminent physician, took a still different position.

"In effect," he wrote, "all that the document does is return to the patient, with an air of largess, some of the rights hospitals have previously stolen from him. It is the thief lecturing his victim on self-protection — i.e., the hospital instructs the patient to make sure that the hospital treats him according to the rules of decency and law to which he is entitled" (2).

The position we take is that continuing improvement of patient care is desirable not only for the well-being of those receiving care but, ultimately, for the whole of society. With the acceptance of this manifesto on patients' rights, the hospital association made a public declaration of intentions, motives or views to recognize the patient's dignity as a human being and to defend his rights. But there are problems and difficulties in protecting these rights.

In spite of this country's massive expenditures to operate one of the world's finest systems of medical technology, laboratories, and research facilities, there is dissatisfaction today with the quality of patient care. In 1973, Americans spent 8 percent of the gross national product on health, compared to 5½ percent in 1962. Health expenditures have risen at the rate of 10 percent annually compared to the 6 to 7 percent growth rate of the rest of the economy (3). More than 80 billion dollars a year are now spent for health care, the third largest industry in this country (4).

A distinguished physician expressed his discontent in this way:

A stay in a hospital exposes an individual to a condition of passivity and impotence unparalleled in adult life, this side of prison. You are dressed in an uncomfortable garment, leaving you exposed and ludicrous; told when you must sleep and when you must rise; informed of what you may eat and when you have to eat it; notified as to when you can have visitors, who they shall be, and how long they can stay. You are discussed in the third person in your presence as though you were some idiot child or inanimate object. If you are unfortunate enough to have an interesting case, you will be presented to a group of strangers who may take the invasion of your privacy as their privilege. Your chart, at the foot of the bed, will contain all the vital information that you would seem to be entitled to have; yet, should you attempt to examine it, you will be treated like a prepubescent caught with a copy of *Portnoy's Complaint*.

Some of this may be necessary for health and some for convenience, but most of it is simply the inevitable result of an authoritative person dealing with people who unquestionably accept his authority.

Hospital regulations are endured by a patient conditioned to seeing his physician as a benevolent father in whose reassuring presence he is prepared to play the role

Reprinted from the *American Journal of Nursing* 77:867, 1977.

of the child. Beyond this, however, more serious rights are violated under the numbing atmosphere of the same paternalism (5).

Another summarized his position in these words:

A paradox of the modern hospital is that while it has developed as a mecca of modern medicine and symbol of medical power, it is increasingly becoming the object of dissatisfaction and sharp criticism because the main emphasis of the hospital is medical technology and administrative bureaucracy (6).

Medical technology is largely the effort of a team of highly specialized members whose focus is on diagnosis, treatment, and research of disease. Cure, the desired outcome of scientific medicine, is reported by Fuchs as occurring in only about 10 percent of patients seen by the average physician (7). Eighty percent of patients have functional or self-limiting illnesses, responsive primarily to the warmth, interest, and compassion of any talented healer with or without medical education. The remaining 10 percent have incurable illness.

Even though most patients say they want to be cured, above all they want someone who cares. It may well be that the demand for caring in this impersonal, urbanized society is greater than ever before.

Thus, the dilemma of the modern hospital. People come to it seeking care and cure. Instead, they receive treatment in institutions large enough to afford the latest technology and a staff ratio of 10 to 12 professionals and nonprofessionals per physician. The physician, as medical team leader, focuses on diagnosis, treatment, the training of other physicians and staff, and research. As a consequence, consideration of patients' interests tends to be sporadic and discontinuous. Therefore, "if genuine humanization is to occur, there must be fundamental change in the structure of the modern hospital. The patient must have a more active and meaningful place in the scheme of things" (8).

The opposite, however, tends to be the case. As a consequence of illness, dependency, and lack of knowledge surrounding diagnosis and treatment, patients are in a position of powerlessness. Physicians, on the other hand, by virtue of their expertise and status possess considerable power. Ultimately, physicians are fully supported by the trappings of legitimacy.

Kelman discusses the legitimate use of power in any social or political system on the basis of criteria defining its application, its users, its limits, and the procedures of redress in response to its abuse:

(a) Those who exercise power and those over whom it is exercised must constitute a community, sharing common values and norms. (b) These norms must include some rules that define the limits within which the power holder must operate — the domain of behavior over which he is entitled to exercise his control, the circumstances under which he may use his power, and the manner in which he may use it; he can be held accountable whenever he violates these rules by going beyond the permissible limits of his power. (c) The person over whom power is exercised may have recourse to mechanisms (such as courts, an ombudsman, public agencies, ethics committees) through which he can question challenge, or complain about the way power is being exercised over him, and he must have the assurance that these mechanisms are not stacked against him; in short, he must have some countervailing power that enables him to protect and defend his own interests in the face of demand from the authorities (9).

The AHA statement on patients' rights undoubtedly was a response to public demand for power to protect the quality of patient care, or, possibly, a response to increasing litigation and costs of malpractice insurance, or to the rise of consumerism.

We contend that the manifesto possesses inherent difficulties and problems in providing for the implementation of patients' rights. Each is discussed in order.

### 1. The right to considerate and respectful care.

Perhaps the most glaring examples of violation of this right are exposures by the press of the conditions in homes for the aged, hospitals for the mentally ill, and institutions for the retarded. One example is enough. The New York Times on Thursday, December 12, 1974, reported:

Dr. Herbert J. Grossman, director of the Illinois State Pediatric Institute conceded on cross-examination in Federal Court in Brooklyn that he might have said after a December, 1972, visit [to Willowbrook Developmental Center for the retarded] that 'the place ought to be bombed.' He had said at that time that there had been 'oppressive overcrowding' and that 'one's feet stuck to the floor' in filth from urine and feces (10).

The dilemma is how to choose between the cost of life-long care and custody of a dependent, virtually nonproductive group and the needs of others.

How much of the economic resources of this society are we willing to allocate to health services? What are the health care priorities in relation to all other local, regional, and national concerns and interests?

Given private resources, what relative value does society place on caring for the mentally ill and retarded in relation to such other options as preventive services and health promotion for mothers and children?

At issue is the question: Is medical care a right or is it a privilege?

Fuchs partly explains right versus privilege from an economic point of view: resources are finite and scarcer than human wants; have alternative uses, such as building of an atomic submarine instead of a hospital; people attach significantly different values to wants; and human behavior, such as overeating and smoking, denies the prime value of health (11). Moreover, there is the view of those who believe that minimum care is a right for all, but those who can afford it have a right to purchase additional superior care (12, 13).

The right to "considerate and respectful care," therefore, may become a consequence of prevailing political and economic factors. Certainly the poor, the aged, the mentally ill, and the retarded are without the economic means to secure superior care and often are without the political leverage needed to secure conditions which support considerate and respectful care.

2. **The patient has the right to obtain from his physician complete current information concerning his diagnosis, treatment and prognosis in terms he can be reasonably expected to understand. When it is not medically advisable to give such information to the patient, the information should be made available to an appropriate person in his behalf. He has the right to know, by name, the physician responsible for coordinating his care.**

How much information does a patient need from a physician to satisfy the requirements for "complete" current information regarding diagnosis, treatment, and prognosis?

What if the physician is unacquainted with the complete, current information in terms of recent research and experimental drugs?

This right raises many questions. What does this right do to the traditional hesitation of physicians to share complete and current diagnostic information with patients?

How much time and effort is a physician expected to give toward completing a patient's education regarding his diagnosis, the treatment and the prognosis?

3. **The patient has the right to receive from his physician information necessary to give informed consent prior to the start of any procedure and/or treatment. Except in emergencies, such information for informed consent should include, but not necessarily be limited to, the specific procedures and/or treatment, the medically significant risks involved and the probable duration of incapacitation. Where medically significant alternatives for care or treatment exist, or when the patient requests information concerning medical alternatives, the patient has the right to know the name of the person responsible for the procedures and/or treatment.**

This is one of the most controversial and complex provisions of the bill and a basis for lawsuits. For example, a 19-year-old patient who had an exploratory laminectomy was left paraplegic. The patient claimed the surgeon failed to inform him of the risk of paralysis. Had the surgeon done so, the patient asserted, he would have refused the elective procedure (14). In another instance, a woman developed a vesicovaginal fistula following an elective hysterectomy. She also claimed she would have refused the surgery had she been informed of the risk of this complication (14).

In still another instance, a 54-year-old schoolteacher, wife, and mother of five, consented to surgery when a diagnosis of malignancy of the jawbone was made. She said she was unaware of the extent of the surgery: removal of the whole lower jaw and two-thirds of the tongue, followed by x-ray to the vocal cords. This treatment left her disfigured and without speech (15).

Were the risks and the extent and duration of incapacitation in these patients disclosed or were these factors unknown or minimized? To what extent should a physician be responsible for determining the extent of a patient's understanding in order to fulfill the terms of this right? In short, what are the reasonable limits of informed consent?

The debate among physicians about full disclosure, or "consent by terror" as it is termed by its critics, continues among hospital staff, law courts, and in the literature. Traditionally, the courts uphold a physician who disclosed risk only to the extent commonly practiced in his specialty and in his locale (16).

A recent study reported the reactions of patients to a general disclosure of complications which might arise out of angiographic procedures. The consent form explained the arteriography technique and the possible complications of clotting with the need for surgery or medication; allergic reaction to the dye; and the possibility of paralysis, bleeding, or pseudoaneurysm with a need for correction by surgery; and, finally, the statistical probability of a serious complication compared to the probability of injury in a car accident. The form asked for reactions to this information, and an opportunity was provided for discussion with a physician before the giving or withholding of consent.

In this study, over 80 percent of patients responding answered affirmatively to the question "Did you appreciate receiving this information?"; less than one-third of the patients were made uncomfortable by the information; less than 15 percent desired further information regarding complications; and only 4 patients out of 232 refused the procedure because of the information given about techniques and risks (17).

**4. The patient has the right to refuse treatment to the extent permitted by law, and to be informed of the medical consequences of his action.**

This provision clearly states the right of a person to reject violation of his body. This right is supported by the Academy of Sciences in its position paper of 1973 indicating that people may refuse the use of heroic or extraordinary measures to prolong life under all conditions (18).

This right also underlies those life-and-death dilemmas of competent adults who, because of religious beliefs, refuse to accept such life-saving measures as transfusions of whole blood, or refuse proxy consent for such measures on their children.

In the case of children, the courts have not universally supported a hospital's attempt to assume temporary guardianship in order to perform needed treatment.

**5. The patient has the right of every consideration of his privacy concerning his own medical care program. Case discussion, consultation, examination and treatment are confidential and should be treated discreetly. Those not directly involved in his care must have the permission of the patient to be present.**

This is an important right easily overlooked in large teaching hospitals when the learning needs of medical students or the urgencies of medical treatment are assumed to take priority over a patient's right to privacy. Needless exposure of a patient's body in teaching conferences, repeated examinations by students, and interviews in a room full of strangers, only some of whom are involved in the patient's care, carried on without the patient's consent, are all violations of this right.

Guaranteeing this right also would prevent facts about the present or past life of a patient from being shared with the family or employer without the patient's specific consent. A patient may have successfully hidden a history of venereal disease, epilepsy, psychiatric illness, alcoholism, or hypertension. Should it now be revealed to concerned persons without consent?

6. **The patient has the right to expect that all communications and records pertaining to his case should be treated as confidential.**

This statement suggests that confidentiality should be preserved under all conditions. Increasingly, hospitals require a written release from a patient before allowing reports, x-ray films, or other communications to be transmitted to another hospital, physician, or insurance company.

Should confidentiality regarding a diagnosis be maintained at the risk of causing harm to others? For example, should a person with incompletely controlled epilepsy, or an elderly person with cataracts, or a person with severe cardiac disease be permitted to drive cars on public highways?

7. **The patient has the right to expect that within its capacity, a hospital must make reasonable response to the request of a patient for services. The hospital must provide evaluation, service and/or referrals as indicated by the urgency of the case. When medically permissible, a patient may be transferred to another facility only after he has received complete information and explanation concerning the needs for and alternative to such a transfer. The institution to which the patient is to be transferred must, first, have accepted the patient for transfer.**

A dilemma arises in this provision from "a reasonable response to the request . . . for services." A pattern of nurse staffing may seem reasonable when the majority of patients are moderately ill and ambulatory but inadequate when several critically ill patients in need of highly skilled, intensive, professional care are admitted to the hospital.

To what extent can a hospital provide, and a patient pay, for a pattern of overstaffing to cover unpredictable periods of great demand for skilled nursing and medical services? Conversely, are not anxious, frightened, ambulatory preoperative patients' requests for explanation and reassurance from professional nurses and interns (both busy with seriously ill patients) reasonable requests which ought to be met?

Difficulties can also arise from the terms "provide evaluation, service and/or referrals as indicated by the urgency of the case." Does this mean that medical, nursing, social service, and other forms of evaluation, service and/or referrals are provided at the request of the patient and the family or at the initiative of the hospital? Who makes this decision and what factors enter into it? Is a patient's ability to pay one of the factors?

8. **The patient has the right to obtain information as to any relationship of his hospital to other health care and educational institutions insofar as his care is concerned. The patient has the right to obtain information as to the existence of any professional relationships among individuals, by name, who are treating him.**

An implication here is that patients do not have unconditional trust in physicians and hospitals. It may well be to a patient's advantage, even though construed as doubt about hospital standards, to inquire about the credentials of the health care and educational institutions affiliated with the care-providing hospitals.

Teaching hospitals by definition add training and research to their care-and-cure functions. These added functions may provide distinct advantages to patients with complex diagnostic and treatment problems. On the other hand, they may be distinctly disadvantageous to patients facing intractable or terminal illness who seek only loving and tender care.

9. **The patient has the right to be advised if the hospital proposes to engage in or perform human experimentation affecting his care or treatment. The patient has the right to refuse to participate in such project.**

Here, the dilemma is between objectivity in scientific research with double-blind studies and a patient's right to know.

Patients in such studies are unidentified either to themselves or to care-providers, thereby avoiding bias which would influence the results of experiments.

Such studies provide patients with a paradox: forgo the research and a chance of being helped or harmed by experimental medications or acquiesce and face a 50 percent chance of being deprived of benefits.

The issue here is that all patients assigned to the placebo category are deprived of the benefits of the drug.

There are several notable examples of research in which scientific objectivity *versus* violation of patients' rights is clearly documented.

The most recent instance is one in which the U.S. Army reportedly gave nearly 600 soldiers lysergic acid diethylamide (LSD) without telling them before or after giving the drug or doing follow-up studies. The researcher's explanation was that the telling would have biased the research (19).

10. **The patient has the right to expect reasonable continuity of care. He has the right to know in advance what appointment times and physicians are available and where. The patient has the right to expect that the hospital will provide a mechanism whereby he is informed by his physician or a delegate of the physician of the patient's continuing health care requirements following discharge.**

Responsibility for continuity of care is stated as a joint responsibility of physicians and hospitals. In practice, however, providing continuity of care requires a high degree of coordination among hospital staff, knowledge of community health agencies, and effective use of resources. If the health of patients is to be maintained at a maximum level of functioning, it may be necessary for a range of community care providers, in addition to physicians, to be mobilized.

Narrowly defined, it is the illness needs of the sick that physicians meet. Those in other disciplines, in contrast, seek to promote health and to maintain health.

It could be argued that this tenth provision fails to acknowledge patients as active collaborators in planning their health care requirements and continuity of care.

A patient's access to his own hospital record bears on the issue of continuity of care. The record serves as a basis for planning care and as a document for recording and reviewing the course of the patient's illness, treatment, and care. It contains valuable health teaching data: physical examination, laboratory tests, findings of diagnostic procedures and surgery, and response to medications.

A major objection to permitting patients to have access to their records is that they lack understanding and may be alarmed if the record is not explained.

Those who rebut this charge view open records as a way to ensure that patients receive information and education about their own health and illness and have continuity of care as they move or change physicians.

Only six states* now allow patients direct access to their records under specified conditions. Studies in Vermont indicate that patients' cooperation was stimulated and their anxiety reduced without adverse effects when they were routinely given a complete copy of their medical records (20).

**11. The patient has the right to examine and receive an explanation of his bills regardless of source of payment.**

Insufficient hospital manpower often prevents patients from receiving the type of explanation needed to understand accounting procedures. Many patients on Medicare find the small payment for which they are responsible an overwhelming burden, and they need help.

A patient's reaction to a hospital bill may well represent reactions to underlying feelings of gratitude about recovery, or ill-concealed feelings of frustration and rage concerning the diagnosis, lack of cure, or lack of care. In any case, the right to examine and receive an explanation of the bill should be supported by offering it well in advance of discharge.

**12. The patient has the right to know what hospital rules and regulations apply to his conduct as a patient.**

Upon admission to a hospital, the sick person is involved in a new role and an often bewildering network of obligations. He is obliged to seek help and to actively cooperate with those offering help.

This contradicts the prevailing notion of many professionals that a patient is essentially passive, based on the social practice of excusing him from performing his customary role, such as that of wage-earner. Further, the hospital environment encourages passivity and dependency. This does not, however, disqualify patients from active participation in their care. If a patient is approached as a person who has the right to accept or reject the care plan, the probabilities are high that the care will be more effective and satisfactions increased.

Health may be maintained longer if the patient sees himself as an autonomous agent ultimately responsible for his own health and well being. Hospital rules and regulations tend to emphasize such factors as visiting hours, the use of liquor and tobacco, the use of television and radio sets, meal hours, and the care of valuables. Increasingly, hospitals are employing patient care coordinators or patients' rights advocates with whom patients may communicate directly if they seek redress or fulfillment of the Bill of Rights.

There are serious problems to confront in implementing the provisions of the Bill of Rights. Clearly, situations involving the autonomy of an ill person seeking help and assuming the role of patient in a large technologically advanced hospital manned by an array of experts who focus on diagnosis, treatment, training, and research are

---

*"... Massachusetts, Wisconsin, New Jersey, Louisiana, Mississippi, and Connecticut patients may have access to their charts under various circumstances." From Annas, G. J., *The Rights of Hospital Patients.* New York, Avon Books, 1975.

highly complex. Such situations are fraught with possibilities for losing sight of the patient's basic rights. Each of the provisions in the Statement is beset with problems. The resolution of these problems rests on the willingness of society to provide the quality of care and treatment that a patient is entitled to receive according to the provisions of the Patient's Bill of Rights.

## REFERENCES

1. American Hospital Association. *Statement on a Patient's Bill of Rights.* Chicago, The Association, 1973, pp. 2–4.
2. Gaylin, W. Editorial: the patient's bill of rights. *Sat. Rev. Sci.* 1:22, Mar. 1973.
3. Fuchs, V. R. *Who Shall Live?* New York, Basic Books, 1975, p. 10.
4. Denenberg, H. S. Health care for all: the wasteland. *The Progressive* 38:15–18, Apr. 1974.
5. Gaylin, *op. cit.*
6. Nuyens, Y. Humanization of the hospital. *Ziekenhus's* (Gravenhage) 1:366–375, August 1971.
7. Fuchs, *op. cit.,* pp. 64–65.
8. Nuyens, *op. cit.*
9. Kelman, H. C. The rights of the subject in social research: an analysis in terms of relative power and legitimacy. *Am. Psychol.* 27:995, Nov. 1972.
10. Kihss, P. Doctor calls Willowbrook 'moderate tragedy' now. *NY Times* Dec. 12, 1974, p. 51.
11. Fuchs, *op. cit.,* pp. 4–5.
12. Moore, C. B. This is medical ethics? *Hastings Center Rep.* (Institute of Society, Ethics and the Life Sciences) 4:1–3, Nov. 1974.
13. Sade, R. M. Medical care as a right: a refutation. *N. Engl. J. Med.* 285:1288–1292, Dec. 2, 1971.
14. Rubsamen, D. C. What every doctor needs to know about changes in informed consent. (Doctor and the law series) *Med. World News* 14:66–67, Feb. 9, 1973.
15. Daly, K. M. Don't wave good-bye. *Am. J. Nurs.* 74:1641, Sept. 1974.
16. Cobbs v. Grant, SF2287, Supreme Court in California, Oct. 27, 1972. *Pacific Reporters* 502 (Second series).
17. Alfidi, R. J. Informed consent: a study of patient reaction. *JAMA* 216:1325–1329, May 24, 1971.
18. The New York Academy of Medicine. *Statement on Measures Employed to Prolong Life in Terminal Illness.* Bull. NY Acad. Med. 49:349–351, Apr. 1973.
19. Treaster, J. B. G.I.'s in test not aware that they received L.S.D. *NY Times* July 24, 1975, pp. 1 and 25.
20. How to reduce patients' anxiety: show them their hospital records. *Med. World News* 16:48, Jan. 13, 1975.

# CHAPTER 40

# The Patient's Bill of Rights and the Hospital Administrator
*Daniel H. Schwartz*

*"The nurse* functions very well at the administrative level" was the comment of a friend of mine about the character of Nurse Ratched, the head psychiatric nurse on the ward of the state mental hospital depicted in the film *One Flew over the Cuckoo's Nest.*

In actual fact this nurse deprives patients of their rights. As the leading character says, she "plays with a stacked deck." She takes patients' cigarettes away from them and rations them because they have been losing them at gambling. She refuses to let patients listen to the World Series because it would interfere with schedules. She claims that the schedules represent the democratic decision of the patients when they do not. She has most patients so thoroughly intimidated that they do not feel they can express their true feelings and opinions.

Yet when the administrative staff of the hospital meets to discuss matters, she comes through as being calm, objective, and rational to her colleagues. By supporting the hospital establishment, she can climb the administrative ladder. I think this portrayal of a nurse exemplifies very well the problem of the hospital administrator in dealing with the rights of patients.

First, the patients are dependent. The degree of their dependence varies. Some patients require total support in order to breathe or to have their hearts beat. Others may require bedpans. Some require feeding.

Second, hospital patients experience loss of identity. This loss varies with their degree of illness and psychological history. Institutionalization, however, deprives all patients of part of their normal personalities. They lose their clothing, the usual routine of daily life, and normal contact with friends and family. They are not themselves.

If Mrs. Smith's normal life is to get up in the morning, make breakfast for her family, wash the dishes, clean the house, take the children to school, care for the younger children during the day, attend an adult education class, prepare dinner, and visit friends with her husband in the evening, she is no longer quite the same Mrs. Smith when she spends a day in the hospital bed being fed by others a menu that is not her usual one, in a room with strangers, seeing only adult members of the family in highly restricted situations, and having anywhere from 30 to 50 hospital staff members come into her room in the course of a few days to do various things that she may or may not understand.

Yet if patients are to have some autonomy, so they are able to be themselves, how does this balance with their medical needs, the purpose for which they are in the hospital?

By entering a hospital a patient is giving up some of his rights. For example, he cannot possibly have as much privacy as he would at home. He is not free, even if he is ambulatory, to leave the hospital for personal reasons unless he is willing to sign himself out "against medical advice." He may not even be permitted to go to the bathroom, although he may physically be able to do so.

The question is how much privacy does a patient have to give up? Should staff members enter the patient's room without knocking, without introducing themselves or their colleagues? Should medical staff members gossip about patients' private lives in the cafeteria? These are matters hard to codify in a bill of rights, but it is clear that while absolute privacy is not possible — even in the outside world, apart from a desert island or a hermit's cave — patients often get less privacy than they could and should in many hospitals.

Administrators deal with staff members who want to do a professional job and often believe they can do it best when patients are docile and submissive, ask few or no questions, and follow orders. In addition, the studies of Kubler-Ross and others show that some nurses and doctors do not like patients to be very ill or to die. One of the first things I was told when I became a hospital administrator was that most nurses believed the ideal patient was a handsome young man who came in for minor surgery and would recover rapidly. Such attitudes are changing with the rapid professionalization of nursing, but "difficult" patients sometimes do not get the care they should.

Some years ago in reviewing the chart of a patient on a tuberculosis ward, I found that a resident had written an order that a patient who was found to have been drinking and had returned two days late from a weekend pass was not to be readmitted to the hospital. This of course was not hospital policy, and I arranged for the chief of our pulmonary division to talk with the resident and explain that despite the patient's uncooperative behavior, we had a responsibility to treat him.

Employees of hospitals as well as the physicians on the attending staff all come to the hospital with many emotions and motivations that make it difficult to organize hospitals so as to treat patients as autonomous persons with rights. As we know, some people apparently go into medicine because they think M.D. stands for MDeity. They expect to have total control over patients and the nursing staff and expect everybody to do as they are told. They assume hospitals are created for their convenience.

The hierarchical structure in which nurses have traditionally functioned is changing, but the degree of their autonomy and responsibility often is not clear, which leads to conflicts that can affect patients. Some nurses are career oriented and expect advancement into nursing administration. In many settings, rewards are far greater for administration in nursing than for bedside care. However, this is beginning to change. Clinical specialization now offers rewards too, and this change may improve the situation for patients.

In nonprofessional hospital categories, such as nurse's aide, diet aide, or housekeeper aide, the work is often thought of as just a job, and the attitude of some staff members in upper levels of the hierarchy does not indicate decent and respectful treatment of them, which in turn may affect patient care.

I once walked into a four-bedded patients' room and found a pile of feces in the middle of the floor. When I asked the nurse why it had not been cleaned up she informed me that she had called the housekeeping department 20 minutes before but they still had not turned up. The nurse did not see this as a situation she should deal with. Changing such attitudes is not a simple matter.

## STAFF PROBLEMS AND CARE

Other emotions and problems that interfere with treating patients properly are the conflicts in the personal lives of the staff that staff members sometimes take out on patients because patients are helpless and dependent. The anger and hostility generated by the oppression of various groups in our society can also create difficulties in patient care. If a patient treats a member of a minority group disrespectfully or even uses a racial epithet, it may be very difficult for that staff member to respect the patient.

The focus of house staff and medical students on advancing their careers can also interfere with treating patients as people. Patients often complain that house staff members and medical students come into their rooms and talk about them as if they were objects with whom the aspiring physicians had no human relationship.

Academically oriented faculty can place a focus on conferences to the detriment of patient care. For example, a fight developed over an x-ray that one physician wished to use for care of the patient while another wished to keep it available for a forthcoming conference. Such a problem could be solved by small copies of x-rays in patients' charts, but the cost forces us to solve this and other such problems through a better understanding and agreement on priorities, by human relations methods rather than by technology. Despite all of these difficulties in a teaching hospital, the quality of patient care is often higher in teaching hospitals than in non-teaching hospitals. Whether it is more humane is a question.

Many hospitals have an automatic rule that bed rails must be up for all patients over 65 at all times. Many patients over 65 suffer from organic brain syndrome and have limited periods of lucidity. The purpose of the bed rails is to prevent falls out of beds. I did a study about a dozen years ago of falls out of beds and found that an equal number of falls occurred whether bed rails were up or down, and falls were more serious from beds in which rails were up.

In discussing reasons with nurses and physicians, I found that many patients felt that the bed rails represented a prisonlike image of confinement and a challenge to overcome. Therefore, in some instances bed rails were the cause of falls rather than a preventative. As a result, in our hospital bed rails are put up only when a physician writes an order that they should be put up. Nurses are authorized to use their judgment in putting up rails pending the issuance of the order.

## CARE AND CONSENT

One of the most difficult problems in patient rights deals with informed consent. Some physicians believe that if all risks of procedures are explained to a patient, the patient would refuse treatment to his detriment. Physicians sometimes believe they have a right to insist on treatment even when a patient declines it to his obvious detriment; failure to inform patients adequately is part of the "the doctor knows best" syndrome. Patients, however, still have a right to refuse treatment.

The question of the competence of the patient to give consent also arises. Legally only a court can decide whether a patient is incompetent, even though many patients are incompetent in the opinion of our psychiatrists. Regular court procedures in hundreds of cases would be an administrative nightmare and not be in the patients' interests. In such instances we try to get consent from both the patient and his nearest relative. Yet frequently I find physicians believe that only the consent of the relative is needed.

A recent study by an open heart surgeon on consents showed that most patients remembered very little of the consent interview (which had been tape recorded at the time), and one patient even denied that there even had been any such interview, although a half-hour tape recording of the interview was done with the patient's consent. These findings raise questions about psychological mechanisms that may block the effectiveness of consent procedures, to say nothing of questions of retrograde amnesia following a traumatic event like open heart surgery during which blood is circulated through the brain and the rest of the body by a pump oxygenator instead of the heart.

In order to deal with the poor attitudes and lack of commitment of professionals to patients as people we have extensive in-service training for nurses, unit managers, and other personnel. We work hard at getting house staff members and physicians to refer to patients by name rather than by diagnosis or room. We take stringent disciplinary action in the case of medication errors involving the giving of a medication to a wrong patient, since in terms of process we consider this a far greater breach of the goals we are trying to achieve than miscalculation of dosage or the giving of a medication at the wrong time. We have a committee on attitudinal change that conducts meetings with groups of employees of all categories on all shifts to seek better ways to help these employees deal with patients as people. We believe we are meeting with some success.

In the end what is needed is to find ways of enabling the hospital staff members to attain their personal goals within the framework of the best kinds of relationship with patients as people. This is not an easy process, and hospital administration has the central responsibility. We cannot do the job if we accept the view that Nurse Ratched of *One Flew over the Cuckoo's Nest* functions well in the administrative process. If we believe she does, the administrative process needs to be turned upside down.

# CHAPTER 41

# The Role of the Patient Representative
*June P. Gikuuri*

*What is the cause* of all the interest in patient's rights? Where did it originate and what were the precipitating factors? I believe it was the civil rights movement in the late 1950s and early 1960s. At that time, the consciousness of the American public arose to try and fulfill the commitment of the constitution to "the right to life, liberty and the pursuit of happiness." Presently, the only statement of commitment we have to health care can be found in the 1966 preamble to the Comprehensive Health Planning Act, which states "the fulfillment of our national purpose depends on promoting and assuring the highest level of health," and this issue is what we are concerned with today.

## NEED FOR PATIENT'S BILL OF RIGHTS
What are some reasons that lead us to believe that perhaps patients have been abused, or their rights have been infringed upon? For one thing, many patients have not been told of the medical alternatives for treatment of their illness. For instance, one of the patients who came to my attention had a fractured vertebrae. The physician told the patient that he would have to have corrective surgery. However, when the physicians went on rounds, the patient overheard them discussing alternatives. The orthopedist conducting rounds told the other physicians outside the patient's door that this problem could be treated medically, but because they would soon leave the service, he wanted to give them some experience in surgical procedure and that the patient would be scheduled for surgery. You can imagine how upset this made the patient. His family came to the community board office and asked if the patient had heard the physician correctly. Was it possible that he could be treated medically, in neck traction for a period of weeks as an alternative to surgery? We spoke to the physician and asked him if there was a medical alternative to treating this patient. Reluctantly, the physician answered yes, but the fracture would be better treated surgically. We asked the physician if this information was shared with the patient. This question resulted in some hostility and prickling; eventually, however, it was explained to the patient that there was a medical alternative to the surgical procedure, and he was put into neck traction for six weeks. Luckily, for the patient, he did not have to have the surgical procedure. This is an example of one of the rights that has been infringed upon — the right to know treatment alternatives for a given diagnosis.

## THE PATIENT REPRESENTATIVE
In 1974 The American Hospital Association's Patient's Bill of Rights was approved by the medical board, community board, and administration of my hospital. Through the encouragement of the community board, the first patient representative for the hospital center was hired in that same year. It was my job to handle pa-

tient complaints or requests for services that were being diverted for one reason or another. I attempted to confine myself to problems that cannot be dealt with in any other department. I do not take over any one else's job. I am not a social worker; I am not an administrator. Whenever a problem comes to my attention that should be handled by another department, I route that problem to the appropriate department.

Patient's rights must continue to evolve and grow. A statement of rights that is applicable in 1976 may not be applicable in 1986. We are now concerned with the right to information, the right to medical records, a discharge summary, and the right to health education. Although we think we have passed the early stages of basic human decency and respect, we must constantly implement and monitor them. There must be someone in every institution who monitors these rights. Someone must be responsible in all settings. A patient representative could be an administrator, who along with other duties is responsible for patient rights, or it could be a provider whose primary responsibility is for patient rights. It could even be a volunteer. It depends on the institution, the volume of patients, the type of patient population, and the type of staff available. In my hospital, there is a paid person to implement and monitor patient rights, and this person is paid by the hospital. The person must be able to cross departmental lines and have the authority to carry out recommendations.

## INFORMED CONSENT

One of the largest problem areas of the American Hospital Association's Patient's Bill of Rights is the area of informed consent. Within this sphere is obviously a very large area for potential and actual abuse. A patient has to sign a consent for every procedure that is done, but many times he is not informed of the procedure, the risks involved, and again, the medical alternatives. We have had patients who were given a paper to sign but were not given an explanation as to what the procedure entails. For example: I have had patients and relatives call me and say, "My husband is going for an operation tomorrow and neither he nor I have been informed of what it is. Can someone explain this operation to us?" When I asked a physician about the procedure he intended to do the following day, he answered, "I told the patient that he is having an exploratory laparotomy." I asked, "Is that what you said to the patient?" He said, "Yes. I've told him several times. I don't understand why he doesn't understand what I'm talking about." Here is another area of concern — explanation of the procedure to be performed in language that the patient can reasonably be expected to understand. Physicians are insulated during their premedical and medical schooling, and tend to forget the everyday language that most people use. Of course, a patient should learn the medical terminology specific to his illness, but he certainly should receive explanations in language that he can reasonably be expected to understand. This is the only way a person can give *informed* consent. Consent can be obtained but it may not be *informed*.

Another example: I had a mother come to my office very upset because her son, age nine, was scheduled for surgery the following day. She had signed the consent for surgery but still was not aware of what type of procedure was to be performed. I called the physician and he told me that the patient had diverticulitis and they were going to correct the problem. I asked him if he had explained it to the mother. He answered me, "Yes, I told her that the patient had diverticulitis and we were going to correct it." That was all. The mother said to me, "Yes, that sounds familiar, but I still do not know what that is." I asked the doctor to explain to the mother what this procedure was in language that she could understand. The physician became very upset and couldn't understand why this mother was so retarded that she did not understand the terminology he used. I felt that this mother was in such a nervous state that I opened the dictionary and let her read the definition for herself. She then felt a little more comfortable with understanding the procedure her son was to have. I did not want to relieve the physician of his responsibility so I sent her back to him for an adequate explanation, which was done much to the physician's chagrin.

## HUMAN EXPERIMENTATION

A patient has the right to participate or not to participate in human experimentation and research according to his own information and his own consent. A flagrant example of infringement in human experimentation is the situation in Tuskegee, started in the 1930s with patients who were diagnosed as having syphilis. Treatment was withheld to enable the researchers to observe what would happen to these patients as the syphilis germ invaded the body [1].

Other areas of infringement on patient's rights are experiments done on fetuses and care of the terminally ill who are often unable to protect themselves. How many of us have seen terminally ill cancer patients in whom the physician insists on finding the primary site of the disease, as if finding it will cure the patient? Doctors often continue to draw blood, order x-rays, and persist in putting the patient through all types of rigors in order to advance the knowledge of medical science. Barber reports a study on human experimentation and research and concluded that stricter curbs on medical research were necessary because a significant number of researchers were found to be very permissive or more willing to accept an unsatisfactory risk-benefit ratio [2]. In other words, the risks to patients were higher than the benefits. We also must be concerned with patients who cannot refuse treatment: prisoner, the fetus, the incompetent, the terminally ill, and children.

## PHYSICIAN-PATIENT RELATIONSHIP

We have been dealing with physicians as deities who have promoted superior-inferior relationships between physicians and patients, but we are about to break through that traditional role and develop, instead, cooperative relationships. As we become better educated concerning health and our rights, physicians will be compelled to relate to us in an intelligent cooperative manner.

Another personal experience concerns a patient who was in the coronary unit and signed himself out against medical advice. Everyone spoke to this man about the dangers: the physician, the nurse, and even the psychiatrist. The psychiatrist stated that the patient was mentally competent and could not be held against his wishes. As a last resort, they brought him to my office in a wheelchair. The nurse did not want to leave him. I spoke to the patient and found that he had many emotional burdens and much guilt. Some very tragic things had happened in his life. Because of his guilt, he just could not stay confined in a hospital. He had to be home, he had to be able to cry, and he had to be able to grieve for the many tragic things that had happened. He revealed many things to me that the staff did not know because I took the time to listen to him. I was very fearful about him being home and called him every day afterward to find out how he was doing. Attempts have been made to get him back into the hospital. He feels that he cannot return to the hospital after he signed himself out. I have told him to return at any time as we are here to take care of him. He cannot be forced to accept treatment unless the illness is contagious or unless he is judged mentally incompetent or dangerous to other people or himself.

## THE LAW AND PATIENTS' RIGHTS

In 1975 the New York State Health Department issued the State Hospital Code (chapter V of title 10, section 720.3), which is another patient's rights statement that is law for New York State and must be implemented in all health facilities in the state. The New York City Health and Hospitals Corporation is also formulating another patient's rights statement. Each document is more progressive than the previous one.

In the past when a patient came to the hospital, his rights were suddenly in escrow; they were frozen. The patient could no longer exercise his rights. We are trying to break through this tradition and give the patient back his rights and responsibilities because rights and responsibilities are linked together. A person should not have to abdicate his civil and common law rights in order to obtain quality health care.

## REFERENCES

1. Barber, B.  The ethics of experimentation with human subjects. *Sci. Am.* 234:25, 1976.
2. Linton, O.  American study of lifetime effects of syphilis ends in 1.8 billion dollar lawsuit. *Can. Med. Assoc. J.* 109:410, 1973.

# CHAPTER 42

# Allocation of Scarce Lifesaving Resources and the Right Not to Be Killed
*Jeffrey Blustein*

## SCARCE RESOURCES AND DEATH

The moral problem of allocating scarce medical resources for lifesaving arises not only in allocating dialysis machines and determining who should have organ transplants, even though these are perhaps the more immediate, visible, and clear-cut instances of medical decision making under conditions of scarcity. Scarcity is also a pervasive feature of all of medical practice and imposes difficult moral decisions on primary, hospital, and federal care levels. Victor Fuchs speaks about the broader problem of scarcity:

> . . . resources are scarce in relation to human wants. It is hardly news that we cannot all have everything that we would like to have, but it is worth emphasizing that this basic human condition is not to be attributed to "the system," or to some conspiracy, but to the parsimony of nature in providing mankind with the resources needed to satisfy human wants . . . . Some advances in technology (e.g., automated laboratories) make it possible to carry out current activities with fewer resources, but others open up new demands (e.g., for renal dialysis or organ transplants) that put further strains on resources. Moreover, our time, the ultimate scarce resource, becomes more valuable the more productive we become [1, p. 4].

The kind of scarcity Fuchs alludes to is certainly responsible for more deaths than the dramatic organ transplant or dialysis cases. But it is easy to overlook the broader picture, because in the dramatic individual cases a relatively clear-cut decision is usually made by a small number of people (hospital committees), and this decision, if negative, will immediately and almost certainly lead to someone's death. Nevertheless, it is a mistake to confine ourselves exclusively to these situations. Even where the causal connections leading to death are complex and possibly obscure and even where a death is not the immediate and clear-cut outcome of someone's (or some small group's) decision, people might be dying as a result of individual and collective decisions. We cannot evade, or try to evade, responsibility for these deaths simply by pointing to their remoteness and uncertainty.

The opposite to overlooking the broader picture would be to claim that there is nothing morally problematic about doing something to prevent all of these deaths. In some cases, to be sure, greater expenditures to forestall further deaths can be accomplished without drastic alterations in the rest of the medical system. But in other cases, a reallocation of funds would entail tremendous alterations elsewhere in the system. For example, providing emergency care without regard to economic circumstances would impose tremendous fiscal problems on nonproprietary hospitals, taxpayers, and the insured. Solutions to particular scarcity problems can and ought to be ranked according to the difficulty (fiscal, professional, space, or manpower) of instituting them both within the medical system and within the broader context of a number of competing social policies. Some goals are, for all practical purposes, unachievable at a particular time. Others pose less insurmountable problems.

Whenever we adopt a policy to distribute scarce resources in a particular way (which we must do since the resources are not unlimited), and when this decision results in some people's deaths, and those who die could have survived if we had chosen a different policy, we are deciding to kill these people. We are not just allowing some to die as a result of the "tragic," but unavoidable, scarcity of our medical resources. This analysis construes killing fairly broadly. That is, a person may be *killed* even though he is no worse off for not receiving a lifesaving scarce resource than he would have been had there been no lifesaving resources available at all. Alternatively, killing might be understood more narrowly to cover only those cases where one person's death results from another's causal influence, *and* the former would not have died if the latter had no causal influence or had done other acts.

Where a therapeutic procedure is available but not widely enough to help all who are qualified to receive it, failure to help everyone who needs it kills those who must go without the procedure. In terms of basic research, the conclusion must be more tentative, but it is no less sobering. Where basic research has not been satisfactorily completed, or not even started perhaps, the people whom we do not help might have been saved. Some are dying under one policy choice who might have survived under another policy. In either case, whether we are considering scarcity on the level of the actually received health care or on the level of basic research, it is not possible to assert that we have simply *done nothing* and that this is why people are dying. The deaths of particular people are or might be the result of action, action for which we must take responsibility.

## RIGHTS AND SCARCITY

Even if we do kill people by making certain decisions, we might not be violating a nonvacuous right not to be killed. I take it as noncontroversial that people ordinarily have a right not to be killed. But is not the value of rights contingent upon the availability of resources for respecting those rights? And when resources are scarce do not the corresponding "rights" become quite valueless?

Moral *rights,* it has often been observed, should be distinguished from moral *ideals.* What divides a right from an ideal has to do with the possibility of respecting the right. When one has a right to x, at least if it is a valuable right, one can be supplied, on demand, with x, or be allowed to enjoy or do x. One does not have a valuable right to x when other people are incapable of respecting his claim to enjoy, or do, or get x. A moral ideal, on the other hand, obligates us to work for that state of affairs in which rights can be respected that now cannot. Not that the moral duty to realize a moral ideal is necessarily any less stringent than the duty to respect rights established once the ideal is realized. It is just that it is a duty of a different sort.

At the heart of the matter, of course, is the question: What is achievable now and what is not? Consider, for example, a lifeboat situation in which there are five people but only enough food and water to save four. Let us also assume that someone's death is indeed unavoidable, since no help is in sight. Does it follow that survival on the lifeboat is not a right to be respected now? Not at all. *Survival for everyone* is an ideal, and lifeboat passengers serve this ideal to the extent that they first try every-thing short of killing one of the passengers before doing so. But *survival for each* is a right of each, and it is a valuable right for each (at least until selection of the victim) even though not everyone can be saved. Here we must distinguish between a collective ideal and an individual right.

Scarcity is sometimes (as on the lifeboat) such that it prevents everyone from getting a good at the same time, and there is a distributional problem on the level of the good to be distributed. But at other times, there is no distributional problem on this level, simply because there is not enough of a particular good to benefit anyone. Enjoyment of the good is an ideal for everyone. The difference between these two cases might show up in a difference in the *criteria* for distributing the scarce resources. Utilitarian considerations may be the relevant criteria in the second kind of case, where choices must be made on the macro level, but in the first kind of case, where choices must be made on the micro level, perhaps other considerations ought to be brought to bear.

## SCARCE RESOURCES AND THE RIGHT NOT TO BE KILLED

Let us now turn to the following problem. Do we, by placing medical research and health care where we do on the scale of priorities, violate the rights of persons not to be killed? I think the answer is sometimes clearly yes. It may be that we cannot save everyone who could be saved by a particular therapy, either because of shortage of the therapy or because the basic research has not been done. Survival of all is here an ideal, perhaps even an unrealizable one. But where a successful therapy does exist and is only in short supply, failure to increase the supply, with the consequent deaths of some who could have been helped, does violate their (valuable) rights not to be killed. In terms of basic research, this violation is not so clear. Failure to devote more resources to basic research violates the rights of persons not to be killed, only on the supposition that with enough money, time, and manpower, research *will* yield a lifesaving treatment or cure in time to save those persons. But the problem is that we cannot always know how successful or how costly research will be until we do it, and thus we cannot always know if we are killing people by not devoting more resources to basic research until we actually devote more resources to it.

It would be a mistake, however, to confine ourselves exclusively to research and the improvement of health care. After all, these factors are perhaps no more inti-mately related to life itself than other instrumental goods. Transportation, for exam-ple, affects the ability to earn a living, and adequate welfare benefits affect the ability to subsist. The rights of persons not to be killed can be violated in a number of ways, and a variety of policy choices can be made outside of medicine to prevent such violations.

## EXCUSABLE KILLING: THE DOCTRINE OF NECESSITY

Even where we do violate the rights of some not to be killed, it is not always inex-cusable. The right not to be killed can be outweighed by other moral considerations.

There are two such considerations: self-defense and necessity. Both can excuse killing. In the case of allocating scarce lifesaving medical resources, self-defense seems, intuitively, inapplicable. The persons who are killed are quite innocent of any wrongdoing, and though self-defense does cover response to persons not fully responsible for their actions, children and the mentally ill, for example, these persons are not clearly innocent of wrongdoing. One may, under certain circumstances, kill a madman in self-defense, but on a lifeboat, for example, where all are equally innocent of wrongdoing and no one is attacking another, self-defense would not excuse the killing of some to save others. Here the excuse of necessity seems to be the appropriate consideration. Necessity excuses killing in cases where the need to kill is dictated more by the spatiotemporal *context* in which the persons involved find themselves, rather than by the actions of these *persons* themselves.

To be more exact, a state of necessity exists when and only when the following conditions are met:

1. The lives are limited to a given situation in place and time.
2. It is reasonable to believe that in this situation, the lives of some — if not all — will be lost.
3. Some lives can be preserved only by killing others [2].

Necessity excuses killings in such states. In the first condition, the persons are involved in a closed system in space and time. It is the confinement to this situation that necessitates the killing. In the second, not all people in this closed system can survive; some deaths are unavoidable. And in the third, there is interdependency between the lives of those involved in the situation; the deaths of some can save others.

Necessity will not excuse *all* killings in states of necessity. On a lifeboat with five passengers only four of whom can be fed, with no reasonable prospect of outside help, a state of necessity exists. But there are any number of alternatives to save passengers: One, two, three, or four can be killed. Necessity will not excuse killing any more than is necessary to save the maximum number of lives. If it did, individuals' chances for survival would not be maximized, and it is unlikely that those in the state of necessity could agree to such a policy. It is important to note further that necessity can excuse killing only when all three conditions defining a state of necessity exist. Thus one can be excused on grounds of necessity for killing some people to save others only if some deaths are unavoidable to begin with. Finally, necessity ought not to be invoked to excuse killings on the ground that they will lead to the greater welfare of those remaining or to the preservation of lives of greater quality. Rather it can be invoked only when the killings are performed in order to save the *lives* of others. In other words, the doctrine of necessity does not tell us *whom* to kill; it only tells us that some persons may be killed.

## APPLYING THE DOCTRINE OF NECESSITY

Now to what extent, and in what way, does the doctrine of necessity excuse the killings that result from a particular choice concerning the allocation of limited lifesaving medical resources? Here I can only indicate in a very rough way some of the relevant considerations. In order for the doctrine of necessity to excuse killings:

1. The closed spatiotemporal system must be defined. In doing so, we must be reasonably certain that the place and time in which the lives are situated does not itself embody some morally arbitrary discrimination.
2. We must establish that no matter which policy is adopted and no matter how the limited resources are distributed, some deaths are indeed unavoidable. Only if some deaths are unavoidable could killings be excusable. Here is an issue of enormous complexity. How much, for example, of the national budget each year goes to finance projects of questionable value? Ought there to be cutbacks in certain areas so that we can free more resources to combat causes of death? How much can be hoped for in the direction of eliminating pollution, automobile accidents, fatal diseases? Would some deaths really be unavoidable if we reordered our priorities, even given the present scarcity of resources?
3. We must establish that some lives can be preserved only by making choices (on the federal, hospital, and primary care levels) that kill others. Here again we confront, among other things, the issue of priorities, not only within the medical sphere, but between it and other spheres as well.

Even within the above parameters, however, there are a number of policies for allocating scarce resources that *might* be reasonable and justifiable. Just within the medical sphere, for example, priority may be given to preventative medicine or to crisis intervention; to the improvement of the best available health care through basic research or the improvement of the actual quality of received health care through support for such services as rural clinics and new medical schools. And then there are a variety of scarcity policies for allocating resources between medical needs (where life is in danger) and other social needs (where life is also jeopardized). But what about the treatment of and research on mild but widespread diseases versus treatment of and research on rare and fatal diseases? It would seem that the doctrine of necessity requires that all life-threatening diseases be researched and treated before resources are devoted to nonlife-threatening diseases. One practical consequence of this policy might be that physicians should concentrate their efforts on taking care of morbidity and that paramedical professionals should be trained to take over care of nonfatal diseases. To justify a different policy (and sometimes we might want to do just this), principles other than that of necessity must be invoked. In any case, each of the above scarcity policies might be a policy of killing some and helping others, and each demands the most serious consideration before its adoption.

REFERENCES

1. Fuchs, V. *Who Shall Live?* New York: Basic Books, 1974.
2. Shuchman, P. Ethics and the problem of necessity. *Temple Law Quarterly* 39:278, 1966.

# The Medical Structure and the Problem of Alienation

# CHAPTER 43

# Alienation and Medicine
*Bernard J. Bergen, Jacob J. Lindenthal, and Claudewell S. Thomas*

## THE PHENOMENON OF ALIENATION IN MEDICINE
Patienthood involves regression while physicianhood involves authority. Patienthood involves dependence while physicianhood involves independence. Patienthood involves passivity while physicianhood involves active manipulation. Illness is a mystery to the patient, and hence what the physician does is frequently mystical. Adults in the population frequently become patients; they rarely become physicians. The relationship between the two often deteriorates into a type of war: While the physician endeavors to manipulate his patient, the physician is also an object of manipulation by the patient. While never before in history has the general population been in more of a position to appreciate what is involved in physicianhood, never before have both parties been as estranged from one another as they are today. Yet the blaming of any one of the parties is as fruitless as blaming one of the sexes for the divorce rate.

We are frequently sensitized to the sense of alienation that takes place in patients. Occasionally, we learn of the alienation physicians experience in their relationships with patients. We are only beginning to learn of the alienation that takes place between physicians and the institution of medicine itself [7, 9]. The very ego ideal to which the physician aspires has undoubtedly contributed greatly to the contemporary alienation within the medical institution. The ideal physician is portrayed as possessing the qualities of an Eagle Scout and accordingly should be trustworthy, loyal, helpful, friendly, courteous, kind, obedient, cheerful, thrifty, brave, clean, and reverent. Since science, which grounds modern medicine, is the religion of the twentieth century, physician's deviations from the highest ethical norms are frequently viewed with dismay and are experienced as an alienating form of disillusionment.

There is considerable validity to Nisbett's argument that we live in an age of the decline of authority [10]; hence it is no mystery when we learn that medicine no longer commands the respect that it once did. An increasing proportion of the population is beginning to understand that decreases in mortality and control of morbidity have been a consequence more of hygienic measures fostered by public health than the result of the use of the hypodermic needle. Medicine has not advanced greatly since 1960, if we employ morbidity and mortality as indices. People are now appreciating that prenatal nutrition, as well as nutrition throughout life, proper shelter, cleaner air, and proper sewage systems can probably have greater efficacy in the alleviation of suffering than can the work of the physician. Today people understand that going to the physician and following his directions are insufficient. They return to a hostile environment that continues to embarrass their organisms. The physician finds himself impotent to change this situation and experiences himself alienated from the ideal of how he wants his patients to view him.

Just a few years ago, going to a doctor involved seeing a physician and perhaps the nurse. Today one goes to see a physician and finds himself divided between myriad individuals who are assigned to help him, many of whom perform rather sophisticated services. The physician is as dependent upon them as he once was on his nurse-receptionist alone. He can no longer take credit for the improvement of patients. It is now a matter of the team that has helped the patient, and the physician's sense of mastery and workmanship is largely gone. The physician is now often in the role of a coordinator of discrete facts and bits of information; only too rarely is he an advocate of a human patient with feelings, needs, and aspirations [11]. As a consequence, all too often "cases" fail to come alive in a way that both comprehension and the physician's own sense of being a live whole person are experienced to the utmost.

## A VIEW OF THE NATURE OF THE PROBLEM

The alienated physician is more than ever a reason to attempt to grasp the nature of the problem that is conveyed by the familiar accusations that medicine neglects the "whole" patient and his "psychosocial needs." This problem has become the open scandal of medicine, the essence of which lies in an absence of attention to the human experience of illness as opposed to its biophysiology.

Efforts to reverse this scandal have not been successful. Psychiatry, for example, allocated, under the division of labor in medicine, the role of being responsible for the experiential dimension of being ill, has not had a profound effect on the overall practices or attitudes in medicine [3]. Indeed, psychiatry itself is now drifting in a direction that neglects problems of human experience.

How, then, can we grasp the full dimensions of the scandal of medicine's inattention to the human experience of illness, which seems so intractable and is one of the roots of the physician's own sense of alienation? We can understand this problem only if we put into relief what tends to remain unspoken about the scandal of medicine, namely, that medicine not only treats the patient's experience of being ill as irrelevant, but also treats that experience as no less than a threat to its position of authority. In effect, when medicine discounts the patient's experience of his illness, it is discounting what the patient signifies as the meaning of the reality of his suffering for himself.

This is the full dimension of the scandal of medicine. In a sense, medicine's demand that laymen concede to it total authority over the reality of their suffering has never really been invisible. We can see the demand operate in any number of spheres and see, as well, the threatening tension involved by the existence of the layman's own experience of the reality of his suffering. Over 20 years ago Balint used the concept of the "apostolic function" to characterize what regulates the doctor's everyday relationship with his patient. "It was almost," Balint wrote, "as if every doctor had revealed knowledge of what was right and what was wrong for patients to expect and to endure, and further, as if he had a sacred duty to convert to his faith all the ignorant and unbelieving among his patients" [2].

Perhaps one of the most visible instances occurs at the point where medicine encounters parallel systems of health care indigenous to other cultures or subcultures [4]. Under these circumstances, scientific medicine often insists upon successfully converting an alien philosophy as its price for involvement. Medicine's encounter with women is a case in point. Sheila Kitzinger has described how medicine demands that a woman giving birth understand her experience as "loathsome, frightening and repugnant" — in effect, as a suffering from an illness undergoing cure [8]. It is a demand that receives its density in all of the intimidating technological apparatus that surrounds the woman; in the very look and touch of those attending her; and above all, in the domination of a technological language of medical procedure that invokes a rule of silence over whatever need she may feel to give voice to the intimacy of her experiences that cannot be conveyed in the remorseless objectivity of the language of medicine.

Nowhere, perhaps, can medicine's invocation of a rule of silence in the name of its authority over the reality of suffering be more clearly seen than in the sphere of the care of the dying patient. It is in this sphere that the threat of the patient's experience, as Phillippe Aries has shown us, appears to medicine in the form of a painful embarrassment — an acute and frightening discomfort [1]. The hospital, in the rules of order that are reflected in its very concreteness — in its very structure and the texture of its ambience — is the invocation of a rule of silence over the patient's threatening expression of his own experience of the reality of dying. The reality of dying, for medicine, must stay contained within the scenario that it organizes and orchestrates, and which the dying patient is expected to enact. This scenario, research tells us again and again, demands that the patient die silently and without fuss — "discreetly," to use Aries' telling phrase. This demand, Aries has shown us, cannot be totally attributed to a reflexive reaction to the fear of dying. What medicine reflects, rather, is that which haunts the modern societies of the Western world — a crushing devaluation of each human's sense of his own individuality.

The hospital, symbolic of modern medicine, is the space in which the institution of medicine lives out most tellingly the control of individual experience. Medicine cannot be conceived of in terms in which it often likes to characterize itself — as an enterprise devoted to healing that wants to operate above the social scene, but which is often harrassed by social processes that impede its mission. Medicine participates fully in the processes of the larger society; what is transacted across the space between the doctor and his patient is a microcosm of what is transacted across the multiple spaces between ourselves and the authority that constitutes the fabric of society. What is there that is threatening about the patient speaking about his experience of suffering? To say it is to open a discourse, at the heart of which is the demand that the other engage the subject's suffering on the ground of his subjective sense of what is real to him about his own suffering. It is the demand of that subject who lies behind the personal pronoun "I" that the other, who claims to be the one who wants to help the subject surpass his suffering, engage that suffering by honoring the assertion of the "I" that it is the "I" who is the source and origin of the values that give a meaning to his life. In sum, it is an assertion from the sufferer that it is his values, grounded in his subjectivity, that gives to the other the authority he has and who, in light of the values he chooses, can withdraw the other's authority over him.

It has become harder and harder in the Western world to speak about the "I" to all those institutions, educational, political, and economic, that lay claim to be the agencies that will help us surpass our suffering. Their demand is that we speak only about our "bowels," our "feet," and our "heart," which all patients recognize in the inner recesses of their being is not speaking at all. It is rather the act of displacing their suffering from the subjective realm, where it derives its meaning from personal values that can give substance to an individual's life, to an objective realm, where its meaning is assigned to it by values hidden in a reality that is capable of being understood only by those whose expertise stamps them with authority. Patients, in effect, deliver themselves to the physician as a set of mysterious signs whose reality only the physician can see, interpret and manipulate in a guardianship over the patient's life. This phenomenon is what Ivan Illich points to when he characterizes modern medicine as a mission to "defend us against evil, premature and unnatural deaths" [6]. It is a mission that medicine carries out with a single-minded devotion, as if the baseline from which its mission can be measured is objectively known and visible to everyone to see: the definition of what constitutes the good, fully ripened, and natural life. Such a baseline is no less than a definition of "health" given a positive content rather than being defined "merely" as the absence of illnesses and diseases. We do not know what constitutes "evil," premature, or unnatural deaths outside of our own values. No one has been able to show that such definitions can exist at all in the realm of fact as opposed to the realm of value. It is a mission, however, for which medicine finds its raison d'etre, not in the realm of reason, but in the mission that organized society itself around the eighteenth century: the great theme of mastering and controlling the forces of nature. As Foucault has shown us, medicine became "modern" when disease became seen as a manifestation of natural forces that display themselves completely and totally within the "confined space of the body" to the "informed gaze" of the physician [5]. He is the authority trained to know the reality of the patients' suffering, which exhausts itself in what he can see going on in the "space of the body." There, in that space, lies both the origin of suffering and its historical course, which it is the physician's capacity and calling to control and master. For the patient to speak is for him to demand that the mode in which he experiences his illness for himself must be accounted for, not as defining the techniques that medicine uses on him, but as defining the sole source and origin of that "baseline" of the "good life" which confers all techniques used on him with an authority. The patient who speaks is thus also the patient who poses a threat to the privileged authority of the physician, which rests on the privileged view of the reality of the patient's experience. It is a privilege the physician insists he holds and, indeed, often believes it is his sacred mission to hold, even though his own alienation tells him it is an almost insupportable burden for him to carry.

To be a patient then is not simply to have one's experience discounted; it is also to fall under a rule of silence about one's values as that which gives authority to everything, and therefore can withdraw it from everything. But what social institution today does not claim a "privileged authority" based on a "privileged" view of the reality of our experience? What social institution today does not place us under a rule of silence in the name of the "sacred mission" to which it is dedicated? This is why the scandal of medicine can be seen in its fullest dimensions only when medicine itself is seen as fully implicated in the social problem of our time: alienation.

To say this, however, is to take away the two views of medicine: Medicine is some sort of villain in the timeless drama of the good guys against the bad; or the view more prevalent within medicine, to the effect that its neglect of the "total patient" and his "psychosocial needs" is a function of its necessary preoccupation with the psychobiological realities of disease. In the final analysis, the scandal of medicine is the scandal of ourselves. Most of us cannot say that we have had a gag in our mouths for over 200 years. For most of us, if we do not speak about our values that can give and withhold others' authority over us, it is because we forswear speaking in favor of the ancient dream that by our becoming an object for another, the other will surely deliver us from suffering. Each physician, by embracing an ego ideal of what he ought to be, seems to be struggling to play his proper role in that ancient dream. But we are beginning to see now that no one can participate in a relationship marked by an inhuman silence without experiencing himself as less than human — as alienated from himself.

## A CALL TO THE FUTURE

Although medicine participates with all of the social institutions of our time in the problem of alienation, it is possible to accord it a kind of privileged position on two grounds. First, it is where we encounter with clarity and stark primitiveness the forces within and without us that play into our experience of alienation — that keep us and motivate us to remain under a rule of silence in the presence of the other's claim of authority over us. Second, and related to this point, medicine can be thought of as the cutting edge for change in our time. It can be thought of as the place where human beings, some of whom lay claim to wanting from others a legitimate authority on the basis of their specialized learning and skills, can begin to look each other in the face and thereby begin to open the possibility of speaking to each other. One wonders whether freedom can support such a burden, for gone would be the mystique of physicianhood and the regressive dependency of patienthood.

The real disservice of medical intervention in terms of adding to the burden of alienation comes about when we all do not share the same value sense about technological medicine and its benefits as do its purveyors. The issue of scientific medicine converting a strange philosophy as the price of making its contribution to the physical and emotional welfare of the holders of that philosophy is one of the crucial issues of our time [4]. As Garrison and Thomas point out, the existence of parallel systems of health care to the established one as well as the failure to bridge these systems contributes to the emotional disarray and estrangement of persons who participate in those subcultural systems but are also dependent on the technological services of established medicine. From the perspective of a need to "bridge" cultures it would appear that our health care systems work best for those who belong to or who accept the dominant culture, not only in terms of accepting the "Rightness of Technological Health Care," but also in terms of the correlates of the capacity to conform to the dominant culture, for example, belonging to the labor force (between 18 and 65), speaking English, being middle class or having middle-class aspirations, being white and educated.

The inner-city dweller, now predominantly black or Spanish speaking, has been confronted for the past 20 years or more by the disappearance of the primary physician and has turned, increasingly, to the emergency rooms of the large hospitals, where the alienating effects of the technologies of medicine are added to by discourtesy, insolence, and indifference. We are discovering once again, as was the case with drug addiction, that societal malaise does not remain confined to the ghetto but radiates swiftly to other locales. Insolence, discourtesy, and indifference, albeit not terribly widespread in the profession, when coupled with a profit-motivated health care inflation, have sent malpractice costs soaring. It is no great credit to the ego ideals embraced by medicine that this is most likely to bring about the "lost reform" of the American polity — namely, national health insurance and ultimately a national health service. Yet neither a national health insurance nor a national health service will diminish the progression of modern medicine toward its own sense of alienation. The physician himself must confront the meaning of his own alienation — the failure of his own ideals and his most secret dreams. Only out of this self-confrontation can he integrate his role of authority into that which human beings truly owe each other. He hardly stands alone, of course, in the need to begin this task of curing himself.

REFERENCES

1. Aries, P.  Death Inside Out. In P. Steinfels and R. Veatch (eds.), *Death Inside Out.* New York: Harper & Row, 1974.
2. Balint, M.  *The Doctor, His Patient and the Illness.* New York: International Universities Press, 1957.
3. Bergen, B. J.  Psychosomatic knowledge and the role of the physician. *Int. J. Psychiatry Med.* 5:431, 1974.
4. Garrison, V., and Thomas, C.  *A General Systems View of Community Mental Health.* New York: Brunner/Mazel, 1975.
5. Foucault, M.  *The Birth of the Clinic.* New York: Pantheon Books, 1973.
6. Illich, I.  The Political Uses of Natural Death. In P. Steinfels and R. Veatch (eds.), *Death Inside Out.* New York: Harper & Row, 1974. P. 25.
7. Ingelfinger, F.  The physician's contribution to the health system. *N. Engl. J. Med.* 295:565, 1976.
8. Kitzinger, S.  The Woman on the Delivery Table. In M. Laing (ed.), *Woman on Woman.* New York: Harper & Row, 1972.
9. Menke, W. G.  Medical identity: Change and conflict in professional roles. *J. Med. Educ.* 46:58, 1971.
10. Nisbett, R.  *Twilight of Authority.* London: Oxford University Press, 1977.
11. Schoolman, H. M.  The role of the physician as a patient advocate. *N. Engl. J. Med.* 296:103, 1977.

# CHAPTER 44

# Sex Differences in Adaptation to Exploitative Society
*Helen Block Lewis*

*The thesis* that civilized society violates human nature and therefore abridges human rights has been with us at least since the Enlightenment. Rousseau mourned the lost "noble savage"; Marx's alienated man and Freud's sexually repressed neurotic are two examples of the injuries that civilization is thought to inflict upon its members. For Rousseau and Freud the conflict between civilization and the individual was inherent in all societies. For Marx the conflict was inherent only in exploitative or profit-driven society. Feminist thinkers in both traditions have specifically detailed the ways in which society's patriarchal values oppress women. My thesis, very briefly stated, is that our exploitative society injures the two sexes differently. It not only creates different distortions of gender identity in men and women, but it also produces different forms of mental illness [7].

The following is a formula for understanding sex differences that has served me as a kind of shorthand statement of the important variables: Sex differences in human behavior (including gender identity differences) are the result of sex differences in chromosomal endowment (XX and XY on the twenty-third pair of chromosomes) times the social inferiority of women (an offshoot of an exploitative social system) – and this interaction must be considered as the numerator over a common denominator: the affectionate or social nature of the human species of both sexes.

Let me address the common denominator first. I base it on accumulating evidence from anthropologists' studies of the difference between ourselves and nonhuman primates and on the relatively recent developments in the study of human infancy. Anthropologists, at least some of them, tell us that the most profound evolutionary change from nonhuman primates to ourselves is the appearance of human culture as our species' adaptation to its environment. And human culture is unique in the scope it gives to affectionate nurturance of the young during a prolonged childhood. As a result, the power of moral prescriptions, that is, internalized affectionate bonds, to govern human behavior from birth to death is also unique to our species.

The emergence of morally prescriptive human culture has been understood as the outcome of the Great Apes' descent from the trees. As Elaine Morgan [9] has shown in her delightful book, *The Descent of Woman,* scientific speculation about the evolution of human culture has itself been dominated by androcentric, that is, male-oriented thinking. The image of Tarzan, the Mighty Hunter, coming down out of the trees has been the model against which human beings' upright posture, opposable thumb, tool using, language, and advanced cerebral cortex have been described. Morgan suggests that human evolutionary advance was also governed by the survival needs of a hypothetical female (primate) ancestor trying to nurture her young during the millions of years of Pliocene drought. Morgan's thesis reminds us that human

This paper is a revised version of a paper prepared for a conference on Genes and Gender in January 1977 and is included in the book *Genes and Gender*, edited by Betty Rosoff, Ph.D., and Ethel Tobach, Ph.D., and published by the Gordian Press, Inc., Staten Island, New York.

culture is an enormous evolutionary advance over primate society, not only because the Mighty Hunter has grown more intelligent and mightier, but because human culture also allows so much greater scope for affectionateness as a force governing human behavior.

As one example, there is a very great difference between nonhuman primates and ourselves in the amount of affectionateness that characterizes sexual behavior. Human sexual behavior is very different from primates' sexual behavior in that sex life is no longer governed by the estrus cycle in the female — that is, by a built-in biological clock. Note again, the significant evolutionary change has occurred in the female. With estrus absent, human beings have lost the prepotent hormonal stimulus to copulation that all other mammals have. In contrast, however, human sexual behavior is unique in that affectionate behavior — kissing, playfulness, touching — have all been incorporated into lovemaking. Among primates, these affectionate behaviors are quite separate from the act of copulation. Compared with primate sex, human sex is replete with revivals of affectionate intimacies that children shared with their parents. It is also unique in the incest taboo — a universal moral prohibition against intercourse between parent and child.

It is worth a brief digression to remark on how not only androcentric thinking but also the Weltanschauung of an exploitative society influences our categories of thought about human culture. We are told that the human infant is the most helpless of all species: completely at the mercy of its parents who, in turn, represent the superior forces of an (assumedly) hostile society. That the human infant is physically helpless is perfectly true. But that its encounters with the world are necessarily with hostile forces does not follow as a given. There are powerful but *benign social* forces operating in every known culture to facilitate the growth of the infant. And although the human infant is physically helpless, it is socially powerful — able as Bowlby [2] has shown, to evoke attachment, that is, affectionate caretaking in its parents of both sexes. Recent work in infant psychology has indeed emphasized the hypothesis that the human infant is, to quote Harriet Rheingold [11], "social by biological origin." Our greatest advance over our nonhuman primate cousins lies precisely in the tremendous increase in the impact of nurturant, affectionate, and moral forces on human behavior. Human infants grow up in an ecology in which moral forces exercise the greatest power over their lives.

Human culture, alas, is also Janus-faced. For reasons that no one really understands, the history of civilized humanity has involved the exploitation of the many by a few, and within this fact about civilization, it must also be said that men have been the exploiters and women subordinate to men. It should be noted that exploitation, warfare, and the subjugation of women are all practiced in the name of the culture's morality. Among nonliterate peoples, there are some cultures that are nonexploitative and in which the sexes are also, correlatively, of equal status. These nonexploitative cultures are few in number: Arapesh, Zuni, and the ancient Kung peoples are among the most famous. Their very existence, however, tells us that exploitativeness is not intrinsic in human nature but is the result of historical forces not yet understood. The existence of these cultures also suggests the correctness of Engels's idea that women's and children's exploitation is a subcategory of generalized exploitative relations within a society.

Why it is that men and not women are the exploiters and the warriors in societies where exploitation and warfare exist is a related, unsolved historical question. Simone de Beauvoir [3] specifically raises this question when she disagrees with Engels that division of labor and the development of technology were responsible for women's subjugation. Why, she asks, should not an originally friendly relation between the sexes have survived the development of technology? Her answer once again reveals that, even in so sophisticated a thinker, there is an acceptance of the concept that "domination" of one self over another is basic to human kind. de Beauvoir assumes that men have a need for "transcendence," while women need only be "immanent." She bases this assumption on the notion that the self develops out of interaction with the "other," and she takes for granted that this interaction is basically hostile, requiring the male self to dominate the other (female).

I do not know the answer to de Beauvoir's profound question, but I suggest that her answer, which assumes a male need for "transcendence," is rooted in understandable ignorance of relatively recent work on the psychology of infants. In this field we learn that human beings are, above all, uniquely social creatures. In the interaction between infants and their caretakers it is not domination and hostility but affection that supports the infant's optimal growth.

Space does not permit more than a brief sample of the evidence for the social nature of human infants. The three-month-old social smile, for example, seems to be a biological given [4]. In fact, Eibl-Eibesfeldt tells us that blind and deaf thalidomide babies, born without arms to touch their mothers, still smile at three months, like normal infants.

There is also the suggestion that girl infants may have an edge over boy infants in their social responses [5]. Two- to three-day-old girl infants, for example, cry more often than newborn boys in response to the sound of other newborns crying [12]. As another example, there is evidence that the mother-infant interaction is smoother and easier for girl babies than for boys [10]. At least two factors may conjoin to produce such results: Mothers have different attitudes toward infant girls and boys, especially in a society that fosters the myth of male superiority, *and* boy and girl babies may bring something slightly different into the mother-infant interaction.

As to the second factor, I am aware that the notion of genetically determined differences between the sexes is not fashionable, especially since the differences are often so interpreted as to promote the subjugation of women. But it also seems to me useless to make the mistake of ignoring genetics, just because their input has been distorted. I used to believe that it was impossible in the present climate of women's social inferiority to obtain any meaningful results about genetically determined behavior differences between the sexes, just as it *is* impossible to obtain any meaningful findings about the genetic determination of black-white differences in intelligence. But the analogy between sex differences and black-white differences in intelligence is false. The reason is this: For differences between blacks and whites, there are no cleanly differentiated gene pools. But when it comes to the difference between having XX or XY as the twenty-third pair of chromosomes, the difference is clean and powerful. (Not that males and females cannot have the same genes derived from the other 22 pairs of chromosomes.) But the presence of XX or XY results in the enormous difference between a reproductive system that is equipped to bear children and nurse them in contrast to a system with only a penis and testicles.

I turn back now to my thesis that differences between the sexes are a product of genetics in interaction with the social inequality of women over the affectionate nature of the species. That an exploitative social order injures humanity is not a novel idea: Marx spoke of the alienation that capitalism breeds in its workers. Erich Fromm has developed in detail some of the characteristic distortions that a profit system inflicts upon the people reared in it. In another tradition, Rousseau was struck by the intrinsic psychic injury that civilization inflicts upon its members; the loss of "pitie" or sympathy of one human being for another. (This notion of Rousseau's was most influential in Levi-Strauss's thinking.) Freud supposed that all social orders, exploitative and nonexploitative, were built upon the suppression of sexuality — a term he used loosely to include affectionateness and nurturance as well.

My thesis is that the injury that an exploitative society inflicts upon human beings is that it requires them to suppress, renounce, or repress their affectionateness, thus creating a profound internal conflict. An exploitative society requires people to treat other people as if they were things, and as a result, it severely taxes the superego of both sexes. My thesis is also that an exploitative society has a different, although equally severe, impact of injury upon the two sexes.

The working model I have developed to understand psychic injury focuses upon the superego. The internalized conflict between exploitative and affectionate values catches men and women at different points in their acculturation and along differing routes of internalization of values, so that the superegos of men and women tend to operate in different modes. Let me emphasize that the differing modes represent equally developed ethical standards. Specifically, women are more prone to the shame of "loss of love" and to the shame of social inferiority, while men are more prone to guilt for the more frequent transgressions against their own affectionate natures that the exploitative world requires of them.

Let us look more closely at the picture for women first. Girls, who may already have some kind of edge in sociability over boys, are trained and encouraged into affectionate, nurturant roles. Their gender identity involves only the emulation of their nurturant mothers. But by the time they are two years old, when their gender identity is well established, they discover that affectionateness is not really a useful commodity in an exploitative world. On the contrary, it is a handicap that brings women into "dependent" relationships in which the other, not the self, is the center of the world. Moreover, women, affectionate as they may be, are second-class citizens in the world of power. One consequence is that women have a terrible sense of loss when the others around whom their lives have been built no longer require nurturing. Their sense of loss and helplessness when they lose their nurturant occupation is also often intensified because they have no other occupation and frequently no gainful employment. So it is no surprise that women should be prone to the shame of loss of love and to the shame of second-class citizenship in the world. It is also a well-established fact that women are two to three times more prone to depression and hysteria than men. As I have shown in my book, *Shame and Guilt in*

*Neurosis* [6], the affinity between the shame experience and both depression and hysteria is quite close.

It seems to me in keeping with the fact that women do not give up their affectionateness, only devalue it, that the clinical picture in depression and hysteria involves no bizarre distortions of human behavior. The symptoms in both depression and hysteria tend to be common and mundane, understandable to all of us as a depressed mood or aches and pains or both.

The demographic characteristics of depression can also be seen to fit the picture I am describing. Depression cuts across class lines; if anything, there is some slight tendency for depression to be associated with affluence. That depression is a disorder resulting from a failure of high ideals of devotion to others was neatly demonstrated in a study by Pauline Bart [1]. Bart predicted that, on the basis of a strong Jewish tradition of wifely and motherly devotion to the family as a center of their lives, middle-aged Jewish women should be more prone to depression than other ethnic groups in which this particular tradition is not so strong. In an empirical study done in California, Bart confirmed her hypothesis. As she puts it, you do not have to be Jewish to be a Jewish mother (and get depressed) but it helps!

For boys the internalized conflict between the affectionate natures and the exploitative demands of the social order in which they are reared assumes a different pattern. Boys are not encouraged to develop their sociability. On the contrary, by the time their gender identity is established at age two, they are aware that they are expected to be aggressive. Their gender identity itself involves extracting from experience with a caretaker of the opposite sex that they are different from her and more like their fathers who are more distant. They are expected, before they are six, to "renounce" their affection for their mothers and their identifications with her – and to become like their not only distant but more aggressive fathers. As one psychologist [8] put it, boys have to solve a problem in person identification, not just take a lesson, the way girls can, in person emulation. At the same time, the culture sets limits on just how aggressive, that is, just how guilty, men may be. No wonder that boys, when they grow up, are not only more prone to problems of gender identity but also to all kinds of obsessional and compulsive disorders, including addictions and sexual perversions. The phenomenological affinity between guilt and obsessive and compulsive states is quite close. Men are also more prone than women to schizophrenia, especially of the paranoid type – in which projection of guilt is an outstanding characteristic.

Schizophrenia of all forms has entirely different demographic and clinical characteristics compared to depression. It is strongly associated with class. If you are poor, black, and male, your chances of falling ill of schizophrenia are much greater than if you are rich, white, and female. These facts seem to me to reflect the direct, early attack that our exploitative cultural order makes on men's sociability (perhaps going along with men's slightly lesser degree of innate sociability). And the facts go along with the chronically high unemployment rates that make so many men face an apparent failure to achieve success in the world of work. Schizophrenia seems to me to be a response to the demand that our exploitative social order puts more directly upon men than women to function in the world as if aggression and not affectionateness were the stuff of which they are made. The greater guilt that results from men's more extensive suppression of their affectionate natures brings with it the clinical picture of bizarre distortions of gender identity, of the self, and of the world that characterizes schizophrenia, that is, out-and-out social withdrawal into madness.

## REFERENCES

1. Bart, P. Depression in Middle-aged Women. In V. Gornick and B. Moran (eds.), *Woman in Sexist Society*. New York: New American Library, 1971.
2. Bowlby, J. *Attachment and Loss* (vol. 1). New York: Basic Books, 1969.
3. de Beauvoir, S. *The Second Sex*. New York: Knopf, 1957.
4. Eibl-Eibesfeldt, I. *Love and Hate*. New York: Schocken Books, 1974.
5. Korner, A. Methodological Considerations in Studying Sex Differences in the Behavioral Functioning of Newborns. In R. Friedman, R. Richart, and R. Van de Wiele (eds.), *Sex Differences in Behavior*. New York: Wiley, 1974.
6. Lewis, H. B. *Shame and Guilt in Neurosis*. New York: International Universities Press, 1971.
7. Lewis, H. B. *Psychic War in Men and Women*. New York: New York University Press, 1976.
8. Lynn, D. Sex role and parental identification. *Child Dev.* 33:554, 1962.
9. Morgan, E. *The Descent of Woman*. New York: Bantam Books, 1972.
10. Moss, H. Early Sex Differences and Mother-Infant Interaction. In R. Friedman, R. Richart, and R. Van de Wiele (eds.), *Sex Differences in Behavior*. New York: Wiley, 1974.
11. Rheingold, H. The Social and Socializing Infant. In D. Goslin (ed.), *Handbook of Socialization Theory and Research*. Chicago: Rand-McNally, 1969.
12. Simner, M. Newborn response to the cry of another infant. *Developmental Psychology* 5:136, 1971.

# CHAPTER 45

# Nursing and the Humanities: An Approach to Humanistic Issues in Health Care
*Sally Gadow*

*A fundamental premise* of nursing practice is that the individual has the right to receive acknowledgment and affirmation as a human being — a human being whose experience is so unique and complex that each individual transcends the categories and quantifications of science. The recipient of health care thus has a right to more than superbly scientific and technically perfect treatment: the right to humanistic care.

This right is addressed by the humanities within the health care professions. The humanities are concerned not only with understanding as fully as possible the depth and scope of human experience, which they share with the sciences, but they are also concerned with creating or uncovering more human meanings for experience. Thus the humanities are ultimately engaged in formulating ideals. By "ideal" or "more human meaning" is meant a way of understanding an experience that enhances the value of the experience for an individual and enables that person to formulate a meaning that uniquely expresses his or her relation to the experience. The "right to humanistic health care" is accordingly the right to receive assistance from health professionals in determining the personal value and meaning of one's illness or disability.

One way of addressing such a right is to identify the humanistic issues within one of the health professions and to outline a program in which those issues are elucidated by the humanities. In the discussion to follow, I outline such a program for integrating the humanities and nursing.

A program that attempts to address humanistic issues in nursing requires an initial identification and even systematization of those issues. Haphazard selection of problems and topics on the basis of current movements, popular themes, or recent judicial decisions will inevitably ignore some issues and overemphasize others.

At least two possibilities for more methodic delineation of issues are available, if we think of the humanities and nursing as the two disciplines that the proposed program seeks to unite. One method is to identify and organize issues from the side of the humanities, that is, according to the three distinct approaches to human experience that the humanities represent: the historical, the aesthetic, and the philosophical. The second possibility is to delineate issues from the side of nursing and then determine which of the several humanities approaches offers the clearest elucidation.

Since current humanities teaching in health professional education tends to use the first method, that is, to address those issues that fall within the purview of the

The original version of this paper, Humanistic Issues at the Interface of Nursing and the Community, appeared in *Connecticut Medicine* 41:357, 1977. It was first presented before the members of the Four-State Consortium for Nursing and the Humanities for the project, "Nursing and the Humanities: A Public Dialogue," which convened at the University of Connecticut Health Center, Farmington, Connecticut, on March 4–5, 1977.

particular humanities disciplines that happen to be represented by the faculty, I shall briefly consider that method first. (My bias will be evident, for I shall devote the greater portion of my discussion to the second method, which seems the more promising for our purposes.)

In looking at nursing from the three humanities perspectives that I distinguished — the historical, the aesthetic, and the philosophical — we can identify different areas of concern. The historical approach focuses upon questions about the origin and evolution of present values and coming changes in nursing. Thus the use of historical studies would be especially relevant in addressing issues, for example, around the involvement of nurses in modifying traditional health care systems. The aesthetic approach, with its focus upon the unique, nonquantifiable dimensions of human experience, would appropriately address such issues as the subjective and expressly individual aspects of the experience of illness, as portrayed, for example, in literature. Finally, the philosophical approach, often the only one employed in humanities teaching in health care settings, obviously addresses issues of the moral permissibility of certain practices, as well as the more difficult issues of, not right and wrong, but permissible versus ideal ways of caring for persons. Beyond these ethical issues, philosophy is also concerned with questions about the nature of humanness and the fundamental forms of human relating; thus it would address, for example, the issue of which particular form of relation constitutes the basis for the nurse-patient relation.

I turn now to the second method, that of delineating issues from the side of nursing, rather than via the humanities approaches. I propose this method as the more useful tool for identifying the so-called humanistic issues, and I do so with the following rationale. There is no division in nursing (or in any human services profession) between humanistic and nonhumanistic issues. The principal elements of the nursing process are persons, and thus the human, unique, nontechnical, value dimensions are by definition an essential aspect of the practice of nursing. There is, in other words, no element of nursing that is without humanistic issues. In seeking to identify those issues, then, a natural method seems to be to lift out *all* of the fundamental issues in nursing, on the assumption that all of them are in principle humanistic. The next step would be to explore the ways in which each of those issues can be most appropriately addressed and elucidated by one or more of the three humanities approaches.

Since I have asserted that the important components of the practice of nursing are the human beings involved, a natural organization of issues suggests itself, based upon the various interrelations of those persons. The issues would thus cluster in this way, emerging around these relations:

1. *The nurse and the patient* (I use the term "patient" simply from custom, rather than as a resolution in advance of the important issues reflected in the range of terms now in use, from "patient" to "client" and "consumer.")
2. *The nurse and other health professionals*
3. *Nurses and nursing*
4. *The nurse and the community*

I would now like to propose formulations of some of the specific issues within each of these categories and formulations of particular humanities approaches that might address those issues.

## THE NURSE AND THE PATIENT
### Defining the Nurse-Patient Relation
What are the historical and the present forms of the nurse-patient relation and what may the ideal relation be, to which both nurses and the public would agree and aspire? The possibilities are not unfamiliar to anyone who has provided or received nursing care: the nurse as parent surrogate, healer, physician surrogate, contracted clinician, patient advocate, health educator, or provider of some unique and yet undefined type of care falling within none of the stereotypic nurse-patient relations, or finally, as my students always insist, a grand and unique synthesis of all of the above.

One approach to this issue is a phenomenological inquiry into the experience of illness, to elucidate the distinction between the body as an external object and the body as a lived interiority. The implications of that distinction for nursing are becoming apparent. To cite one example: Studies by nurse researcher Dr. Jean Johnson have shown that postoperative patients were affected more positively (in terms of less analgesia, shorter hospital stay, etc.) when given preoperative information about the subjective sensations they might experience after surgery than when given a description of the objective phenomenon they would undergo, namely, the surgical procedure [1]. The question of whether nursing should ideally address the individual as a lived subjective body, an organ-system object-body, or a unity of the two surely has implications for the nature of the nurse-patient relation.

A second philosophical concern with the body and illness is the issue of the value of suffering, or the religion of health. Existentialists maintain that we are free to determine the meaning that any experience will have for us, no matter how anguished or unfree we feel. Nietzsche makes a still stronger claim that pain is a positive, even necessary, means of finding one's own way, of creating oneself, and the sympathy of those who rush to comfort us only devalues our experience. The question then of whether suffering is a defect of human existence or a uniquely human possibility for self-determination is an issue that nursing must address. If suffering is an experience the meaning of which — to be a truly human meaning — must be decided by each individual (unencumbered by the health professions' negative value of illness), the task of the nurse then becomes infinitely complex, in assisting each patient in an individual manner to establish the particular meaning that his or her suffering is to have.

In addition to the approaches of phenomenology and existential philosophy, the perspective of history is essential in helping clarify the extent to which the traditional role of the nurse as steadfastly caring and minimally technical has been a function of the level of technology involved in health care. Without question, the role of the modern nurse encompasses far more knowledge and skill concerning equipment and apparatus than has ever before been the case. Does the advancement of technology in nursing signify a necessary concomitant: a decline in the customary caring that has characterized nursing? Is nursing faced with a decision between preserving its commitment to *caring,* on the one hand, and following the progress of medical science with more technically intricate *curing,* on the other hand?

*Ethical Problems Unique to the Nurse-Patient Relation: Are There Any?*
What are the moral implications of the particular type of relation that exists between nurse and patient? Are there ethical decisions that nurses should be prohibited from making, just as there are clinical or therapeutic decisions they are not permitted to make? Or, on the contrary, do nurses have ethical responsibilities to patients that even surpass those of other professionals less intimately and continuously involved with the patient? A philosophical analysis of such ethical issues would have to address the fundamental nature of the relation between a person who needs care and one who assumes responsibility for rendering that care. For example, does the patient have the right to receive from the nurse information withheld by other professionals? Does the nurse, if acting as parent surrogate, have the right to act paternalistically when the patient does not make "healthy" decisions? Does the nurse, if acting in contractual partnership with the patient, have the right to withdraw care when the patient refuses to assume responsibility for his or her health? Does the nurse, if acting as healer, analogous to the physician, have the right to cultivate the placebo effect that is thought to accompany all of the actions of a healer, even to the point of patent deception? Or does the nurse, if acting as patient advocate, have an obligation to protect the patient from every erosion of human dignity and value, including deceptions in the name of health? Finally, if nursing is a synthesis of all of these roles, how are values reconciled when the roles conflict, as they often clearly do?

## THE NURSE AND OTHER HEALTH PROFESSIONALS
The relation between nurses and their colleagues in health care has become a pressing issue because of at least two developments: the emergence of new health professions with which the traditional professions must articulate and the increasing autonomy of midlevel practitioners as physicians become willing to delegate more of their tasks and authority. The issues for nursing can be formulated in the following ways: (1) What does the increasing number of independent nurse practitioners signify for the future of nursing? Is nursing practice striving to become another version of medical practice, rather than an alternative and complementary form of care? Or are nurse practitioners seeking to offer a true alternative to the traditional medical practitioner by providing the same general care but in a different fashion? If so, what is that difference? And why is an alternative thought to be necessary? (2) A second issue is the changing relation between nurse and physician in their roles as members of the health care team (as distinct from their roles as independent practitioners). Nurses are becoming increasingly responsible and critical concerning the quality of care patients receive from their physicians. Nurses challenge orders to resuscitate patients, orders not to resuscitate patients, and at times simply make

decisions directly contrary to orders that they think unjustified. The nurse is of course professionally obligated to challenge a medically unsafe or unsound order. Is there a comparable obligation to refuse an order that is ethically indefensible in the nurse's view? Does the physician's ethical as well as medical opinion have more authority than that of the nurse (especially when the latter is in agreement with the expressed view of the patient)? Should the nurse, in other words, be legally required to honor an order, such as the directive to continue intravenous fluids for a terminal patient — or to discontinue them — when the order is based solely on the physician's moral view of the case in question and rests entirely on values, as many decisions in health care inevitably do?

Here the disciplines of history, law, and moral philosophy are all needed in addressing such an involved issue as the legal, ethical, and professional divisions — past, present, and future — between nurses and other health care providers. One approach that might prove interesting as well as bring issues quickly to the surface would be an interprofessional debate of a hypothetical clause in California's "right to die" legislation allowing the nurse to assent to a patient's request for discontinuing extraordinary life-sustaining measures when the attending physician refuses to do so.

## NURSES AND NURSING: THE RELATION OF NURSES TO THEMSELVES AND THEIR PROFESSION

### Images of Women Healers in Art and Literature

Because 97 percent of nurses are women and because many of the issues around the nurse-patient and the nurse-physician relation are entangled with issues of women's history, as evidenced by the cliché of the nurse as mother of the patient and wife of the physician, it would seem crucial to address explicitly and directly the relation between women and nursing. One way of clarifying that relation is by examining images of women and women healers in literature and the visual arts. Such an inquiry would constitute a historical as well as an aesthetic approach by documenting the images of women that cultures have expressed through their artists, images that not only are reflected in art and literature but also may in part be created by artists and writers.

### Women in the History of Health Care

Another way of clarifying the relation between women and nursing is to examine historically the situation of women in the health sector, which would entail analyzing not women as such, but the socioeconomic and political systems that generate and perpetuate the present situation of women. If the traditional role division within the family is accepted as part of the basis for the sex hierarchy in the health sector, critical historical studies of the family are required for elucidating the so-called family structure of the health team, with its male leader, female auxiliary, and devoted dependents. On the other hand, since the study of history offers us interpretations rather than facts, it would be simplistic to conclude that the same social and economic needs that seem to have delegated the role of nurturing support to women in the family are also the reason why nursing has been traditionally a role of "motherly care." If that were the case, we would have to predict that the liberation of women will mean the liberation of nursing, but it is not clear that the caring functions of the nurse are entirely analogous to those of the mother, or that — even if they are — nurses would want to discard them for more "liberated" functions.

### Holistic Nursing: Whole Nurse or Whole Patient?

It is often maintained that optimal healing depends upon the interaction of both patient and professional as whole human beings. This view is then translated into concern for the patient as a unique psychobiological whole that is self-responsible and self-healing. What is lost in that translation is the professional. To what extent is the whole person of the healer — in this case the nurse — a necessary corollary for optimal healing of the whole patient? In recent years, nurses have been acutely concerned with achieving full professional stature as health care providers. Is the commitment to professionalism antithetical to the concern for holistic, or whole person, healing of the patient if that in any way requires whole person involvement of the nurse? The distinction between personal and professional involvement has been kept carefully sharp, ostensibly for the protection of the patient. Patients now complain that their care is too impersonal. Do they mean too professional? Or are they simply responding to the movement of humanistic health care, which I interpret as the movement to soften the distinction between personal and professional, as the practitioner becomes more fully, subjectively as well as objectively, involved, more human, while the patient becomes more knowledgable, assuming responsibilities for health and decision making formerly the sacred domain of the professional?

This issue, I propose, is not best addressed by the humanistic psychologies, although they can surely be credited with much of the personalization of the professional. The issue requires a sound and new conceptual framework within which to develop means for unifying and transcending the once contradictory relation between professional and personal. What seems to be needed is, among other things, a more ample and complex, and thus more human, concept of the body. Efforts at rehumanizing health care cannot survive confinement to a concept of body that belonged to medicine when it was hoped to be an exact science and the practitioner a scientist — or, in the case of the nurse, a research assistant — whose objectivity was inviolate, if at times untherapeutic. The task of addressing the problem belongs to philosophy and art, philosophy providing a conceptual reexamination and a phenomenological rediscovery of the whole person as body, and the arts providing aesthetic, nonanalytic access to human experience through images of the body that the artist portrays. A single work with the impact of Degas's tender painting of a girl-woman sitting unclothed on a bed, bewildered, wondering, can in itself make the point that one's experience and image of one's body are so intimate and inward, and often excruciatingly fragile, that the height of dehumanization would be an objective, "professional" treatment of the body as an external object, a clinical entity.

## THE NURSE AND THE COMMUNITY

### Community Control: Nursing and a National Health Service

The insulation of the health industry from the needs of the community is slowly diminishing. Consumerism, plus the public consciousness raising with respect to health and health care, means that all of the health professions will be held increasingly accountable for the distribution of practitioners, both in geography and in specialty, and for their professional training, particularly as it concerns the values of human dignity, quality care, and nondiscrimination. The issue for nursing is thus the extent to which the profession can and should share with the public the responsibility for adjusting the maldistribution of health workers. A related issue is whether that adjustment should extend to the very foundation of health care, namely, professional education, with the possible effect that the levels within the professional hierarchy would be equalized by educating health workers together in health team schools, with a basic curriculum for all categories of practitioners and with control of admissions and curriculum exercised by community health boards. If this type of health service assured more equitable and rational distribution of care, what should the position of nursing be, particularly if public participation threatened to democratize health services to the point of eliminating distinctions between professions, or if the public were more interested in warm and personalized care such as nursing auxiliary persons sometimes supply than in the theoretical and technical mastery that the professional nurse possesses?

### Nursing Values and Aging: Stereotype or Ideals?

In a nursing class recently on the subject of midlife crisis, I heard students express almost unanimously a reluctance to contemplate middle age because it is the prelude to aging, and aging seemed to them little better than dying. Their attitude, simply reflecting the prevalent social negativity toward aging, is not likely to change as a function of nursing education. Studies have shown that nursing students' stereotypic attitudes toward the aged are changed, if at all, by prolonged contact with elderly patients. The issue of how to change attitudes is, of course, the concern of educational psychology. The humanities must address the more fundamental issues: What images and ideals of the aged *should* health professionals be educated to have, in place of the negative stereotype of deterioration and decay? Are there more positive and freeing, and thus more human, meanings of aging that nursing and the humanities together can propose as alternatives to the concept of aging as a disease for which the cause and the cure have not yet been found?

The issue of more human meanings of aging is one for which nursing and the humanities are ideally suited to integrate their particular approaches to human experience. Historical studies are needed as a basis for understanding alternative attitudes of other cultures toward aging, as well as the origins of present values — and disvalues — of aging in this society. Against that background, an integration of the philosophical and aesthetic approaches can provide an interweaving of conceptual ideals and concrete images that express the positive meanings of aging as a human value. To this fabric of meanings that the humanities weave, the nurse brings experiential meanings of aging that evolve from sustained, intimate care of the aged, those human beings who as individuals defy generalization and succeed where educational strategies fail in humanizing the stereotypic attitudes of young professionals. Presumably, education fails where the aging themselves succeed because the latter not only present but also embody the human meanings of their experience, while education tries at best to neutralize negative values without offering vital, compelling alternatives. It is certain that together the humanities and nursing can articulate such alternatives.

REFERENCE
1. Johnson, J.  Unpublished paper. Johns Hopkins University School of Health Services, Baltimore, Maryland, February 1977.

# CHAPTER 46

# The Moral Situation in Nursing
*Catherine P. Murphy*

CRITERION FOR A MORAL SITUATION

In order to have a moral situation, Barry Chazan [5] holds that several criteria must be met. First, the individual must be faced with a human confrontation or a conflict between human needs or the welfare of others and the need to choose between alternative behaviors or actions. Second, the choice made must be guided by moral principles that are universal prescriptions for behavior, embodying some theory of justification. Third, the choice must be guided by a process of weighing reasons, and the decision must be freely and consciously chosen. Last, this choice is affected by feelings brought by the individual, and caused by the situation, and the particular context of the situation.

Nurses meet the first criterion, for they are constantly confronted with conflict between human needs in which they are faced with the dilemma of choosing alternative actions. For example, nurses are faced with all the inherent moral quandaries associated with life and death situations, truth telling and informal consent; behavior control and drug research on human subjects; and when their professional services become a scarce commodity, they must practice triage in the allocation of care in understaffed situations. Because of the very nature of their work, nurses in health care institutions are constantly confronted with moral conflict. Since nursing decisions are concerned with human life and they directly affect the welfare of other human beings, they ought to be measured by ethical or moral standards.

The second criterion for the moral situation focuses our attention on the role of moral principles in guiding choices to be made. Philosopher R. M. Hare advises us that a moral principle is a statement that conveys some directive or prescriptive force [18]. He emphasizes a moral principle as a weaker form of a prescriptive statement, because it should not directly cause or command certain behavior but rather suggest or propose certain behavior that the agent himself chooses in guiding his actions. Since the moral principle should not be a directive made by an authority figure, it becomes then an internal suggestion that the agent himself uses in guiding his behavior.

The nurse's role involves inherent conflict between two moral principles or normative components of her role. First, a nurse is morally obligated to recognize the right of the patient as an individual, or to put it in nursing language, she has a commitment to meet the individual needs of her patient. On the other hand, as an employee in a health care institution, the nurse is subordinate to the administration and, hence, must uphold the utilitarian goals of the institution: the greatest good for the greatest number. The consequence is that a patient must be ready to sacrifice his individual rights or needs for the greater good of all other patients in the institution. Since hospital administration has for all too long controlled both the educational and practice settings for nurses, they have also controlled the socialization process in nursing. To make sure that their utilitarian goals were not interfered with by dissension, hospitals saw to it that nurses received "proper" moral indoctrination by stressing the virtues of loyalty, duty, subservience, and blind obedience to authority.

The nursing values of unquestioning obedience to authority and dedication to duty that were rooted in the profession's military and religious tradition have led to the "ours is not to reason why" approach to nursing practice. As late as 1973, the American Nurses' Association president, Rosamond Gabrielson, noted that we were still struggling with our "born in the Church — bred in the Army" authoritarian heritage [13]. Nursing's Victorian values of unthinking obedience to authority figures and fixed rules and regulations are dysfunctional for the newly defined role of the nurse as a morally responsible agent who is an advocate for the patient and guardian of his rights. Each moral situation requires an independent moral judgment on the part of a moral agent, and a nurse must be able to engage in moral reasoning that is based on moral values and principles that are separate from institutional norms and authority. Nursing's antiquated values and moral principles must be examined in the light of modern standards, for the concept of professional autonomy is lost in an environment governed by rigid authoritarianism. Autonomous behavior on the part of a professional nurse should be autonomy within the context of disciplined moral reasoning and action. Nurses need to direct their use of newfound freedom with real values and responsible moral judgments.

Since Chazan's third criterion of weighing choices and having freedom to choose among alternatives is closely tied to his fourth criterion in the moral situation in nursing, it seems appropriate to consider them together. While an increasing number of professional nurses practice outside of hospitals, nursing by and large is still practiced in hospitals. For this reason, my consideration of situational contexts in the moral situation is limited to this type of setting.

## CONTEXT OF THE MORAL SITUATION IN NURSING

The particular context or moral atmosphere in which the moral situation takes place in nursing is such that it limits a nurse's ability to perform and make moral decisions. Organizational theorist Herbert Simon maintains that in order to analyze the effect of environment on the decision-making process, one must study the organization's "anatomy" and "physiology" [38]. The anatomy of an organization can be found in the distribution and allocation of decision-making functions, while its physiology can be viewed by analyzing the processes whereby an organization influences individual members by supplying their decisions with organizational premises. Organizations control centrality of decision making by preventing those low in rank from being able to weigh competing considerations and by requiring that employees accept conclusions reached by members higher in organizational rank. This control is accomplished by imposing general rules to limit discretion or by actually taking the decision-making function out of the hands of the employee. The organization further structures the employee's decision-making environment by control of the communication system and by internalizing values and decision rules through training and indoctrination [32].

Hospital physicians are in a staff position and not usually employed or subordinated to administrative authority. Even though outside of the administrative chain of command, they have authority to issue orders directly to nurses. Nursing staff are in a line position, which makes them responsible to the administrative hierarchy as well as to physicians. This dual chain of command creates conflict and severely limits the decision-making role of hospital nurses because they are subordinated to two lines of authority at one time. A range of empirical studies shows that physicians are central to the decision-making process and nurses are relegated to the role of implementing their decisions, for in effect, the physician's orders determine the framework for all the care given [8, 14, 15, 22, 29].

The network of communication in the hospital hierarchy allows the physician to move freely throughout the institution while the nurse must communicate through "proper" bureaucratic channels. The pattern of communication between nurses and physicians has been portrayed as the nurse-doctor game by psychiatrist Leonard Stein [39]. The rules of the game are that the nurse must appear passive, and if she should be bold enough to offer information or recommendations, it must be done in a manner that makes it look as if it was initiated by the physician. Making a suggestion to a doctor can be equal to insulting or belittling him, and the penalties for failure to play the game can be severe.

What are the consequences of such an atmosphere and how does it affect the attitudes of nurses and the quality of the nurse-patient relationship? A host of studies focusing on the dynamics of ethical decision making, such as those by Lawrence Kohlberg and others, clearly shows that the moral reasoning of an individual is affected by the moral atmosphere [25]. Studies by nurse researchers Bueker, Jarratt, Simms, Harrington, and Theis support this finding, since they conclude that the social organization of the hospital affects the value orientation and role performance of the nurse [4, 19, 22, 37]. In one recent study on the moral reasoning of nurses, the findings suggest that most of the participants were at a conventional level of reasoning that stresses obedience to authority and the need for harmonious relationships with institutions and authority figures even when patients' rights were being violated [33].

## VALUES NURSES BRING TO THE MORAL SITUATION

In addition to the situational factors that limit conscious, freely chosen alternatives for action and the ways in which the context of the moral situation can affect the manner in which a nurse values or chooses certain behaviors, what are the attitudes, values, and predispositions that nurses bring to the moral situation that also affect their choices? Emotional needs, internalization of role stereotype, social class, educational training, and sex affect the attitudes and behavior of nurses as organizational employees. Dominant-submissive patterns of behavior are frequently seen in the personalities of employees in bureaucratic organizations. Sociologists Bensman and Rosenberg believe that this pattern of behavior is a factor in attracting employees to begin with, and it is subsequently reinforced with continued employment, for the precise and inflexible rules of organizations offer a sense of psychological security for many employees [3].

Some researchers suggest that a theory of match fits an individual's mode of moral reasoning or ideology to his working environment [11, 20, 26]. Robert Hogan's research on moral reasoning claims that individuals are self-selected to occupations on the basis of their personality and moral reasoning. He found that individuals who place high value on rules and codified procedures in regulating human affairs were drawn to occupations that preserved and defended conventional social institutions, while individuals who considered rules a hindrance were drawn to occupations that promoted social change [20].

Moral reasoning and related personality characteristics may be a preselection factor in choosing nursing as a career. The sex-linked nurse stereotype that embodies obedience, caring, warmth, and nurturance has been the image that is constantly reinforced in books, games, and the media [10, 35, 36]. In the past, studies have indicated that high school and college students who reject nursing as a career do not feel that nursing offers opportunity for self-realization, originality, or creativity [12, 28]. Claire Fagin contends that this image of nursing often discourages "the career minded, more aggressive women from choosing nursing" [10]. Moral reasoning may also affect student success in a school of nursing. Research shows that there is a strong relationship between selectivity in recruitment and decrease in need for subsequent employee control in organizations. If personalities are shaped to fit organizational roles, there is much less need to socialize and control employee performance [9].

The educational environment in nursing has been portrayed as being rigid and authoritarian, and it is said to reflect teaching styles that impart subservience and appeal to authority for approval [2, 16, 21, 31, 34]. The student-teacher relationship has been compared to that of recruit and drill sergeant in the military, with faculty rigidity inhibiting independent action on the part of the student [39]. Research findings indicate that nursing students perceived that there was an absence of justice in their educational environment, that they did not feel free to express their opinions, and that faculty failed to treat them as autonomous individuals [34, 40, 41]. In a longitudinal study of nursing students, Olesen and Whittaker found that if nursing students did not conform and reflect the nursing value of "modesty" and "humility," they were branded as "overconfident" and considered "potentially unsafe." The dropout students who scored higher on nonauthoritarianism than the more successful remaining students were unable or unwilling to cultivate the required demeanor [34]. It seems that faculty in nursing education who act as recruiting agents for the profession place high emphasis on the conformist attitudes of the profession and, consequently, weed out students whose values do not match the conventional values of the educational and practice environments.

Since the traditional role of the nurse has been one of subservience and unquestioning loyalty and obedience to authority, it only stands to reason that the conventional mode of moral reasoning is functional for the role that the nurse has been forced to play in the health care system. Nursing literature and research suggest that there is a selection bias in nursing administration in favor of nurses who are subservient and loyal to the institution and its authority figures [1, 6, 17, 27, 30, 43], and there is a strong possibility that mode of moral reasoning may be functional for achieving success in nursing practice.

Nursing has been predominantly a female profession, and recent attention has focused on the difficulties encountered by nurses as members of a female profession. Researchers in one study on moral reasoning suggested that attainment of autonomous morality was more difficult for women because it involved conflict with the traditionally defined feminine role of dependence and irresponsibility [17]. Nurse researcher Cleland holds that nursing's lack of autonomy is directly attributed to the social position of women in society and is a result of the socialization process of females [7]. Sociologist Strauss maintains that the practice of nursing reflects the feminine virtues of "responsibility, motherliness, femininity, purity, service, and efficient housekeeping, while it has omitted the 'political (equal rights) reformer' themes" [42].

## NEED TO CHANGE THE MORAL SITUATION IN NURSING

I maintain that it is necessary to have nurses engage in moral reasoning that is separate from institutional norms and authority. If a nurse is oriented toward the utilitarian moral principles of the institution, she is not capable of reducing the welfare and claims of a group of people to the welfare and claims of patients as individuals. When faced with a situation of moral conflict, she will weigh alternative solutions in terms of the consequences for the social order of the group or institution rather than in terms of justice to the individual patient. Justice becomes a principle for social order rather than for personal moral choice, and the consequence of being unable to engage in moral reasoning that weighs competing claims of each individual in moral situations results in unthinking obedience that renders a nurse helpless when faced by conflicts of duty [25]. The nurse, then, submits to the edict of superiors and the fixed rules and regulations of the institution. The "collective morality" of this kind of reasoning has been destructive to nurses' individual morality, for it has led to loss of personal integrity and accountability and has permitted nurses to absolve themselves of misdeeds by placing blame and responsibility on others. Since this type of reasoning is directed toward maintaining rules and the existing social order for its own sake, it cannot provide rational guides to social change and the creation of new norms in the health care system. Nurses cannot be expected to be agents of social change if they do not go beyond this level of reasoning.

While it would not be wise to do away completely with all rules and authority in nursing, there is a need to develop a more reflective and selective approach toward dealing with them. Nurses have had much responsibility with very little authority and virtually no accountability. A crisis of authority is under way in society and health care today, and nursing needs to begin to think about the basic nature and function of authority itself. We must begin to reconsider the whole meaning of authority in order to develop a new locus of authority, namely, authority that stems from the patient and not the health care hierarchy. Nurses must begin to think of their moral and political authority as coming from the client and not from the health care bureaucracy.

## WAYS TO IMPROVE THE MORAL SITUATION IN NURSING

Throughout history the nurse has been the humanist in health care. Her role as the only logical advocate for the patient and guardian of his rights must be made legitimate. In order to prepare the nurse for the role of patient advocate, I propose a marriage between nursing and philosophy to increase the nurse's knowledge and expertise in ethical decision making. Nurses must be engaged in moral thinking, and they must be guided through, and made aware of, the moral situation and its components [5]. We must become more and more aware of the forces that shape our values, and we must look at the values of our patients as well as our employers.

In order to prepare nurses to be moral agents, we must create educational and service environments that allow nurses to act as moral agents. Studies concerned with the moral atmosphere of educational environments have indicated that students respond best to a combination of moral reasoning, moral action, and institutional rules as a relatively unified whole in relation to their mode of moral reasoning [23]. The potential of a moral atmosphere to stimulate moral reasoning depends upon its environmental level of justice and the extent to which it provides role-taking opportunities by allowing individuals to share in responsibilities and the decision-making process [24]. The educational environment in nursing must have a high level of institutional justice that is based on democracy and fairness and recognizes the dignity and rights of the learner. Students should be encouraged to enhance their role-taking abilities by becoming actively involved in the social and moral functioning of their school and practice environments. In the teaching of ethics and in confronting situations that require ethical reasoning in nursing education, the following criteria are necessary for stimulation of cognitive conflict and moral reasoning: There must be exposure to situations that present moral conflict and contradiction for the learner's current moral structure; interchange and free dialogue between teacher and student; and moral reflection that involves identity questioning, commitment, and consistency between moral reasoning and action [24].

The current practices in nursing service, wherein value decisions are formulated by those in authority and nurse subordinates perform the technical aspects of labor with unquestioning obedience and lack of responsibility for consequences of actions, must be abandoned.

Service agencies must therefore:

1. Change the justice structure of their institutions so that nurses can share in decision making and responsibility and be free to make moral choices, thereby creating a moral atmosphere where there is commitment and consistency between moral reasoning and action.
2. Institute courses in ethical reasoning through in-service education for all nurses.
3. Expose new graduates to role models and working environments that reflect and

support principled, autonomous moral reasoning, so that cognitive awareness of moral principles can be developed into a commitment to their ethical application in the practice of nursing.

## REFERENCES

1. Anderson, R. M.  Activity preferences and leadership behavior of head nurses: Part I. *Nurs. Res.* 13:239, 1964.
2. Baumgart, A. J.  Are nurses ready for teamwork? *J. N. Y. State Nurses Assn.* 3:4, 1972.
3. Bensman, J., and Rosenberg, B.  The Meaning of Work in a Bureaucratic Society. In A. Etzioni (ed.), *Readings on Modern Organizations*. Englewood Cliffs, N.J.: Prentice-Hall, 1969.
4. Bueker, K.  A Study of the Social Organization of a Large Mental Hospital. Doctoral dissertation, The American University, 1969.
5. Chazan, B. I.  The Moral Situation. In J. F. Soltis and B. I. Chazan (eds.), *Moral Education*. New York: Teachers College Press, 1973. Pp. 39–49.
6. Christman, L. B.  Nursing leadership – style and substance. *Am. J. Nurs.* 67:2091, 1967.
7. Cleland, V.  Sex discrimination: Nursing's most pervasive problem. *Am. J. Nurs.* 71:1542, 1971.
8. Coser, R. L.  Authority and Decision-Making in a Hospital: A Comparative Analysis. In A. Etzioni (ed.), *Readings on Modern Organization*. Englewood Cliffs, N.J.: Prentice-Hall, 1969.
9. Etzioni, A.  *Modern Organizations*. Englewood Cliffs, N.J.: Prentice-Hall, 1964. P. 87.
10. Fagin, C. M.  Professional nursing – the problems of women in microcosm. *J. N. Y. State Nurses Assn.* 2:7, 1971.
11. Fontana, A., and Noel, B.  Moral reasoning in the university. *J. Pers. Soc. Psychol.* 27:419, 1973.
12. Frank, E. D.  Images of Nursing Among College Freshmen Women in New Orleans. Doctoral dissertation, Columbia University, 1969.
13. Gabrielson, R. C.  A message from the American Nurses' Association. *J. N.Y. State Nurses Assn.* 4:26, 1973.
14. Georgopoulos, B. S., and Mann, F. C.  *The Community General Hospital*. New York: Macmillan, 1962.
15. Goss, M. E. W.  Influence and authority among physicians in an out-patient clinic. *American Sociological Review* 26:39, 1961.
16. Group, T. M., and Roberts, J. I.  Exorcising the ghosts of the Crimea. *Nurs. Outlook* 22:368, 1974.
17. Haan, N., Smith, B., and Block, J.  Moral reasoning of young adults: Political-social behavior, family background, and personality correlates. *J. Pers. Soc. Psychol.* 10:183, 1968.
18. Hare, R. M.  *The Language of Morals*. New York: Oxford University Press, 1964.
19. Harrington, H. A., and Theis, E. C.  Institutional factors perceived by baccalaureate graduates as influencing their performances as staff nurses. *Nurs. Res.* 17:228, 1968.
20. Hogan, R.  Moral Development and the Structure of Personality. In D. De Palma

and J. Foley (eds.), *Moral Development: Current Theory and Research.* New York: Wiley, 1975.

21. Jacox, A.   Professional socialization of nurses. *J. N.Y. State Nurses Assn.* 4:6, 1973.

22. Jarratt, V. R.   A Study of Conceptions of Autonomous Nursing Actions Appropriate for the Staff Nurse Role. Doctoral dissertation, University of Texas, 1967.

23. Kohlberg, L.   Moral Stages and Moralization: The Cognitive-Developmental Approach. In T. Lackona (ed.), *Moral Development and Behavior.* New York: Holt, Rinehart and Winston, 1976.

24. Kohlberg, L.   The cognitive-developmental approach to moral education. *Phi Delta Kappan* 56:670, 1975.

25. Kohlberg, L.   From Is to Ought: How to Commit the Naturalistic Fallacy and Get Away with It in the Study of Moral Development. In T. Mischel (ed.), *Cognitive Development and Epistemology.* New York: Academic, 1971.

26. Kohlberg, L.   Stage and Sequence: The Cognitive-Developmental Approach to Socialization. In D. A. Goslin (ed.), *Handbook of Socialization Theory and Research.* Chicago: Rand-McNally, 1969.

27. Kramer, M., McDonnell, C., and Reed, J.   Self-actualization and role adaptation of baccalaureate degree nurses. *Nurs. Res.* 21:111, 1972.

28. Lande, S.   Factors Related to Perception of Nurses and Nursing and Selection or Rejection of Nursing as a Career Among Seniors in Roman Catholic High Schools in New York City. Doctoral dissertation, Columbia University, 1964.

29. Lefton, M., Dinitz, S., and Pasamonick, B.   Decision-making in a mental hospital. *American Sociological Review* 24:822, 1959.

30. Lieberman, B. P.   The role of the nursing supervisor in implementing the new definition of nursing practice. *J. N.Y. State Nurses Assn.* 4:39, 1973.

31. Mauksch, I. G.   Let's listen to the students. *Nurs. Outlook* 20:103, 1972.

32. Mouzelis, N. P.   *Organisation and Bureaucracy.* Chicago: Aldine, 1967. P. 127.

33. Murphy, C. P.   Levels of Moral Reasoning in a Selected Group of Nursing Practitioners. Doctoral dissertation, Teachers College, Columbia University, 1976.

34. Olesen, V. L., and Whittaker, E. W.   *The Silent Dialogue.* San Francisco: Jossey-Bass, 1968.

35. Richter, L., and Richter, E.   Nurses in fiction. *Am. J. Nurs.* 74:1280, 1974.

36. Schorr, T. M.   Is that name necessary? *Am. J. Nurs.* 74:235, 1974.

37. Simms, L. A.   The Hospital Staff Nurse Position as Viewed by Baccalaureate Graduates in Nursing. Doctoral dissertation, Teachers College, Columbia University, 1963.

38. Simon, H. A.   *Administrative Behavior.* New York: Free Press, 1976. P. 220.

39. Stein, L.   The doctor-nurse game. *Am. J. Nurs.* 68:101, 1971.

40. Stein, R. F.   The student nurse: A study of needs, roles and conflict. Part II. *Nurs. Res.* 18:433, 1969.

41. Stone, J. C., and Green, J. L.   The impact of a professional baccalaureate degree program. *Nurs. Res.* 24:287, 1975.

42. Strauss, A.   The Structure and Ideology of American Nursing: An Interpretation. In F. Davis (ed.), *The Nursing Profession: Five Sociological Essays.* New York: Wiley, 1966.

43. Williamson, J. A.   The conflict-producing role of the professionally socialized nurse-faculty member. *Nurs. Forum* 11:357, 1972.

# CHAPTER 47

# The Human Rights of Patients, Nurses, and Other Health Professionals
*Bertram Bandman*

*If a physician* prescribes a drug that is contraindicated in the nurse's judgment and if she finds support for her position from other health professionals and in the New York State Nurse Practice Act, what should she do? In a conflict between a nurse and a physician, who if either of them has the right to decide? Or if a patient refuses a blood transfusion, such as is evident in cases of Jehovah's Witnesses, or if a patient refuses a procedure or medication, which both the physician and the nurse consider essential to life, who if anyone has a right to decide? Does a nurse researcher have a right to determine the limits of a subject's right to refuse? Do nurses and patients have rights? Human rights? What kinds of rights?

In this paper, I shall try to examine what it means to say that patients and nurses and related health professionals have human rights and legal and institutional rights. First, I shall indicate some things that have been said about the meaning of rights and especially about human rights. Second, I shall try to show why patients and nurses have human rights and that in addition nurses and other health professionals have a further role that is best characterized in terms other than rights.

## WHAT ARE RIGHTS? WHAT ARE HUMAN RIGHTS?
### What Are Rights?
Rights have variously been defined as powers, needs [21], interests, claims, valid or justified claims [15], and entitlements [22, 32]. Each of these definitions has strengths and difficulties. Following H. J. McCloskey [22] and Elsie Bandman, in part, I shall refer to rights as *just entitlements* for making effective claims and demands.

### What Are Human Rights?
Following in part the work of some recent philosophers [11, 15, 17, 38], I think we may identify a human right as a right that is a moral right of fundamental or "paramount importance" [11], essential to "a decent and fulfilling human life" [30], one that is more important than any other right and is shared equally by all human beings [15, pp. 84–85; 25].

Human rights are important because they are "independent of legal or institutional norms" [25], and I would add, accordingly transcend all other rights. Human rights are so important that they annul, cancel, and override all other rights that conflict with them. Huckleberry Finn has a right to free his friend Jim, no matter what the legal norms of the South are at that time. According to one writer, "human rights serve as an independent standard of political criticism and justification" [25].

A brief version of this paper appeared in the *American Journal of Nursing* 78:84, 1978. Copyright January 1978, the American Journal of Nursing Company. Reproduced in part with permission from the *American Journal of Nursing,* Vol. 78, No. 1.

There are, I suggest, three conditions for a right of importance, namely, a human right.

### Freedom

H. L. A. Hart has well stated that if there are any rights at all, there is at least the equal right to be free [17]. The right to be free is the right to one's *domain,* whether it be one's body, one's life, one's property, or one's privacy. Jeanne Berthold calls this "the right to self-determination" [8, pp. 516–517]. It is the area of one's life over which, as Joel Feinberg aptly puts it, one is "the boss" [14].

John Holt brings out this feature of a right with an example, which applies to any human right: "When the law gives me the right to vote, it is not saying I must vote. It only says that if I choose to vote, it will act against anyone who tries to prevent me." In saying I have a right to vote, the law does not, however, say, "what I must or shall do" [19]. To have a right means you do not have to do what you have a right to do. And "if I have a right to do X," according to Joel Feinberg, "I cannot also have a duty to refrain from doing X" [15, p. 58].

To have a right is also to be free not to use it and also to be immune to a charge of wrongdoing for using or not using it. To exercise one's right may be unwise or foolish, but one is never wrong to exercise one's rights.

The right to be free for nurses, physicians, and patients alike includes the right not to be brainwashed, lied to, kept ignorant, deceived, tricked, involuntarily or unknowingly given drugs or medication, put to sleep, or otherwise unjustifiably coerced, where harm to one's self or others is not involved. The right to be free and autonomous is a necessary condition of a human right and is absolute in the sense that one cannot have these things done and still be free to live a "decent and fulfilling life" as a human being [30].

### Rights Imply Duties

A second condition is that rights imply achievable obligations on others [7, 11, 12, 33], obligations that are of the utmost importance [11] to everyone's equal right "to a decent and fulfilling human life" [25]. As Feinberg points out, "rights are necessarily the grounds of other people's duties [15, p. 58].

### Rights Presuppose Justice

Freedom and duties are not sufficient to characterize and orient a human right. A right of "paramount importance," like a human right, "is something of which no one may be deprived without a grave affront to justice" [11]. According to St. Augustine, rights flow from justice. There are no rights without justice [29].

Considerations of impartiality and equality are requirements of justice, which includes everyone's equal right to "a decent and fulfilling life" and to the means for achieving such a life. Justice as equality means that everyone is treated in a similar way in similar situations [25]. Justice as equality applied to health care practice means that everyone is justly entitled or has a right to a similar quality of health care.

Justice in connection with human rights means not only that the older political rights, sometimes called freedom or "option rights," are necessary, such as the right to vote, worship freely, and not to be tortured, but also the newer social and economic rights. These rights are known as the rights to well-being, sometimes called rights of assistance or recipience or welfare rights [3, 16]. I prefer to call them need fulfillment rights.* These rights include the right to social security, education, and health care. A just conception of rights provides both for option rights and for need fulfillment rights.

### Difficulties with Option and Need Fulfillment Rights

One difficulty with the older option rights has been that they were too narrow and exclusive and of benefit to only a few. Large estates and "the high fences" Robert Frost once wrote about no longer "make good neighbors."

A difficulty, however, with the newer rights of recipience or need fulfillment is the *thinning out of rights*. One finds claims for the rights of physicians, nurses, patients, mental patients, cardiac patients, in addition to the rights of policemen, the poor, the aged, Italians, blacks, women, garment workers, the comatose, the gay, prisoners, fetuses, unborn future generations, animals, and trees. The declaration and demand for more and more social, economic, educational, ecological, and health care rights of all kinds can only mean a declining possibility of imposing correlative obligations.

In order to have a right, people have to accept comparable duties implied by the rights they claim. The expansion of claims to rights can have the danger of trivializing and making less valuable each right, since all rights cannot possibly be fulfilled. The thinning out of rights of need fulfillment is also notable in the role of rights in nursing care. The American Hospital Association's Patient's Bill of Rights promises more than anyone can deliver. For example, the right to complete current diagnostic information along with most other rights cited makes promises that exceed the possibility of implementation.

A solution to both difficulties (of rights being too exclusive and too numerous) is to extend the older rights to include all human beings, recognize basic need fulfillment rights, divide rights into equally small shares as Jesus did with the loaves, but not promise rights that go beyond what human beings can achieve. On this view, human rights are limited to the most essential necessities of life that constitute an equal "floor" of support below which no one falls [25].

There are no rights of importance, no human rights, without freedom, however small the domain has to be cut to make room for everyone's equal right to it, including the freedom to eat and freedom from avoidable pain and disease. Nor are there any rights of value unless others pay the price of that freedom by providing for that freedom in the form of duties [3]. Without correlative duties imposed on and accepted by others, rights are empty. Third, without justice to orient and regulate the relation and distribution between freedom and duties on a rational basis, rights are blind.

## HUMAN RIGHTS OF PATIENTS AND CORRESPONDING DUTIES OF HEALTH PROFESSIONALS

### The Human Rights of Patients as Option Rights and Need Fulfillment Rights

A human right to life includes not only the option right to be left alone, but as Virginia Held eloquently puts it, "it also includes being able to acquire what one

---

*Joel Feinberg uses this term but in a different connection in Rights, Duties and Claims, *American Philosophical Quarterly* 3:139, 1966.

needs to live" [18]. To have any rights of importance is to have not only the equal right to be free, but also the right to "an equally decent and fulfilling human life" and to whatever means are necessary to achieve such a life, including appropriate health care.

The human right to live this sort of life makes the deprivation of one's health care rights "a grave affront to justice" [11]. The Tuskegee syphilis experiment [9], the involuntary Willowbrook Hepatitis program, the use of 600 United States Army servicemen for an LSD experiment without their knowledge [4], and Nurse Ratchet in *One Flew over the Cuckoo's Nest* telling Murphy to take his pills because they are good for him and refusing to tell him why — these are all grave affronts to justice [11]. They violate patients' human rights in health care [4].

### The Human Rights of Health Professionals and Patients

Health professionals, including nurses and physicians, have the equal human right to a "decent and fulfilling human life" along with patients.

Claire Fagin cites the human rights of nurses, which include the right to be heard, the right to participate freely, the right to satisfaction, and the right to question or doubt [13]. These rights also include nurses' rights to their beliefs, to advocate patients' health care rights, to refuse to give treatments they believe are contraindicated, the right to challenge, and the right to fulfill oneself professionally and as persons.

### Patients' Interests and Rights Paramount

Elizabeth Carnegie ties the nurse's role to the protection of patients' rights. She says, "Nurses have always known that the patient is the chief consideration" [9]. And Carnegie quotes Minnie Goodnow, who said in the early part of this century that "the patient is the main thing — the reason for it all — the unit — the one chief consideration, the one [to] whose welfare all else must be subordinated" [9]. Carnegie concludes, "Patients' rights have always come first with nurses" [9].

If patients' rights are uppermost, nurses may advocate patients' rights quite effectively, it seems, without having to have special rights [6], rights that would imply others' obligations, presumably patients' obligations and, on occasion at least, be in direct collision with patients' rights. If nurses and physicians serve the human rights of patients, there is no reason to accord nurses and physicians special rights, at least when these rights conflict with patients' rights. One of the functions of rights oriented by justice is to protect the powerless, not the powerful, and in the relationship of the patient to the health professional, the patient is relatively powerless.

Attributing rights to nurses and physicians also adds to the population explosion of rights and a resulting *thinning out* of everyone's rights. Such a thinning out generates needless conflicts between patients' and nurses' rights, with a resulting loss to patients' rights, where it even becomes possible for the investigator to "usurp the individual's right to freedom of choice" [8, p. 518]. If the patients' interests are paramount, the role of health professionals would seem to be to serve those interests and not be in a position to usurp them.

*Nurses and Other Health Professionals Have Duties But No Special Rights Against Patients*

To have a right is to have the right to be free to do or not do what the right provides. To have a right is also to have a right not to. The odd thing about nurses' rights is that if nurses are free, nurses can choose not to exercise their rights. For example, if a health professional has a right to stop a patient from bleeding to death could not the health professional also say, "I have a right to stop this bleeding, but as with my right to vote or to play golf, which I'm also free not to do, I'll exercise my right and not stop the patient from bleeding to death"?

The statement that "if I have a right to do X, then I cannot also have a duty" either to do or "to refrain from doing X" [15] also spotlights a difficulty for those who champion nurses' and physicians' (special) rights. This difficulty is that, first, the so-called right is *to* prescribe and apply appropriate treatment, and second, it is not to permit medications and drugs that are contraindicated in place of X. In the first case this right is a duty *to do,* to prescribe, and to treat, and in the second case it is a duty to refrain, cancel, or refuse. If X is believed to be contraindicated, a health professional has a duty, not a right, to refrain from administering or prescribing it.

The absence of a duty or the absence of a duty to refrain from doing something is not applicable to the so-called special rights of health professionals. If they fail to do or omit to do what they say they have a special right to do or not to do, but which it is their duty to do or refrain from doing, then they are liable, culpable, and open to the charge of wrongdoing and malpractice. Hence their duty is hardly a "right" that one is free to exercise with immunity to the charge of wrongdoing.*

If patients have a right and thus a choice, others then have a duty and no choice. If health professionals refuse to carry out their duties, they become liable to malpractice. For example, if a nurse is in a conflict with a physician, believing that a medication is contraindicated, is it the nurse's right or her duty to refuse to give the medicine to the patient? If, as I think, the nurse has a duty to refuse, she cannot have a right to decide whether to refuse; but neither does the physician have a right to decide; he, too, clearly has a duty not to do what is contraindicated.

I turn next to the correlation between rights and duties. The right to refuse to be a research subject implies the duties others have to forebear from research. Yet one finds that "the nurse has . . . the right to conduct research" [1, p. 105; 8; 26]. Even Jeanne Berthold weakens her protection of patients' and subjects' rights by asking, "When is it the responsibility of the investigator to usurp the individual's right to freedom of choice?" [8, p. 521]. Recalling Cranston's remark, "a human right is something the deprivation of which is a grave affront to justice" [11], we can interpret an investigator's responsibility to mean that it can never be to usurp an indivi-

---

*For those who like logical symbolism, by a familiar *modus tollens* argument, one can derive the following conclusion:

1. If I have a right to prescribe $X_1$ (or refuse $X_2$), then I cannot also have a duty to prescribe $X_1$ (or refuse $X_2$, whichever the case may be).
2. *But I have a duty to prescribe $X_1$ (or refuse $X_2$).*
3. Therefore, I cannot have a right to prescribe $X_1$ (or refuse $X_2$, whichever the case may be).

dual's right to freedom of choice, for this usurpation would be a gross violation of a human right.

If nurses have a right to "usurp the individual's right" for purposes of research, this right means that others, presumably subjects, have a corresponding duty to submit to such research. This assumption would seem to be a patently absurd consequence of patients' or subjects' rights to refuse to participate in research projects.

Ernest Nagel similarly suggested a difficulty in ascribing to presently living geneticists the right to plan future generations [24] ; for presumably the right to plan future generations would imply correlative obligations on future generations not yet born to provide for and accept the geneticists' right to plan.

If it is a nurse's right to administer blood or give a medication or participate in sterilizing a patient or do research, it is then a patient's corresponding duty not to interfere or refuse and to submit and be a willing recipient to the experiment or transfusion or medication or sterilization. If nurses and physicians have rights, presumably patients have duties, an absurd consequence in attributing special rights to nurses and physicians that conflict with patients' rights.

If, however, a patient has a right to refuse X (medication, sterilization, or blood transfusion), but the health professional also has a right to administer X, in a conflict of rights, one right or the other gives way. Either the patient receives X or not. But we cannot have two rights simultaneously, at least not two conflicting rights that generate contradictory obligations, a patient's right to refuse and the right of health professionals to administer X. But if rights imply duties, including patients' rights to refuse X, then health professionals have a duty to protect the patients' rights of refusal. Otherwise, there are no real rights to refuse.

However, if patients have rights, others have duties toward right-holders; and with respect to the rights of patients, health professionals cannot have rights that coincide with patients' rights without colliding with patients' rights.

If rights collide and some rights are to be satisfied, some rights will need to be set aside or vacated in favor of others. In a conflict, patients' human rights, being the fundamentally important rights, override, cancel, and annul the special rights of nurses and physicians. Accordingly, it appears to be an anomaly to ascribe specialized role-derived rights to health professionals that conflict or may conflict with patients' fundamentally important human rights, rights that are important enough to override or cancel all other rights.

## NURSES AND PHYSICIANS HAVE DUTIES AND PRIVILEGES
### Rights, Duties, and Privileges

What is the special role of nurses in relation to patients? I believe nurses and physicians have no special rights, only privileges [2] . Privileges are "carved out" of rights, as Feinberg points out [15] . To have a privilege, to cite a second metaphor Feinberg adopts from more imaginative writers, is to have a key to a lock that belongs to a right-holder. If the right-holder is unable to unlock the gate, he or she may extend the privilege of using the key to another person — an advocate, one might say.

If patients' interests and rights are of "paramount importance" to nursing, as Carnegie [19], Porter [27], Berthold [8], and others contend [10, 20, 23, 28], and if nurses regard themselves as protecting patients' rights as advocates, then nurses have the duty and also the privilege of acting on behalf of patients, but not the right to do so. However, having the duty does not give nurses the right or freedom to decide whether or not to act on patients' behalf. Nurses have no choice in the matter, and hence they could not possibly have a right in the matter.

For a patient to have general and special human rights in his or her role as patient is to have rights that are fundamentally important, and they are more important than any other legal or institutional, role-derived rights that may be regarded as privileges instead. At the top of the scale of importance, we refer to patients' human rights. In marked contrast and at a subordinate level of the scale, we may refer to an instrumental privilege, a means of doing something for someone.

In this juxtaposition of rights and privileges, we move away from the rights of nurses and physicians and relocate their roles as clearly subordinate and instrumental to that of patients — akin to Hume's dethroning of reason and relocating it as "the slave of the passions." The role of health professionals, being responsible and having duties to patients, is to serve the health needs, interests, and human rights of patients.

What is a privilege? According to Feinberg's citation in *Black's Law Dictionary,* a privilege is "an exceptional or extraordinary power or exemption," and one that is revocable [15, p. 56].

Nurses and physicians are privileged. They are "permitted to carry keys to the lock on the gate" of a patient's domain of freedom, when this domain includes a patient's body. To shift to a woodcutting metaphor: Privileges are "carved out" of rights at the right-holder's request and at the right-holder's pleasure, assuming the right-holder is informed and the consent is voluntary. The physician and the nurse do not "carve out" a privilege without the consent and pleasure of the right-holder. The hand may be the nurse's or physician's, but the decision is the patient's.

A physician's privilege of operating on a patient depends on a patient signing a consent form. The physician operates only with the *consent* of the patient (except in an emergency). This privilege is recognized in law as a patient's right to decide what happens in and to his or her body [2; 5, pp. 95–96]. However, a privilege, unlike a right, "lacks guarantee . . . and can be withheld or withdrawn at one's pleasure" [15, p. 57]. A license gives a person a privilege, not a right. A physician's privilege, unlike a right, implies no correlative duty on the part of the patient, and the patient may withhold or withdraw the privilege at any time, just as a subject or patient may terminate an experiment at any time.

If a right invests a right-holder with the authority to hold others responsible [2], in a patient-physician or patient-nurse relation, the physicians and nurses are responsible to their patients. If, moreover, those who are held responsible are accorded privileges, which by definition are revocable, physicians and nurses have the privilege of serving their patients. Consequently, in a rights-privilege relationship, patients have the right to treatment, and physicians and nurses have the privilege rather than the right to treat patients.

*The Subordination of Health Professionals' Privileges to Patients' Human Rights*
However, within this restriction, one can easily "carve out" the privilege in place of
the special right which Fagin, Peplau, Elsie Bandman, Catherine Murphy, and
others understandably wish to confer on nurses, the privilege,* for example, of not
participating in a medical practice, such as electroshock therapy, that, in a nurse's
professional judgment, is contraindicated. In addition, the sort of positive right
Fagin wishes to confer on nurses can also be expressed as a privilege, such as the
privilege of "diagnosing and treating human responses" in accordance with the
New York State Nurse Practice Act, for example.

Rights single out the most important values, needs, and desires of a society. Rights
hold others accountable for carrying out the duties thus implied. On this basis,
rights are clearly attributable to patients, and duties, responsibilities, and privileges
thus created from such rights belong to health professionals.

Since human rights, being fundamentally important, are independent of and
transcend other legal and institutional rights, they are of greater importance than
other rights, including the so-called special role-derived rights of health professionals.
One way to mark off this distinction is to refer to patients as having rights and
health professionals as having corresponding duties and privileges.

*A Limited Case for Special Rights Against Other Health Professionals*
There may, nevertheless, be a small but vital area in which nurses have special rights.
The role of involved advocacy may sometimes make some rights voluntarily transferable. In the law, a will manifests the transferability of rights. Transferability may
also occur in common law, for example, when a foster parent becomes deeply and
beneficially involved with a child. The foster parent may gain the esteem and confidence of that child to the point where the child develops a deep sense of trust in
the foster parent. This relationship may be due to the fact that the foster parent
has made evident that he or she is disposed to protect or provide for the child's
*vital rational interests.* In such relationships, a foster parent may, in the eyes of the
child, act on the child's behalf. What may have begun as a privilege may end up as
a special right, somewhat as a relationship between friends or lovers may eventuate
in marriage.

The relationship in the nurse-patient relation is analogous. Deep involvement
with each other may render rights transferable. In this sense, nurses have special
rights to help and sustain their patients. It is in this sense, too, that nurses may have
special rights *against* other health professionals. Nurses have such rights whenever
they are demonstrably in the *right* and others are in the *wrong.* (The etymological
connection between *having a right* and being *right* or between *rights* and what is
morally or legally right is not − nor was it ever − unintended [9, p. 442] .) Nurses
gain or earn special rights through training, competence, and experience along with
their natural sympathy and understanding, and often their judgments and beliefs
coincide with rationally justified beliefs about what serves the vital interests of
their patients.

*Insofar as a nurse's role does not conflict with a patient's, but rather with the role of a physi-
cian or another nurse, a nurse's special human rights of challenging and doubting may also come
into play.

Although, according to this account, nurses do not have special rights against patients, they may very well have special rights against other health professionals. The converse, too, is true, that is, other health professionals may have special role-derived rights against nurses. But no health professionals have special role-derived rights against patients' *human rights,* and the determination as to who in a given situation has special rights depends on the extent to which a given health professional is serving the *vital rational interests* of patients. That is why, to return to the case cited at the outset, a nurse does have a right against a physician whenever a physician's order is contraindicated. But it could be the other way; that is, the physician could be right and the nurse wrong.

For a nurse correctly to claim that a physician's order is contraindicated, however, is the *exception* rather than the *rule.* As Dorothy Nayer has pointed out to me, quite correctly I believe, the paradigm of physician-nurse-patient relationships is that the physician is ordinarily recognized to "give the orders" for the patient. The physician is acknowledged legally, historically, and traditionally, and on sound grounds, to *know* more than other health professionals; and he or she is consequently considered the most essential link in the restoration of health.

The physician is primarily and ordinarily responsible for the patient, and the physician ordinarily gives the orders, *unless* they are contraindicated, either because the physician does not really know the patient's condition or medication, or is negligent or overworked, or demonstrably has an interest other than the patient uppermost in mind. In such cases — but it would seem to apply only in such cases — the nurse not only has the legal right to refuse to do what the physician orders, but also a moral basis for exercising a role as an advocate for the patient.*

Thus the nurse ordinarily respects the contractual relation between the physician and the patient *unless* there is a clear-cut case of the physician not putting the patient's interest first. A physician gives the orders *unless* these orders are contraindicated.

If the physician makes an error, who best safeguards the patient's health interests? The family is not always present. The patient may be too sick. The persons who are naturally most suitable to safeguard the patients' interests are those who give continuous care, namely, nurses. Thus a defeasible clause provides protection of the patients' rights and interests if physicians fail to provide for their patients' rights and interests in health care. The special rights nurses would have against physicians would be provided by the defeasible provisions of contracts that physicians have with their patients.

The principle governing the attribution of special rights is the consistency of role-derived rights with the general human rights from which such special rights are logically derivable. And the test of consistency between special role-derived rights and general human rights is, it seems, quite effectively put by Carnegie, Minnie Goodnow, and others, namely, that the patient comes first [10].

---

*The notion of "unless" clauses in contractual relations is known in law and morality as "defeasibility"; it consists in the elucidation of those conditions that would "defeat" a contract. It was first given philosophical prominence by H. L. A. Hart in "The Ascription of Responsibility and Rights," in A. Flew (ed.), *Logic and Language* (Oxford: Blackwell, 1952).

Since, unlike Claire Fagin, I see no real difference between a right *against* and a right *to,* I also see a case for a nurse's special role-derived right *to* do things on behalf of the rational vital interests of her patients, providing, however, that such rights never collide with the patients' human interests and rights and have been knowingly and voluntarily transferred by the patients to the nurse.

CONCLUSION

Patients and health professionals share equally in having fundamentally important human rights. But to distinguish and subordinate the specialized role of health professionals from patients, patients have special health care rights and health professionals have the privilege and the duty of correcting each other's mistakes whenever indicated. Finally, if there are instances in which rights are transferable, such as when deep attachments develop, nurses may gain special rights to provide for and protect their patients' rational vital interests and even exercise these rights against other health professionals.

The reason I think it is nevertheless rare that nurses have special rights is that to have such rights is to know and to be in a position systematically to care for the rational vital interests of patients. But who is ever really in a position to know or care that much for someone else?

In any event, the test of the privileges and special role-derived rights of health professionals is the consistency of privileges and rights with the fundamentally and overridingly important human rights of patients. For without human rights, which imply freedom, duties, and justice, there are no rights at all.

REFERENCES

1. American Nurses Association. The Nurse in Research: ANA Guidelines on Ethical Values. New York: American Nurses Association, 1968. (Also in *Nurs. Res.* 17:104, 1968.)
2. Bandman, B. Nurses have no rights. *Am. J. Nurs.* 78:1, 1978.
3. Bandman, B. Some legal, moral and intellectual rights of children. *Educational Theory* 27:169, 1977.
4. Bandman, E. L., and Bandman, B. Rights are not automatic. *Am. J. Nurs.* 77:5, 1977.
5. Bandman, B., and Bandman, E. L. Rights, Justice and Euthanasia. In M. Kohl (ed.), *Beneficent Euthanasia.* Buffalo, N.Y.: Prometheus, 1975. Pp. 96–98.
6. Bandman, E. L. The Rights of Nurses and Patients: A Case for Advocacy. This volume, Chap. 48.
7. Benn, S., and Peters, R. S. *The Principles of Political Thought.* New York: Collier Books, 1959. Pp. 102–103.
8. Berthold, J. S. Advancement of science and technology while maintaining human rights and values. *Nurs. Res.* 18:514, 1969.
9. Brandt, R. B. *Ethical Theory.* Englewood Cliffs, N.J.: Prentice-Hall, 1959. Pp. 436–442.
10. Carnegie, M. E. The patient's bill of rights and the nurse. *Nurs. Clin. North Am.* 9:557, 1974.

11. Cranston, M.  Human Rights, Real and Supposed. In D. Raphael (ed.), *Political Theory and the Rights of Man.* Bloomington: Indiana University Press, 1967. Pp. 49–52.
12. Cranston, M.  *What Are Human Rights?* New York: Basic Books, 1962. P. 41.
13. Fagin, C.  Nurse's rights. *Am. J. Nurs.* 75:82, 1975.
14. Feinberg, J.  Voluntary Euthanasia and the "Inalienable Right to Life." Paper presented at Bioethics and Human Rights Conference at Long Island University, Brooklyn, N.Y., April 9, 1976, and forthcoming in *Philosophy and Public Affairs,* winter 1978.
15. Feinberg, J.  *Social Philosophy.* Englewood Cliffs, N.J.: Prentice-Hall, 1973.
16. Golding, M.  Towards a theory of human rights. *Monist* 52:4, 1968.
17. Hart, H. L. A.  Are There Any Natural Rights? In A. I. Melden (ed.), *Human Rights.* Belmont, Calif.: Wadsworth, 1970.
18. Held, V.  Abortion and Rights to Life. This volume, Chap. 11.
19. Holt, J.  Why Not a Bill of Rights for Children? In B. and G. Gross (eds.), *The Children's Rights' Movement.* Garden City, N.Y.: Anchor Books, 1977. P. 321.
20. Kelly, L.  The patient's right to know. *Nurs. Outlook* 24:26, 1976.
21. McBride, A.  Can family life survive? *Am. J. Nurs.* 75:1651, 1975.
22. McCloskey, H. J.  Rights. *Phil. Quart.* 15:118, 1965.
23. Murphy, C. P.  The Moral Situation in Nursing. This volume, Chap. 46.
24. Nagel, E.  Comments on the Presentations of Drs. Ehrman and Lappé. This volume, Chap. 9.
25. Nickel, J.  Are Social and Economic Rights Real Human Rights? Paper presented at Society for Philosophy and Public Affairs, City University of New York, Graduate Center, New York City, March 1977.
26. Notter, L.  Protecting the rights of research subjects. *Nurs. Res.* 18:483, 1969.
27. Porter, K.  Patient's rights and nurse's responsibilities. *Hospitals* 17:102, 1973.
28. Quinn, N., and Somers, A.  The patient's bill of rights: A significant aspect of the consumer revolution. *Nurs. Outlook* 22:240, 1974.
29. St. Augustine.  *The City of God* (translated by G. Walsh et al.). Garden City, N.Y.: Image Books, 1958. Pp. 468–472.
30. Scheffler, S.  Natural rights, equality and the minimal state. *Can. J. Phil.* 6:64, 1976.
31. Schlotfeldt, R.  Can we bring order out of the chaos of nursing education? *Am. J. Nurs.* 76:105, 1976.
32. Wasserstrom, R.  Rights, Human Rights and Racial Discrimination. In A. I. Melden (ed.), *Human Rights.* Belmont, Calif.: Wadsworth, 1970. Pp. 96–110.
33. Williams, P.  Rights and the Alleged Right of Innocents to Be Killed. This volume, Chap. 18.

# CHAPTER 48

# The Rights of Nurses and Patients: A Case for Advocacy
*Elsie L. Bandman*

*In this paper,* I propose that nurses have special rights because they have special responsibilities in matters affecting the health of people in their care. I suggest that nursing needs to demonstrate its claim to special rights through advocacy of patients' rights on matters of health and health care delivery.

What are the special rights to which nurses lay claim? Most could be subsumed under two broad rubrics: the right to independent judgment or autonomy and the right to participate in making decisions affecting client health and, ultimately, the health of the members of society.

The current New York State Nurse Practice Act appears to provide considerable scope for independence of judgment. It permits nurse "diagnosis" as "discrimination between physical and psychosocial signs and symptoms essential to effective execution and management of the nurse regimen . . . distinct from a medical diagnosis" [11]. It permits "treating" as the performance of measures central to the implementation of a nursing regimen as well as the carrying out of "any prescribed medical regimen" [11]. It defines the practice of professional nursing as "diagnosing and treating human responses to actual or potential health problems, through such services as case finding, health teaching, health counseling and provision of care supportive to or restorative of life and well-being . . ."[11].

Care that is "supportive or restorative to life and well-being" has been given traditionally by nurses to people in nearly every health agency and hospital in this country. Nurses maintain continuity of the caring and curing functions of hospitals 24 hours of every day. The professional ministrations of the nurse are indispensable both to the implementation of the medical regimen and to ameliorating the psychosocial needs of the client faced with discomfort, possibly mutilation, fear, and uncertainty. A relationship of the nurse and patient is a logical consequence of the necessary interaction for implementation of nursing and the medical regimens. It is in practice difficult to separate the two, since the degree of knowledge and sensitivity to human responses with which the nurse carries out prescribed medical regimens may significantly affect the patient's behavior toward his or her illness. Nursing is thus intimately related to the cure function of medicine, and conversely the need for nursing care is recognized as a compelling reason for hospitalization.

Effective implementation of the practice of professional nursing as defined requires a relationship between nurse and patient characterized by concern, trust, scientific knowledge, competence in practice, and skill in human relations. The provision of supportive and restorative care to the patient implies the patient's right to expect that his or her best interests and self-interests will be the primary obligation of the nurse.

Out of this relationship comes the nurse's special rights and duties to the patient. These are, for example, the right to persuade the patient to stay in the hospital instead of signing himself or herself out, to urge him or her to continue with uncom-

fortable but beneficial treatment, such as diet or bathroom restrictions, or to represent the unaffiliated patient's desire to reject further surgery.

William Nelson, a professor of philosophy, says:

. . . Most rights are special rights. Special rights are conditional in certain ways and
are therefore limited in scope. These rights and their correlative obligations arise
out of specific relationships and their content reflects those relationships in the
sense that the latter can plausibly be regarded as justifying the former. An account
of the conditions giving rise to a kind of special right determines the scope of such
rights in a nonarbitrary way [10].

Feinberg defines the doctrine of the moral correlation of rights and duties as the
theory "that acceptance of duties is the price any person must pay in order to have
rights. . . . There can be no rights without duties and that a prior condition for the
acquisition of rights is the ability and willingness to shoulder duties and responsibilities" [7, p. 61]. The general opinion in nursing is that nurses carry an excess of
duties and responsibilities without the rights arising out of their special relationships
of obligations to patients.

If, as the American Nurse's Association's *Code for Nurses* states, "the nurse's
primary commitment is to the clients' care and safety" [2, p. 8], each nurse clearly
has a special right and the correlative obligation arising out of the special nurse-
patient relationship.

What is not so clear is the relationship of rights and obligations between a nurse
and the hospital. The hospital promises and the patient expects safe, decent, and
respectful care. Both the hospital and the patient expect, in fact, that the nurse will
be the promise-keeper. If the nurse is unable to do so because of the limitations of
the situation, such as when intensive care units shut down over weekends in municipal hospitals because of retrenchment policies and patients in these units who
still need highly skilled care are transferred to understaffed units, does a nurse inform a family that staffing is inadequate and that there is need for additional nursing? What are the nurses' special rights and corresponding obligations when nursing
care falls below the level of safety due to inadequate staffing?

Truth telling is conventionally viewed as desirable, especially when truth is deliberately sought by the questioner. Most of us have been reared on such injunctions
as "Always tell the truth" or a variation of it. Despite the widely held belief that
truth telling is "good," it actually is a major ethical issue and a dilemma for hospitals, their employees, and all those engaged in the nurse-patient relationship.

According to the second statement of the "Patient's Bill of Rights," the patient
has a right to obtain from his physician complete current information concerning
his or her diagnoses, treatment, and prognosis in terms he or she can reasonably
understand [1]. Suppose the client does not understand, or misunderstands, because
of fear, anxiety, age, or simply because of unfamiliarity with medical terms. The
*Code for Nurses* provides that the client be fully involved in planning for his or her
care [2, p. 4]. In developing care plans then, should the nurse parrot what the physician said, add other information, make her own response, or respond in terms of
the hospital's legal policies or in terms of the patient's emotional needs? Suppose

the patient is not given full information concerning his or her diagnosis, treatment, and prognosis for seemingly no really good reason. Does the patient have a right to seek that information from his or her nurse?

What then are her rights? Are her rights independent of the physician? Or do they depend upon the nurse's special rights and obligations arising out of the *Code for Nurses,* which states that "the primary commitment is to the client's care and safety" [2, p. 8]? The nurse, I maintain, must face the moral issue of truth telling just as the physician and hospital must. The *Code for Nurses* further states that "the nurse assumes responsibility and accountability for individual nursing judgments and actions" [2, p. 9].

I would like to pose a few of the moral issues regarding truth telling that lie between client and nurse. What words shall be used to tell what truth to which patient at which time? If a patient asks, "Do I have cancer?" following his or her physician's departure, does the nurse fulfill her truth-telling function by saying, "Yes you do," or does she say, "Why don't you ask your physician?" Even these simple words have different meanings for different people. The phrase "ask your doctor" instead of a direct "no" may for some people confirm the suspected diagnosis, since the nurse's words are interpreted as avoidance. The same answer may be interpreted by other clients as an indication of the nurse's complete trust in the physician, the nurse's lack of knowledge, or the nurse's complete indifference.

On the other hand, if the nurse says to the patient, "Yes, you do have cancer," the effect can be equally as varied. Some patients may feel enormously grateful that their burden of ambiguity and outright deceit is lifted and begin to deal with the reality of their situation. Other clients may deny the cancer, and others may be devastated. Whatever the response, is the nurse prepared to commit herself to that client in that special relationship of rights and duties that such a disclosure implies? Is the nurse sufficiently accountable and knowledgable to give an accurate prognosis in terms of the illness, its treatment, and its risks? Is the information really hers to give?

## MODELS OF RELATIONSHIP BETWEEN PHYSICIAN AND PATIENT

One view holds that any and all discussion of medical diagnosis, prognosis, and treatment is the exclusive right of the physician supported by legal definitions of the practice of medicine. Szasz and Hollander [13], however, view the physician-patient relationship as three different modes of interaction influenced by particular technical processes and the social setting. The first model involves a passive patient and an active physician. The analogy is drawn between this model and the infant-parent relationship in which the person acted upon is unable or not expected to contribute. Even though it is practiced, it is more suitable for a state of emergency, anesthesia, coma, or for persons with conceptual deficits.

The second model is widely used in medicine. It involves the patient in a cooperative, but essentially powerless, position seeking advice from the powerful physician, who guides the decisions of these individuals toward what he considers to be the good ends. Here the relationship is seen more like the parent guiding his adolescent child.

The third model postulated is that of mutual participation. It includes the patient as an equal who uses his or her own experience to contribute to a mutually agreed upon treatment program. It is characterized by democratic processes of sharing, identity, respect, and interdependence.

An addition to Szasz and Hollander's three models is the model suggested by Gray, also a physician. He points to the use of various assistants that enable the physician, "a product which is, for a large share of its traditional function, over-trained . . . [to be] primary in responsibility but secondary in contact" [8, p. 28]. He sees this role as a major source of dissociation in the patient-physician relationship, which he considers to be the "most essential feature of medical practice . . . and in danger of being destroyed" [8, p. 27].

Each of the proposed models appears to share a common feature — the exclusivity of the physician's authority for the caring and curing functions of health care. The reality is that this concentration of authority and responsibility inherent in the physician role for caring and curing is supported by laws governing medical practice.

## THE NURSE-PATIENT RELATIONSHIP

Where do these models place the nurse giving care that is "supportive or restorative to life and well-being" in relation to patients bewildered by unexplained treatments, confused by technical terms, ignorant of alternatives, and seeking information regarding diagnosis and proposed therapies? In response to these not infrequent patient situations must the nurse remain silent because of the restraints imposed by the nature of the contractual arrangement between physician and patient?

Various responses can be made that suggest the need for the role of nurse advocate for patients' rights. The concept of defeasibility as developed by H. L. A. Hart is one solution [9, pp. 148–150]. Its use in this connection was suggested to me by B. Bandman [4]. Hart states that there are positive conditions required for the existence of a valid contract: the participation of two parties or more with an offer by one and acceptance of the offer by the other, and including a written memorandum in some cases. These conditions are necessary but not always sufficient, because unless certain conditions can be met, the contract may be defeated. Hart lists a number of conditions that lend themselves to the idea of a contract as a defeasible concept. They include: fraudulent and innocent misrepresentation or nondisclosure of material facts; the use of duress or undue influence, the presence of lunacy or intoxication; contracts made for immoral purposes, or which restrain trade or prevent the course of justice; contracts impossible to perform or prevented through an unexpected change of circumstances; a claim barred by the passage of time. Hart is of the opinion that a "definition of contract requires as necessary conditions that the minds of the parties should be fully informed' and their wills 'free' "[9, p. 150]. He cites Pollock as stating that in order to secure a valid contract and make the consent binding on the giver, "the consent must be true, full and free" [9, p. 150].

The concept of defeasibility gives considerable protection to patients' rights in the event such protection is not provided by physicians. A hospital may issue a statement of patients' rights that the patient or the family may invoke. A patient without family or resourcefulness could well be represented by the nurse as the advocate of the patient's rights [6]. In the last analysis, each health professional is morally and legally responsible for his or her own acts; therefore, Hart's *unless* concept applies to other health professionals as well. For example, tell the truth unless it causes serious danger to the patient. Keep a promise unless fulfilling it will injure the person; follow the medical regimen unless it is harmful to the patient.

The application of Hart's defeasibility concept offers considerable scope for nurse advocacy. It respects the legal framework of medical practice and the physician-patient relationship while honoring the purpose, spirit, and substance of patients' rights. It recognizes the legitimacy of the nurse-patient relationship described in the *Code for Nurses* as "the nurse's primary commitment is to the patients' welfare and safety" [2, p. 4].

Annas and Healey [3] see the focus on patients' rights, which they believe was caused by movement within the legal framework, toward redefinition of relationships previously protected. The process has been to correct imbalances through emphasis on a fair distribution of rights and duties, powers, and obligations. A new self-consciousness concerning basic human rights is abroad and is being expressed in manifesto rights on behalf of the powerless as contained in the American Hospital Association's statement on a Patient's Bill of Rights [1].

Annas and Healey probe the complexities of patients' rights advocacy primarily from the perspective of the patient-physician relationship. The source of concern is, however, primarily recognition, representation, and protection of the patient's personal interests. Decisions made about the patient obviously affect his or her life in significant ways. Therefore, his or her rights and interests as defined by the American Hospital Association must be ensured. Annas and Healey assert that the "traditional doctor-patient relationship" has given way to technology, group practice, research and training commitments, and other "ambiguous identifications of the decision-maker . . . [and] of the person or entity that commands the decision-maker's loyalty" [3].

They view the advocate as representing the patient with all power residing in the patient; the advocate's purpose is to assist the patient to exercise his or her rights. They state four conditions as basic to effective advocacy: (1) access to records and to consultation on behalf of the patient at the request of the patient; (2) membership in hospital committees significant to the quality of patient care; (3) ability to deal with patient complaints through established hospital mechanisms; and (4) access to all patient support services as requested.

The professional nurse, in many instances, is assuming much of this responsibility. Much more could be accomplished by way of improved nursing care. Berthold stated the case in 1969 as one of reluctance to assume the full professional role:

Nurses . . . have accepted, overtly at least, both physicians' and hospital administrators' value systems. Nurses have played their own covert games to circumvent one or the other when they were in conflict or when, in their opinion, a third type of behavior was more appropriate for a specific patient. It seems to me, however, that nursings' social responsibility is more far-reaching than it has been willing to accept, at least to date . . . . Specific treatments and procedures have changed markedly, but the way we structure our system of health care has remained basically unchanged and many administrators do not appear willing even to acknowledge the question as a serious social concern. Many physicians and nurses are neither altering their own basic role concepts, nor involving themselves in relevant dialogue and decision-making either with others to a sufficient degree to result in expeditious and orderly change within the health profession [5].

There is a critical need for a professional level of practice in lieu of the current managerial, task-oriented mode as Sheahan points out:

One may look at nursing's functions and elaborate vocabulary and be deluded into thinking that professional care is being given . . . we say that nurses with professional preparation are able to diagnose or identify nursing problems, decide upon an appropriate course of action, develop and evaluate nursing care plans? Yes, but even when this process is actually operative in hospital nursing, it guarantees only that an orderly process will be followed in determining and . . . delivering nursing care. By focusing attention on the form or steps of the nursing process, we are deflected from attending to the substance of the process. If we look at the latter, we may discover that nursing decisions are generally decisions without substance, all of a minor or operational nature, such as how to explain or carry out a regimen of care that someone else has prescribed or how to keep the patient safe and comfortable between medical therapies [12].

If nursing is to realize its full professional status as a major contributor to meeting the health needs of people, it must move into a position of advocacy for patients' rights in matters of health. This step could revitalize the professional role to that of a fully participating member of the interdisciplinary team now caring for patients.

The framework of a patient's bill of rights has been well established by the American Hospital Association and the National League for Nursing. Full-scale implementation is both desirable and necessary. It would redefine the nurse-patient relationship, the nurse-physician relationship, and the employer-employee relationship of the hospital and professional nurse. Nurses can thus practice within a holistic, humanistic framework with the nursing process used in the resolution of major patient concerns and self-interests.

## REFERENCES

1. American Hospital Association. *Statement on a Patient's Bill of Rights.* Chicago: American Hospital Association, 1975.
2. A.N.A. Committee on Ethical, Legal and Professional Standards. *Code for Nurses with Interpretive Statements.* Kansas City, Mo.: American Nurses Association, 1976.
3. Annas, G. J., and Healey, J. The patient rights advocate. *J. Nurs. Hosp. Adm.* 4:25, 1974.

4. Bandman, B.   The Human Rights of Patients, Nurses, and Other Health Professionals. This volume, Chap. 47.
5. Berthold, J. S.   Advancement of science and technology while maintaining human rights and values. *Nurs. Res.* 18:514, 1969.
6. Carnegie, M. E.   The patient's bill of rights and the nurse. *Nurs. Clin. North Am.* 9:557, 1974.
7. Feinberg, J.   *Social Philosophy.* Englewood Cliffs, N.J.: Prentice-Hall, 1973.
8. Gray, D. E.   Letter to a young friend. *Dialogue* 3:27, 1976.
9. Hart, H. L. A.   The Ascription of Responsibility and Rights. In A. Flew (ed.), *Logic and Language.* Oxford: Blackwell, 1952.
10. Nelson, W. N.   Special rights, general rights and social justice. *Philosophy and Public Affairs* 3:410, 1974.
11. New York State Education Law.   Title VIII, Articles 130–139. Albany: New York State Department of Education, 1976.
12. Sheahan, D.   The Game of the Name: Nurse Professional and Nurse Technician. In M. H. Browning and E. P. Lewis (eds.), *The Expanded Role of the Nurse.* New York: American Journal of Nursing Co., 1973.
13. Szasz, T. S., and Hollander, M. H.   The physician-patient relationship. *Arch. Intern. Med.* 97:585, 1956.

# The Right to Health Care — Alternative Health Care Delivery Systems: Issues in Economics, Control, and Distribution

# CHAPTER 49

# The Right to Health Care: An International Perspective
*Victor W. Sidel*

Over the past few years the slogan, "health care is a human right, not a privilege," has been heard increasingly, not only from those on the political left but from many others as well. As with most slogans, the words are more an evocation or expression of an intuitive feeling than the precise exposition of a philosophical, moral, social, or legal concept. What the slogan does reflect is a deep-seated dissatisfaction with medical care as it exists in the United States, and indeed — albeit for the most part for other reasons — in most countries of the world. The demand for health care as a human right is a reflection of dissatisfaction with the undeniable problems of access to medical care of those who are relatively powerless in American society — the poor, prisoners, the mentally subnormal — and of the problems of access of the poor in other societies. Within the United States some of the problems that are perceived as limiting equitable access — and therefore limiting the practical assertion of the right to health care — are the geographic maldistribution of medical care resources, the emphasis on specialization with its attendant diminution of the resources available for primary medical care and chronic illness, and the institutional, economic, and social barriers to access among specific groups of the population.

Looked at in international perspective, it is clear that relief of certain human needs — the specific ones may differ from society to society — is seen as fundamental, not to be denied under any circumstances, and relief of other needs is seen as a luxury to be granted or denied as social or individual circumstances dictate. Among the needs that are seen as most fundamental, at least in modern industrialized societies, are needs whose relief would appear to be corollaries to the right to life: food, clothing, shelter, protection from assault, and, more controversially, an adequate level of education and of social stimulation and, more controversial still, of free suffrage and of social interaction. Whatever the philosophical basis of a "right," few would intuitively deny that in a society in which, for example, there is plenty of food to go around none should be allowed to starve for lack of it. As health care has come to be seen as more and more efficacious in the protection of health and the prevention of premature death and disability from disease, and as it has become more and more generally available, the need for it has come to be seen by many as fundamental a need — and the right to it therefore as fundamental — as that for food or shelter.

Even for food, however, there have been substantial recent questions raised about the universality of the right to access, when quantities appear to be limited, unless other social conditions — such as reduction of the rate of population growth — are imposed. Since health care is usually viewed as a resource whose supply is limited, this argument, too, has its parallel in the debate about access to health care as a right.

The reactions to the assertion of the right to health care in the United States have taken a number of different forms, in part because of the imprecision of the assertion of the right and in part because of ideological or practical objections to its assertion. One form of response has been to focus on the distinction between the demand for a right to "health care" and the demand for a right to "health." The World Health Organization puts the two rights into the same paragraph in the preamble to its Constitution [19]. This paragraph states the principle: "Health is a state of complete physical, mental and social well-being and not merely the absence of disease or infirmity." The paragraph continues: "The enjoyment of the highest attainable standard of health is one of the fundamental rights of every human being without distinction of race, religion, political belief, economic or social condition." And it concludes: "Governments have a responsibility for the health of their peoples which can be fulfilled only by the provision of adequate health and social measures."

The World Health Assembly, the annual meeting of the delegates of the member nations of the World Health Organization, declared without qualification in 1970 that "the right to health is a fundamental human right." The constitutions of some individual nations also recognize such a right. For example, the Constitution of the German Democratic Republic, adopted in 1968, guarantees to each citizen the "protection of his health and working capacity" [20].

An example of a recent analysis of the two kinds of rights was provided by John S. Millis, former President of Western Reserve University and author of the "Millis Report" on graduate medical education. In an editorial in *The New England Journal of Medicine* in 1970 he states: "Today, there are many voices advancing the idea that health is necessary to life, liberty and the pursuit of happiness, and that, therefore, it should also be regarded as a basic human right" [9]. Millis goes on to argue that a society can guarantee access to health services but not to health itself, since many of the factors that determine a person's health are independent of the health care system and indeed, he alleges, independent of the efforts of society itself "except through intolerable restrictions in personal liberty."

Many critics of a "right to health care" extend the analysis one step further. They allege that since much of personal health, and therefore much of the need for health services, is under personal control, those who do not take appropriate responsibility for the protection of their own health should not be granted an equitable right to socially supported health care services. They ask, for example, whether someone not wearing a seat belt who is injured in an automobile accident or someone who has smoked cigarettes heavily who develops lung cancer is entitled to the same right of access to socially provided health care services as are those who are injured or develop lung cancer without having personally contributed known excess risk factors to the process? In short, if someone by personal actions denies himself his right to health, does he still retain the right to health care if it implies additional costs to others in the society [4]? Clearly, if one is required to "earn" the right to health care, this is the expression of an argument against an implicit right to it.

A second form of argument against the assertion of a right to health care is much more direct and is based on the idea that the rights of some involve duties for others. A right to health care, this argument states, implies onerous duties for health workers, and it is immoral, illegal, or both to demand of doctors and other health care workers that they provide health care services in difficult locations or using methods other than those they would themselves choose to live in or use. This argument was sharply stated in 1971 by Dr. Robert Sade:

The concept of medical care as the patient's right is immoral because it denies the most fundamental of all rights, that of a man to his own life and the freedom of action to support it. Medical care is neither a right nor a privilege: it is a service that is provided by doctors and others to people who wish to purchase it [12].

A third form of the argument against the assertion of the right to health care is based on the allegation that it is economically impossible to provide the "best available" medical care to everyone in a society and that therefore the most that the right can imply is guaranteed access to some "decent minimum" of health care. In an excellent recent article taking this point of view, Charles Fried [3] reaches the following conclusions:

1. To say there is a right to health care does not imply a right to equal access, a right that whatever is available to any shall be available to all.
2. The slogan of equal access to the best health care available is just that, a dangerous slogan which could be translated into reality only if we submitted either to intolerable government controls on medical practice or to a thoroughly unreasonable burden of expense.
3. There is sense to the notion of a right to a decent standard of care for all, dynamically defined, but still not dogmatically equated with the best available.
4. We are far from affording such a standard to many of our citizens and that is profoundly wrong.
5. One of the major sources of the exaggerated demands for equality are the pretensions, inflated claims, inefficiencies, and guild-like, monopolistic practices of the health professions.

These conclusions lead Fried into a number of what he calls "practical proposals." Among them is the assuring of each person of "a certain amount of money to purchase medical services as he chooses." There is no notion in the conclusions and proposals that the right to health care implies equal access and is a step toward equity. In fact, there is an explicit statement that the right to health care has nothing to do with equal access, and there appears to be no consideration of the idea that the ability of some to purchase more or better medical care in the marketplace may, in the face of limited total resources, lead to a denial of the right to adequate health care, not to speak of equivalent or optimal care, to others. In short it appears to me that the argument that the right to health care implies no more than the right to some minimum standard of care while others in the society, whatever their relative need for care, can purchase all they want is in essence a negation of and an argument against the right [8].

Finally, a fourth form of the argument against a right to health care is that the vast majority of professional medical care is useless or dangerous, and since no one should have it, there is no right to it. The viewpoint is stated most explicitly by Ivan Illich, but many other recent analysts state arguments about the "limits of medicine" that can be used to support the same view [6].

Some of these arguments and the responses to them, and other issues related to health care as a right, have been recently discussed in *Ethics and Health Policy* [18]. Let us now turn to the patterns of medical care in some other selected societies and analyze their relevance to a right to health care in the United States.

Although the idea of a community physician, available at public expense to everyone in the community, was apparently known to the ancient Greeks and was introduced into what is now the Soviet Union by, of all people, Tsar Alexander II (as part of the Zemstvo reform of 1861), the modern expression of health care as a right has been largely advanced in the industrialized countries of Europe through social insurance mechanisms rather than through direct provision of services. Germany in 1881 under Bismark and Great Britain in 1911 under Lloyd George introduced methods of social insurance to cover the costs of medical care as a complement to previous inadequate poor law guarantees. The United States is now alone among the industrialized countries of the world in failing to have widely available social insurance for meeting the costs of medical care.

Britain in 1948 took a more momentous step toward making medical care a universal right. Finding that insurance mechanisms alone were insufficient to provide equal access and an adequate standard of health care for all its citizens, the Labor government under Aneurin Bevan introduced the National Health Service. The service, despite a 1974 reorganization, is still basically divided into three parts. The first is a general practitioner service in which each general practitioner has a panel of patients, usually about 2500 and usually consisting of whole families, for which he is paid by the government on a capitation basis and for whom he provides both office care and home visits as needed. The second is a hospital service, staffed by salaried consultants and other hospital personnel, to which access, except in emergencies, is controlled by the general practitioner. The third component is what might be called public health services, including visiting nurses. All these services, with minor exceptions, are provided without cost to the patient at the time of service. The absence of charge and the public control over the distribution and nature of the services has appeared to distribute medical care resources much more equitably throughout the country and to remove many of the barriers to access to the system, although serious problems remain. A conscious attempt to provide socialized medicine in a country in which only certain selected sectors have been socialized, to provide equality in health care in the face of inequities elsewhere in the society, seems to have led to a much closer approach to health care as a right than in most other nonsocialist countries [7, 17].

The pattern of development in the Soviet Union has been quite different. Starting from a far more limited economic base, the Soviet Union in the 1920s moved directly into a salaried, community-based service. Unlike the British attempt to meet all sectors of health care needs as equitably as possible, the Soviets concentrated early on maternal and child care services and on occupational health. The Soviets have also trained large numbers of health workers, both doctors and others, so that they now have a doctor-population ratio almost double that of the United States or Britain. All of them are salaried, and no charge is made for any services. As a result of these efforts, there has been enormous progress in some areas and undoubtedly an enormous increase in equity. On the other hand, the services are highly bureaucratic and appear to pay less attention to certain forms of chronic illness than to emergency and acute illness [2, 5].

China is an example of a much different process of attempting to provide medical care as a right for all [15, 16]. In China medical care, as education and all other human services, is seen less as an end in itself than as a way of transforming society and those who live in it. Health services are deeply embedded in a social matrix. The Chinese health care system illustrates a blend, which appears in many ways to be unique to China, of "self-reliance," responsibility on the part of the individual to protect his own health and that of his neighbors, and of societal responsibility to organize and channel health care activity for local, regional, and nationwide well-being. Health care activity represents a complex mix of decisions made by individuals themselves, although within a common ideological and cultural framework, and decisions made at a series of organized societal levels. At the most local of these societal levels, where in a decentralized society such as China's many of the most important decisions are made, individual and societal decision making are so tightly intertwined as to be almost indistinguishable.

To "serve the people," to work for the good of the society, is the widely stated ethic, expressed in countless signs and posters and in the conversation of all with whom we spoke in China. The health care system has attempted to utilize and promote this commitment to the welfare of the group. The focus on mass participation, mass education, and "mobilizing the masses" to engage in sanitation work, wiping out schistosomiasis, intense efforts in immunization, or in birth control programs can all be seen as part of a broader societal view. Whether one's child is immunized against polio and whether one is treated for one's illness becomes the formal concern of the local community, whose stake in the immunization program is emphasized over its importance to the individual and whose share in the treatment program is communally based [14].

Eighty percent of China's people live in its rural areas, and one of the most important steps toward equity in health service has been the shift of resources from the cities into the countryside. In the last few years, about one and one-half million barefoot doctors have been trained. They receive a preliminary course of training for three to six months conducted by commune health centers, county hospitals, or mobile medical teams. After a period of practice, refresher courses are given from time to time, so that they are able to take on more and more responsibilities. After several years of service, some of them are selected by the communes to receive regular medical education in medical colleges; they then return to their own communes to continue providing medical care.

In rural areas a cooperative medical service system has been adopted. The funds of the cooperative medical service come partly from the peasants themselves and partly from the public welfare fund of the production brigades, and in case of necessity the service receives a subsidy from the government. With the cooperative medical service, medical care is given free at the time of service except for a small registration fee. Should a patient be sent to the county hospital or any other hospital, all medical fees are paid from the cooperative funds. In urban areas, and in some instances in rural areas, there are at times larger direct payments by the patient for care.

There are health workers analogous to the barefoot doctors in the factories and in the urban neighborhoods. They provide preventive medicine, health education, occupational health services, first aid, and limited primary care on the factory floor or in the factory health center. Supervision and continuing education, as well as referral, are provided through the doctors and assistant doctors in the factory or in the neighborhood clinic. The worker doctor, like the barefoot doctor, performs health work part-time while continuing his or her other duties and is paid a salary similar to that of other workers in the factory.

The cities of China are divided into districts of several hundred thousand people; districts are divided into "neighborhoods" or "streets" of about 50,000 people; neighborhoods are divided into "residents' committees" or "lanes" of about 1500; and residents' committees into "groups" of about 100 people. Services are decentralized to the most local level at which they can be given. Residents' committees usually have health stations in which the personnel, local housewives who are called "Red Medical Workers," provide first-level care. They are trained for short periods and supervised by doctors and assistant doctors who work in the clinic or hospital at the neighborhood level; they can refer patients to those facilities or directly to the district general hospital when necessary.

People in the community are mobilized to perform health-related tasks. During the decade after China's liberation in 1949 a large-scale attack was made on illiteracy and on superstition. People were encouraged to build sanitation facilities and to keep their neighborhoods clean, and campaigns were mounted against specific diseases. In all these health campaigns it was repeatedly stressed that health is important not only for the individual's well-being but also for that of the family, the community, and the country as a whole.

During the Cultural Revolution (1966–1969), much reorganization was undertaken in medical education. Higher medical schools began to admit students who had had less schooling than previous entrants but who had the experience of working in factories and in communes; these students are usually selected by the people with whom they work and whom they return to serve. The curriculum has been restructured to place greater emphasis on practical rather than theoretical education, much more training in Chinese medicine has been added, and the length of time was experimentally reduced to about three and one-half years instead of the previous six.

National statistics are not yet available on the current health status of China's population, but recent visitors report a nation of healthy looking, vigorous people. Although much of China is still poorly developed technologically and its people — particularly in the rural areas — work very hard for long hours, there is no evidence of the malnutrition, infectious disease, or other manifestations of ill health that often accompany this level of development. Although changes in health care have certainly played an important role, the improvements in health status are not due to changes in health care alone; improvements in nutrition, sanitation, and living standards are at least as important as the changes in health care.

With experiences in other societies in mind, we can now return to the four arguments against a right to health care analyzed at the beginning of this paper. The first, a form of "blaming the victim," attempts to distinguish between a right to health and a right to health care and argues that people relinquish all or part of their right to health care by failing to practice healthful habits. But most health and illness are socially determined rather than individually determined. I refer not only to the obvious instances of environmental pollutants and other unhealthy societal conditions but also to the fact that most personal health practices are culturally and societally determined. Cigarette smoking, for example, is fostered by seductive advertising and by the easy availability of cigarettes. The percentage of teenage women who smoke is actually increasing while the incidence of lung cancer among women rises rapidly. As Sweden and other countries have shown, social action can change personal habits. For example, the wearing of seat belts and abstinence from alcohol in relation to driving can be societally imposed. It is largely society that — in the words of Norman Bethune — "makes the wounds"; in my view it is the right of every person to have that same society help protect him from the wounds and to help bind him after they are made.

The second argument holds that the right of doctors and other health workers would be abrogated by the maintenance of a right to health care for the patient, since this would imply a "duty" for the doctor. This argument neglects the fact that well over half of the costs of medical education and some 40 percent of the costs of health services in the United States are publicly financed. It also neglects the fact that restrictions on the number of medical students and on licensing create a near monopoly of the profession. The principle should be made clear in the United States, as it has been made clear in other countries, that accepting large quantities of public funds implies a significant obligation to the society. Perhaps morality or legality requires that some of these demands be made of those who enter medical school from now on rather than imposed ex post facto, but from a societal viewpoint the demands surely must be made.

The third argument, a much more subtle one, states that there exists only a right to a minimum standard of health care, not a right to equal access to the "best available." If one believes that a demand for equity and justice is the foundation and the goal of a society, the only way of approaching that goal in an inequitable and unjust society is to demand that those who now have least access will in the future have most access to the best health care that the society can offer. It is not simply a matter of "equality" of access to the best care; the poorest in the United States are also the sickest, and their needs for medical care are far greater than those of the wealthy. Even equality of access to the best care is inherently discriminatory. What we face is a matter of equity, and equity demands redistributive justice, for health care as well as for the other resources of the society.

The fourth argument, that medical care is not worth much and may produce more harm than good, strikes me as the least relevant of the arguments against a right to care. Even if the allegation is true, so long as the vast majority of the rich consider medical care a resource worth having for the protection of their health, it seems inequitable not to provide access to that same care to the poor. The time when those with easy access to the care decide to forego it will in my view be time enough to deny a right to care to those whose access to it is now difficult or denied.

If we are indeed to view medical care as a right in the United States, not to be abridged because of income, geographic region of residence, culture, or other factors, how can that theoretical right be made into a practical reality? Certainly it is a sine qua non that significant financial barriers must be removed between the patient and the medical care system. But this is not enough. Even after the direct payment barriers are removed, substantial barriers to exercising the right will remain. These barriers, as analyzed by Bergner and Yerby, include information barriers, physical and program barriers, and barriers imposed by professional attitudes and training. Even the financial barriers are more complex than can be solved by simply removing the direct charges for medical care; costs of transportation, of baby-sitters, and of lost time from work also act as barriers to equitable access to health care for the poor [1].

The only possible way to overcome these barriers is through the construction of a national health system, as in the United Kingdom and the Soviet Union, but based on our own models. Care must be provided for everyone within defined geographic boundaries, using structured methods for movement into and through the system for receiving care, by a system that forbids the making of profit and the construction of corporate empires in health and that uses the full potential of everyone working in it. But the Chinese experience suggests that much more than even this is needed: that health care, and indeed all response to human needs, must be woven into the fabric of the community itself. Only when each person in the society truly has reason to believe that the best way for his or her personal needs to be met is to help in meeting the needs of everyone in the society, that the best way to raise the standard of living for one's self is to raise the standard of living of the entire society, that one's own right to life, liberty, and pursuit of happiness is entirely dependent on the rights of others to the same level of access to resources and power, can a right to health care or to any other kind of care become a right capable of being fully exercised in the United States.

There is yet one more issue that must be addressed in a paper on an international perspective on the right to health care. It is the broader ethical question of equity and justice among nations, the morality of the people of one country having relatively abundant medical care (not to speak of abundance of food, clothing, shelter, and other necessities of life) while the people of many other countries have little medical care and indeed little of anything but hunger, illness, and despair. The injustice, the immorality, and the ethical bankruptcy of such a situation is to me clear on its face — although Garrett Hardin and others would argue that sharing our substance with others is to destroy ourselves by bringing too many people into the "lifeboat." In my view we destroy ourselves by *not* attempting to help those barely clinging to the gunwales.

That we need to help — that the right to health care or to other basic necessities is meaningless in one country unless it is granted to all — is to me clear, but how to accomplish it is less clear. Certainly we must do the obvious things: stop stealing resources (in the form of trained personnel and profits) from other societies, stop selling inappropriate practices (such as the profitable sale of infant formula by persuading mothers in poor countries to abandon breast-feeding [11]), and start shifting massive amounts of resources in the opposite direction. But direct bilateral transfer of these resources, even if it were politically feasible in the wealthy countries, would only in the short run increase dependence and increase human misery.

Again, some lessons from China may be relevant. China has taken the position that it will not attempt to export its own methods but will attempt to help countries that request China to do those things that the country wishes to do, whether that entails training barefoot doctors or building a railroad. The Canadian effort, through the International Development Research Center, is attempting in similar ways to be guided by the wishes of the people of the recipient countries. Even the World Health Organization, after decades of building "centers of excellence" based on methods from industrialized countries in developing countries that could spend less than a dollar a person per year on all health services, has begun to recognize that such efforts decrease equity and probably decrease overall health. The World Health Organization, as exemplified by recent decisions including the publication of a book entitled, *Health by the People,* is beginning to move in the direction of providing aid that emphasizes self-reliance and mass participation [10]. The United States, by sharing its abundance through international agencies that accept this principle, could be instrumental in helping to make possible the assertion of a right to health care in other countries as well as in the United States.

In summary, international experience suggests that the way to meet the objections that have been raised to the assertion of a right to health care is by building that health care into society in such a way that it is under the control of the people themselves. While this may seem at the moment like an unrealizable dream in a country like the United States, it is only by the assertion of the right and by vigorous efforts to come closer to its practical application that we can fulfill our moral commitment, both to those who need better health care today and to those who will as a result of the control over their own lives be healthier tomorrow [13].

## REFERENCES

1. Bergner, L., and Yerby, A. S. Low income and barriers to use of health services. *N. Engl. J. Med.* 10:541, 1968.
2. Field, M. G. *Soviet Socialized Medicine.* New York: Free Press, 1967.
3. Fried, C. Equality and rights in medical care. *Hastings Report* 6:29, 1976.
4. Fuchs, V. R. *Who Shall Live?* New York: Basic Books, 1974.
5. Hyde, G. *The Soviet Health Service: A Historical and Comparative Study.* London: Lawrence & Wishart, 1974.
6. Illich, I. *Medical Nemesis: The Expropriation of Health.* New York: Pantheon Books, 1976.
7. Levitt, R. *The Reorganised National Health Service.* London: Croom Helm, 1976.
8. Mechanic, D. Rationing health care: Public policy and the medical marketplace. *Hastings Report* 6:34, 1976.
9. Millis, J. S. Wisdom? Health? Can society guarantee them? *N. Engl. J. Med.* 283:260, 1970.

10. Newell, K. W. (ed.). *Health by the People.* Geneva: World Health Organization, 1975.
11. Rensberger, B. Drop in Breast Feeding Causes Problems in Poor Countries. *New York Times,* April 6, 1976.
12. Sade, R. M. Medical care as a right: A refutation. *N. Engl. J. Med.* 285:1288, 1971.
13. Sidel, V. W., and Sidel, R. *A Healthy State: An International Perspective on the Crisis in United States Medical Care.* New York: Pantheon Books, 1977.
14. Sidel, V. W., and Sidel, R. Self-Reliance and the Collective Good: Medicine in China. In R. M. Veatch and R. Branson (eds.), *Ethics and Health Policy.* Cambridge, Mass.: Ballinger, 1976.
15. Sidel, R., and Sidel, V. W. Health Care in the People's Republic of China. In V. Djukanovic and E. P. Mach (eds.), *Alternative Approaches to Meeting Basic Health Needs in Developing Countries* (a joint UNICEF/WHO study). Geneva: World Health Organization, 1975.
16. Sidel, V. W., and Sidel, R. *Serve the People: Observations on Medicine in the People's Republic of China.* Boston: Beacon Press, 1973.
17. Stevens, R. *Medical Practice in Modern England: The Impact of Specialization and State Medicine.* New Haven: Yale University Press, 1966.
18. Veatch, R. M., and Branson, R. (eds.). *Ethics and Health Policy.* Cambridge, Mass.: Ballinger, 1976.
19. World Health Organization. *Constitution.* Geneva: World Health Organization, 1960.
20. World Health Organization. *Health Aspects of Human Rights With Special Reference to Developments in Biology and Medicine.* Geneva: World Health Organization, 1976.

# CHAPTER 50

# Whether National Health Insurance?
*Malcolm MacKay*

*Last year* we spent 8.6 percent of our gross national product on health care. This amount is double the percentage spent 25 years ago. Does this mean we are twice as healthy? Regretfully, the answer is no. Life expectancy has remained nearly constant, causing one commentator to note that "there is no longer any significant relationship between the money spent on health and the longevity of the population" [2]. These facts beg the question: Just what have we gotten for the extra expense?

Some argue that modern medicine has little of real value to offer or, in fact, may be medically and socially destructive [3]. Many vehemently disagree. Others doubt the ultimate value of some modern medical treatment but believe there is a moral obligation to do everything medically possible, regardless of expense or probability of cure, in dealing with the seriously ill.

Medical, economic, and moral considerations come together in the current debate over national health insurance. The threshold question should be: How much health care should every citizen be entitled to receive? Senator Kennedy flew cancer specialists from around the world to Washington to consult on his son's condition. Is everyone entitled to this degree of care? If the answer is yes, what about competing societal demands? All of us, perhaps especially the politicians, have shunned attempts to define how much medical care is enough; such avoidance is no longer excusable. More encouragingly, public concern is focusing on a related question: Why have health costs escalated?

There are many partial answers to this question: increases in labor and equipment costs, technological advances, new drugs, and lack of meaningful health resource planning. Another partial but substantial answer that deserves more attention is the growth of third-party payment mechanisms — private and public — in the last 40 years. Medicare, Medicaid, and private insurance, while making health care available to millions of people who otherwise could not afford it, also have allowed both providers and consumers of health care to become insensitive to the true cost of such treatment.

During the last decade, largely through Medicare and Medicaid, the government has greatly expanded its third-party role in the purchasing of health services. It now accounts for well over 40 percent of the more than $139 billion expended annually for health in the United States, up from 25 percent of a total of $40 billion in 1965. Before Medicare and Medicaid, there were two natural brakes on a hospital's ability to pass along costs: the lack of prepayment coverage for the aged and payments of less than actual cost by the government for care rendered to the medically indigent. Now there are no such cost restraints.

The consequence of removing the consumer of health treatment from the usual process of the marketplace is that providers — particularly doctors — assume the role of both buyers and sellers. Although payments to doctors are, compared to payments to hospitals, a relatively small part of our total health expenditures, doctors determine how much or what treatment should be administered in what setting and by whom. Because the vast majority of doctors are reimbursed on a fee-for-service basis, there is a financial incentive to render treatment, be it necessary or not. Increasing the number of doctors through medical school expansion or by liberalization of licensing requirements for foreign-trained students may do nothing to lower the cost of doctor services and little to improve physician distribution. In fact, licensing more doctors may simply increase society's total health bill through greater use of facilities and more physician salaries.

How can health care costs be contained? Four very different ways, either alone or in combination, suggest themselves. One is socialized medicine, with the government running the hospitals and employing the medical practitioners as civil servants. This method would increase the demand for health care by making it free (or virtually free) to all. However, by underfunding the socialized system, demand could be controlled by long lines at doctors' offices and waits of months and even years for elective surgery. Less medical care for most of the population would be an ironic and regrettable but very possible consequence of a health care system run by the government.

Another way to control costs is through a public utility type of price regulation of doctor and hospital reimbursement. Although simple and direct, many believe this solution to one set of problems may create another set of difficulties. These critics draw an analogy from the failures of price regulation involving other industries. As Roger Noll of the California Institute of Technology put it:

Imposing public utility regulation upon [the] existing system is probably the only step that could be taken to make the industry even more insulated from incentives to be efficient [4].

Martin Feldstein of Harvard is equally adamant that direct regulation of physician fees and hospital payments is inappropriate because there is no "technically correct" way to set the quality of care [1].

A third suggested approach is to regulate the amount of health care resources. In many areas of the country we have overbuilt the hospital system, and this overbuilt system has generated a demand for doctors, staff, and equipment far beyond real needs. Empty beds cost almost as much to maintain as do full beds. More importantly, empty beds and underused medical professionals and equipment have a way of creating utilization. A hospital's high occupancy rate may bear little relation to that hospital's real social usefulness.

Limiting new hospital construction and actually closing hospitals, strict public review of new equipment purchases by hospitals and doctors, and placing limits on the number of new doctors entering the system each year may make eminent sense in terms of the overall public interest. It also, however, can create bad politics. Local communities, hospital worker unions, and professional staffs can create considerable pressure against hospital closures. Local hospital planning and approval agencies often reflect, in the words of Bruce Vladeck of Columbia, "pluralist interest-group representation . . . [which] . . . tend[s] to lead to bargaining, log-rolling, and collusive competition among narrowly-defined special interests, with the interests of the broader general public less well-served" [5]. Nevertheless, better resource allocation, combined with meaningful utilization review, can be achieved if the public cares enough to demand it.

The final suggested solution is very old-fashioned — bring ordinary marketplace incentives to the health care industry. This method would necessitate a restructuring of third-party health payment mechanisms, so that they no longer distort the demand for health service by relieving the buyer and seller of cost considerations. One way to achieve this goal is to require substantial coinsurance or deductibles scaled to income. This plan would make the patient and the doctor more reactive to costs and create something of a marketplace control on these costs. The development of alternative health delivery systems, such as health maintenance organizations where doctors are salaried rather than paid on a fee-for-service basis, and the publishing of drug prices and physician fees help to offer the consumer a choice.

Again, however, political reality intrudes. Publicly funded payment programs, collectively bargained benefit packages, and private insurance have succeeded in varying degrees in eliminating meaningful out-of-pocket health expenses for many Americans. Will many of us vote for politicians who advocate reducing our present health benefits? In addition, just as socialized medicine is the approach that is most progressive in terms of transferring income, deductibles, and coinsurance — like the sales tax — it may be seen as regressive in terms of income distribution.

The Carter Administration is focusing on cost control before it develops a comprehensive national health insurance proposal. How the cost issue is handled may determine the shape of the entire health system for at least a generation. At present, doctor and hospital price regulation and the direct delivery of care by the government are probably the most palatable approaches politically. However, as the public becomes more cognizant of the underlying economic realities, better resource allocation and marketplace incentives may become more acceptable. They deserve serious consideration, if for no other reason than the government's poor record of directly delivering services and regulating prices in nonhealth areas.

REFERENCES
1. Feldstein, M. S. The Feldstein Plan. In R. D. Eilers and S. S. Moyerman (eds.), *National Health Insurance.* Homewood, Ill.: Richard D. Irwin, 1971.
2. Feldstein, M. S. A new approach to national health insurance. *Public Interest* 23:93, 1971.
3. Illich, I. *Medical Nemesis.* New York: Pantheon Books, 1976. P. 13.
4. Noll, R. G. The Consequences of Public Utility Regulation of Hospitals. Paper delivered at the Conference on Regulation in the Health Industry, Institute of Medicine, Washington, D.C., January 7–9, 1974.
5. Vladeck, B. C. Interest-group representation and the HSAs: Health planning and political theory. *Am. J. Public Health* 67:23, 1977.

# CHAPTER 51

# Reflections on the Right to Health Care
*Joseph Margolis*

*Victor Sidel* invokes the slogan, "health care is a human right, not a privilege" [9].
If health *care* is a human right, it is hard to resist considering what share of the
bounty of the affluent West ought rightfully to be applied toward the global fulfillment of that right. The question is not really one of allegiances. As far as I am concerned, it is scandalous that the world goes hungry, forever wars against itself, and
suffers unnecessary disease and death. Rather, the question concerns what can be
said about the claim to a human right to health care. To turn to that question is to
intend to return, by a detour, to the actual plight of actual societies. Rejecting
Sidel's slogan is not primarily a way of repudiating personal or national concern and
responsibility in providing health care internationally (or even nationally) but, more
realistically, a decision regarding tactics and strategies with an eye to improving such
a service.

## THE LANGUAGE OF HUMAN RIGHTS
The language of human rights is extravagant and vacuous at one and the same time.
There is, of course, an importance in claiming that men are endowed with so-called
human, unalienable, or natural rights. The issue is largely historical and dialectical:
an emphasis on a favored form of equality as against some established privilege, and
a constant reminder that humane revisions of political life must be fairly interpret-
able in terms of the fundamental concerns of human endeavor. But the emptiness
of familiar, so-called human rights may be seen at a stroke once it is remembered
that, *even though inalienable,* such rights as that of life and liberty are not thought
to be necessarily violated by capital punishment. It could hardly be supposed for instance that if the right to life and capital punishment were logically incompatible,
that fact would have escaped the attention of the American and French revolution-
aries and their successors. But what is the point of appealing to the *human* right to
life if there are politically legitimate ways of taking life?

In theory, *no one can forfeit his human rights;* he can only forfeit his positive or
legal or political rights, that is, the actual, particular rights accorded and protected
in a particular society. By parity of reasoning, then, to claim that all men have a
natural right to health care tells us nothing about what may go contrary to that alleged right. Presumably, the systematic neglect of one suffering or starving part of
the world by another more fortunate part may be, however unkind, as correct in
terms of honoring human rights as any more generous effort. On the other hand, the
appeal to the idiom of human rights is an extravagance. Since health care delivery
directly and immediately entails large financial and other economic commitments,
insistence on the international (or even national) scope of that right cannot help
but be seen as an indirect claim to the fair distribution of the wealth and facilities
of privileged societies or of privileged groups within a society. It is difficult to suppose that the natural right to health care is simply discovered as a result of a straightforward inquiry of a securely nonpartisan sort. And it is notorious that the claims

of underprivileged societies or communities to the effect that they have a *right* —
let alone an inalienable right — to dispose of the economies of the privileged is both
unrealistic and counterproductive and may even invite anomalous questions about
the further natural rights of effective exploitative communities — for example, the
white settlers of America. The language of natural rights is simply insensitive to the
detailed needs and histories of the societies affected, which is not to deny, of course,
that at certain times it may well prove to be an ideologically effective instrument.

We may collect these considerations more systematically. By human, unalienable,
or natural rights, I understand nothing but those most generic, so-called prudential
concerns attributable to men, underlying all rational efforts to provide a moral justification for political, legal, economic, and related institutions [5]. I do not mean
by this that all rational efforts to justify institutions rest on the theory of natural
rights. I mean, rather, that all such efforts depend on the admission of certain prudential interests: such general interests as life, liberty, property, the pursuit of happiness, dignity of person, security of person. These interests tend to be statistically
prevalent. Also, they are determinable rather than determinate, in a sense rather like
that in which red, green, and yellow are determinate colors, whereas "color" signifies only that *some* color may be specified. In the context of rights, however, there
need be *no* determinate right corresponding to the admission of determinable rights.
That is just the point of citing the compatibility of capital punishment and the
natural right to life, of punishment and liberty, and of comparable positive constraints and any of the other putative natural rights. Determinate *interpretations* of
natural rights are, when institutionalized, the positive or legal or traditional rights
of some society. Being determinate, they may be determinately violated.

In the context of international health care, however, it would be quite unconvinc-
ing to suppose that we could stipulate precisely which determinate set of rights
should be internationally enforced. Certainly, we could not do this on the grounds
of having discovered the relevant natural rights. So-called positive rights, then, are
determinate entitlements authorized by some power competent to enforce their protection. Rights, from this point of view, are legitimate powers, residing in persons,
to posit certain valid claims, or those claims themselves [1, 2]. Broadly speaking,
then, the theory of rights may be said to distinguish rights in three rather different
ways: first, *natural rights,* which are determinable formulations of prevailing prudential interests construed as inherent and inalienable grounds for certain valid
claims on the part of persons; second, *positive rights,* which are determinate entitlements fixed by a particular law or tradition within a society; and third, *general rights,*
which are ideological specifications (or subspecies) of natural rights favored in one
historical context or another — determinable like natural rights, which are their
genera, but formulated in a way to favor enforcing or promulgating certain positive
rights and positive obligations.

A large part of the United Nations Declaration of Human Rights is concerned
with general rights; and the thesis that health care is a human right is, pretty clearly,
a putative instance of what I am terming a general right. So seen, insistence on the
right of health care is the exercise of a partisan instrument — here noted as such,
without prejudice to the substantive issue.

We may state all of this somewhat less formally. All speculation about rational public planning takes for granted that human beings tend to favor certain interests, as of life and liberty. These are the prudential concerns that every policy must make provision for. But there are indefinitely many ways in which such concerns may be managed in different societies and in societies at different times in their development. To call them natural rights is simply to favor one among a number of alternative views of their role in a political doctrine and program. In any case, they are, inevitably, compatible with a great variety of particular institutions — some of which (capital punishment for instance) show that nothing politically significant follows merely from the admission of such rights. Only positive rights count, in the sense that only the particular social arrangements that particular societies understand as a fair way of protecting natural rights are effectively recognized and operative rights. These positive rights, for instance, that one has the right to a trial by jury under certain circumstances, cannot, on the foregoing argument, be directly derived from a set of natural rights. What is needed is an intermediate set of general rights; that is, some frankly partisan conviction about how to interpret natural rights so that particular positive rights will be generally understood as the fair arrangements they are taken to be. General rights, then, provide the flexibility needed in order to reconsider whether actual positive rights do or do not "really" ensure our natural rights, but they also yield a certain inflexibility, both in terms of stubborn convictions about existing rights and in terms of resistance to other ways of thinking about human needs and interests.

These considerations may seem very wide of the mark of what should be provided in the way of international or national health care. But they are actually pointedly pertinent. For, in the absence of some form of moral cognitivism [7], determinate rights can only be justified in relation to some favored ideology and to certain more general and relatively neutral constraints, such as those of consistency and coherence, historical relevance, and plausibility as an interpretation of generic prudential interests. If so, then, regardless of our loyalties, appeal to the natural or general right of health care cannot fail to be the expression of a larger conviction regarding the valid disposition of the wealth, resources, technology, and labor of various privileged communities. It is safe to say that there is no recognized criterion for assigning the entailed obligations neutrally. Hence, talk of the right of health care, however useful, cannot fail to be marked as a maneuver intended to be ideologically persuasive and fruitful: hence, to be assessed in terms of the advantage of alternative idioms.


## THREE IDIOMS

It is important to emphasize that, by and large, there are at least three distinct idioms for formulating the proper moral engagement of mankind. One is the idiom of rights, which entails correlative obligations or duties; for if A has a right, then B is obliged to respect it. But, second, we may opt for the idiom of duties or obligations itself, which can be operative without the provision of any individual rights at all [2]. Third, there are various idioms that eschew rights and obligations altogether, which emphasize, instead, virtues or what is right and wrong or what is good and bad [6]. The advisability of casting the issue of health care in terms of one or another of these *idioms* depends in large part on the objectives to be realized and the susceptibility of the populations to be mobilized.

## HUMAN RIGHTS AND MORAL COGNITIVISM

It is clear that, if the doctrine of natural rights is both vacuous and extravagant, and if, in the absence of a moral cognitivism, the choice of one idiom or another is a matter of political strategy and rhetorical advantage, we should see the question of contributing to international or national health care for what it is: a dialectically engaged debate about what, given the personal and political commitments of the members of our own society, is convincingly congruent with those commitments. Here, the complications are plain: Independent political states and economic communities within states having strikingly unequal facilities and power; conflicting and even inimical ideologies; the absence of international controls realistically linked to facilitating more than the most moderate programs of cooperation, or capable of being managed independently of national rivalries; and, perhaps most important, the absence of a global ideological conviction that would justify the claims of citizens of one part of the world to a share of the medical facilities of another. There is some reason to think that exploiting the idiom of rights may be relatively more useful within the United States than in the international setting. Even in the first case, it is bound to generate an adversary concept of medicine that, in time, may prove counterproductive. It is much more convincing to think of encouraging contributions to a national or international program of health care in terms of recognizing the *medical needs* of various communities and their members than in terms of the inalienable *human right to medical care* and the entailed obligations. Both idioms are linked to our basic prudential concerns. At the present time, the first is less controversial and less ideologically freighted than the second. In fact, it permits a recognition both of obligations and of virtues, and it is not incompatible with a doctrine of rights. The second requires the direct admission of valid claims and is bound to be resisted.

Both Victor Sidel [9] and Malcolm MacKay [4] complicate the issue hopelessly by pretending (opposed to one another as they are) to be in possession of the correct moral orientation of medicine. MacKay holds essentially to a marketplace conception of medical service — fee-for-service; but he inclines uneasily in the direction of "minimum performance standards" [3, 8]. Roughly, then, he rejects a natural or general right to health care but seems prepared to consider some obligations regarding performance standards *once* particular contracts are instituted. Thus seen, he answers his initial question ("Is there a universal entitlement to health care?") by default. Given our analysis, he could have acknowledged a "human" or "general" right to health care and maintained his present position, though it is true at the present time that the use of the idiom of rights is construed as entailing certain priorities and constraints affecting the free market conception. Sidel understands the alternative this way as well. Intent on promoting medical care as a right in the

United States, Sidel stresses that "it is a *sine qua non* [but no more] that significant financial barriers must be removed between the patient and the medical care system"; and he draws loosely on the Chinese achievement in community-based medicine — apparently an effective blend of policies based on the themes of "self-reliance" and "serve the people" — in order to urge that "health care, and indeed all response to human needs, must be woven into the fabric of the community itself." Somehow, the issue has been confused. For one thing, there is no reason to think that the Chinese People's Republic is committed to a natural rights doctrine; second, the collectivism of the Chinese is hardly portable to the United States; and third, it can hardly be denied that health care delivery is "woven into the fabric" of the American community or that, to some extent at least, "significant financial barriers" have been removed between the patient and health care system.

SOME DIFFICULTIES

The real difference between MacKay and Sidel lies elsewhere. MacKay wishes to emphasize *the minima of public responsibility* for health care delivery in a society presumed to function satisfactorily as a market economy. And Sidel wishes to emphasize the public provision of *adequate health care delivery* in a society that, by comparison with collectivistic paradigms, is failing because it is a market economy [10]. Once it is put this way, however, the issue has *some* empirical import. There are also some ironies. Both MacKay and Sidel are primarily concerned with improving health care delivery in the United States. Neither is directly concerned with the provision of health care as an internationally recognized right. And both are committed to a somewhat hidden agenda, namely, the compatibility of a proposed health care system with the adequate functioning of the entire society — where what is to count as "adequate" subtends radically opposed political ideologies. MacKay is committed to an updated version of laissez-faire; and Sidel, to a liberal-oriented socialism. MacKay emphasizes, therefore, the cost to the economy and the threat of inefficiency in intruding considerations that are not directly in accord with the dynamics of a market system. And Sidel emphasizes the priority of certain humane considerations that the economy must serve, even at the price of revising a market orientation. The empirical remainder, the range of comparison, lies essentially with estimating the resilience of a market economy to accommodate the kind of health care improvements that critics like Sidel advocate; the ideological issue, which most assuredly is not open to direct empirical resolution, concerns, all things considered, what point between MacKay's "minima" and Sidel's "adequate" to favor and with what attendant political changes.

My own partisan view is that a free market conception altered to accommodate the admittedly humane concerns that Sidel emphasizes can only be a patchwork, can only be dialectically unpromising in a setting in which corporate life and corporate responsibility dominate our thinking. But that is another way of saying that it is utterly simplistic to speak of managing the health care problem in the United States, at the present time, in terms of choices of global ideologies, alternative rhetorical idioms, the adoption or rejection of a market economy, the application or rejection of the Chinese model, or anything of the sort. These doctrinal disputes permit only the gradual shaping of social conscience, hence of social forces that may one day have to be reckoned with. It is the comparison of actual health care delivery around the world that fixes the empirical margin of realistic dispute within the United States. The perceived need for improvement, here, yields under pressure, to changes that are likelier to subvert the ideological quarrel between MacKay and Sidel than to be led by it.

REFERENCES
1. Bandman, B.  Rights and claims. *Journal of Value Inquiry* 7:206, 1973.
2. Feinberg, J.  The nature and value of rights. *Journal of Value Inquiry* 4:243, 1970.
3. Fried, C.  Equality in rights on medical care. *Hastings Report* 6:29, 1976.
4. MacKay, M.  Whether National Health Insurance? This volume, Chap. 50.
5. Margolis, J.  *Negativities. The Limits of Life.* Columbus, Ohio: Charles Merrill, 1975.
6. Margolis, J.  The conceptual aspects of a patient's bill of rights. *Conn. Med.* 39:582, 1975.
7. Margolis, J.  Moral cognitivism. *Ethics* 80:136, 1975.
8. Nozick, R.  *Anarchy, State and Utopia.* New York: Basic Books, 1974.
9. Sidel, V. W.  The Right to Health Care: An International Perspective. This volume, Chap. 49.
10. Williams, B.  The Idea of Equality. In Peter Laslett and W. G. Runciman (eds.), *Politics and Society* (second series). Oxford: Blackwell, 1962.

# CHAPTER 52

# A Note on Bioethics
*Sidney Morgenbesser*

*Many defend* or seem to defend the thesis that there is a right to health care or perhaps more fully, but still schematically, a fundamental or derived natural right to equal basic health care. They further seem to claim that this right obligates governments and states to institute some form of socialized medicine for all its citizens or perhaps for all its inhabitants. I am sympathetic to a modified version of the conclusion about policy, agnostic about the premise, and skeptical about the argument. The transition from right to governmental obligation may be too abrupt; at best, governments and states may have some form of conditional obligation.

No obligation on a specific state need arise if its members have their health needs met in a morally decent way by private institutions. And though private practice and institutions may fail in a given state, that state — perhaps because of lack of adequate resources — cannot be obligated to institute some meaningful form of health care. Of course, though state A may be too poor to do anything about the health care of its citizens, state B may be rich enough to assure adequate health care both for its citizens and the citizens of A. One may argue that state B is obligated to help A provided that state A is the poorest of states, state B is in a position to help, and health care is the direst of needs. Note the emphasis on offer; state B cannot legislate for members of other states.

It is of course open to those who emphasize the natural right to health care to modify their position and claim that the natural right in question gives rise to a relevant prima facie obligation upon governments. Note, however, it is equally open to many who do not begin with the notion of natural right to reach similar conclusions about policy. Those who speak of natural rights should not dismiss notions of natural obligations and duties, duties and obligations we simply have because of our humanity and which do not arise because of some actions or commitments on our part. And it may be that we have natural obligations that are logically prior to natural rights, obligations toward others that do not arise because of their prior rights. And among such natural duties and obligations may be not only such negative obligations as not to inflict pain and cause suffering but also positive ones to help relieve the sufferer and help heal the sick.

Previously, it may have been believed that we could meet these obligations by relying upon doctors to donate time and services to the ward and clinic and by relying upon relevant others to contribute to the upkeep of hospitals that would be available to all. For a variety of reasons we believe, and I think correctly, that the system of private fulfillment of such obligations cannot work and that socialized medicine is a better way of meeting them. A decent society learns to socialize its members' obligations: Members of a decent society may know that they are obligated to tax themselves for adequate health care for all. In such societies it is not only prudence that motivates commitment to national health care. Members of a decent society do not want to have relative advantages over others in matters of sickness and health.

360

Here then is a case for socialized medicine that has some appeal to me but which I cannot defend here. I add that neither the case that I have suggested nor the one that begins with the notion of natural rights may be the best argument to use in order to justify a socialized health care program to members of our culture. For the best strategy might very well be the one that begins with the ideals of democracy and decency to which our culture seems to be committed and draw the consequences. In morals, as in science, we are in the middle; there may be no road to justification that begins from a point outside the culture and tries to judge it transcendentally. And I have begun from my culture and have seen myself as offering a comment on A Physician's Prayer widely but perhaps doubtfully attributed to Maimonides:

Inspire me with love for my art and for your creatures. Do not allow thirst for profit, ambition for renown and admiration to interfere with my profession. For these are the enemies of truth and can lead me astray in the great task of attending to the welfare of your creatures. Preserve the strength of my body and soul that they may be ever ready to help rich and poor, good and bad, enemy as well as friend. In the sufferer let me see only the human being [1].

One may be agnostic about the thesis that there is a right to health care and partisan of the view that there are moral rights, for example, the right of the patient to be informed, that ought to be respected in medical practice. Similarly, one may be skeptical about the necessity and advisability of some form of socialized medicine and with good reason be committed to the view that our health care system is in need of revision. The literature is filled with complaints about our doctors and hospitals. Doctors are criticized for being insensitive to the rights of patients and nurses, for wanting to continue their old and, to them, comfortable roles of father figure and moral authority; our hospitals are condemned for being too big, too inefficient, and too motivated by market considerations.

The case for changes in our medical system is strong and familiar. The case, once admitted, calls for deep changes in national and local policy toward industry, slums and cities, rural areas; requires shifts of investments from hospitals to preventive medicine. Macro-ethical-medical issues need be considered as well as micro-ethical-medical issues, with emphasis on rights of patients and nurses.

Many hospitals now advise patients of their rights, and some even present patients upon entry to the hospital with a bill specifying their rights. Indeed some have gone further and have provided money for ombudsmen who fight for patient rights; after all patients cannot be expected to picket. It may be useful, however, to remember that patients have a right to waive their rights, and it may not be a bad thing if they do so. Provided there is well-founded trust in the background, there is nothing wrong, I think, if the patient says to the doctor, "Look doctor, do what you think is best, I trust you." It may simply be a misapplication of our democratic ideals to expect or desire that the patient in a hospital always wants to make his or her own decisions. There is a converse problem if we approach the situation from the vantage point of the doctor. Many doctors seem to believe that there are two obligations upon them — to inform the patient and to do everything for the best interests of the patient. But these obligations may conflict; it is not evident which obligation has priority when they do. The patient has a right to be informed and to decide upon the mode of treatment; it is not necessarily wrong if the doctor decides.

Some of the remarks about the rights of patients can be generalized and applied to nurses. For they too have the right to be treated with dignity and respect, and one may add the obligation to treat others similarly. In contradistinction to patients, however, nurses can be expected to press for their rights and the cogency of their legitimate claim that they be assigned tasks that enable them to meaningfully use their talents and training and to make relevant decisions. The same case can be made for all workers in a hospital.

And it is only perhaps because they think the answer obvious that nurses do not specify a decision procedure for deciding which decisions should be made by them and which ones by others. They may very well believe that with relevant information all men and women of goodwill could agree about these matters. Who then are relevant agents of goodwill? The answer all the workers — doctors, nurses, and so on — in the hospitals is not quite the right one. After all, members of the community are affected by hospital practice and they too should be involved in decisions about hospital practice and aim. Should we then say that decisions should be made by local communities?

There is no point continuing to raise these questions that do not admit of easy or even of relevant answers until we are clear about the macro-ethical-medical issues to which I have alluded. Then we may know which decisions about hospitals should be made on a national level, which ones on a local level, and which decisions should be left to hospitals to treat autonomously. And even then we will not be clear about our obligations to citizens of poorer states.

## REFERENCE

1. Fredenwald, H. *Jews and Medicine.* Baltimore: Johns Hopkins, 1944. Vol. 1, p. 28. See the discussion in S. W. Baron, *A Social and Religious History of the Jews* (2nd ed., rev. and enlarged). Philadelphia: Jewish Publication Society of America, 1958. Vol. 8, p. 239.

# CHAPTER 53

# Some Considerations Regarding Ethics and the Right to Health Care
*Michael R. McGarvey*

*It is important,* I believe, in discussing ethical premises of the notion of the right to health care to reinforce a distinction made by Charles Fried [2] between the "Right *to* health care" and "Rights *in* health care."

## RIGHTS IN HEALTH CARE – THE DOCTOR-PATIENT RELATIONSHIP
Rights *in* health care assume that the patient has established contact with a provider of health services. Rights *in* health care derive from the traditional relationship between physician and patient, which, in turn, is based upon a patient's expectations of the physician [1]. Among these expectations are the following: The welfare of the patient comes first; patients are not to be harmed; patients are not to be killed; the professional is competent in the field; and the physician is dedicated to life and health.

Once a relationship is established between a patient and a physician or other health care provider, a covenantal and a presumably therapeutic relationship develops.

The ethics underlying professional behavior in the "doctor-patient relationship," and therefore the patient's rights *in* health care, are traced most comfortably to those theories of normative ethics belonging to the formalist school (deontological theories). Ethicists of this school examine the bonds and relationships between people, groups of people, and institutions. The formalists judge actions and practices on the basis of these relationships, not solely on consequent "out-comes." Lying and betraying a trust would be seen by a formalist as disrupting bonds between people. From this ethical point of view, such actions would not be justified simply to achieve a favorable end result.

## MEDICAL ETHICS
Until rather recently "medical ethics" and the enforcement of professional standards of behavior were left largely to the benign neglect of professional physicians' organizations. Several events of recent years have, however, focused public attention on professional behavior and forced legal involvement: (1) the discovery of ethically questionable examples of experimentation with living human subjects; (2) development of life-system-sustaining mechanical devices such as the respirator; and (3) the development of organ transplant techniques, with associated issues of conflict of interest.

Despite these developments, the nature of the relationship between a given patient and a given provider of health care has remained relatively consistent with tradition. This relationship has, far more often than not, been based upon the expectations noted earlier and their satisfaction.

363

Though the area of rights *in* health care is a field rich with intriguing and important topics for thought and debate, I believe that the issue of the right *to* health care poses more far-reaching, socially significant questions.

## THE RIGHT TO HEALTH CARE

The right *to* health care deals generally with access to health care services and, in a more limited sense, with access to particular types of high technology medical services. Whereas rights *in* health care depend largely upon the behavior of individual providers, the right *to* health care is a function of social policy and is, therefore, largely within the realm of politicians, policymakers, special interest pressure groups, and bureaucrats.

At this point one enters the world of the utilitarian ethicists, economists, and cost-benefit analysts, aligned in varying array with or against physicians, therapeutic nihilists, and advocates of social egalitarianism. The complexities of this slippery terrain have led as distinguished a moral thinker as Paul Ramsey to conclude that the large question of how to choose between medical and other societal priorities is "almost, if not altogether, incorrigible to moral reasoning" [7].

In light of the massive spending on health care services in the United States, which now approaches 9 percent of the gross national product, economists suggest that we have reached, if not far exceeded, the point of diminishing returns on our health care dollar. Fuchs [3] points out that the connection between health and medical care is not nearly as direct or immediate as often believed. He supports his contention by pointing out that health differences are probably far more dependent upon genetic factors, environment, and life-style than upon medical services. He concludes that "We cannot have all the health or all the medical care that we would like to have. We have to choose." Fuchs further asserts that, though economics may assist in the rationality of decision making, it cannot provide the ethical base or value judgments involved.

From a different point of view, professional iconoclast Ivan Illich [5] suggests that the risks of contact with physicians and the medical care system may, in fact, outweigh the benefits to be gained. There is no question that the powerful new medical technology is a sword capable of cutting in both directions — to cure and to kill. And skepticism is not limited to lay observers. In a recent survey, fully 38 percent of nurses indicated that they would not wish to be hospitalized in the institutions in which they worked [4]. Advertisements for a recent popular book by a physician author proclaim in large letters, "Warning: Your doctor may be hazardous to your health."

Despite all of this, people *want* medical care. They *value* health services. Therefore, to the extent that ours is a society based upon principles of freedom, justice, and equality of opportunity, we must attend seriously to those who claim health care — a complex of services essential to the pursuit of life, liberty, and happiness — as a right that should be available to all regardless of the ability to pay.

## AN ETHICAL RATIONALE

Efforts to establish the ethical basis of the claim to health care as a right have looked to standard conceptions of social justice. These conceptions may be summarized as follows: (1) to each according to his desert; (2) to each according to his societal contribution; (3) to each according to his contribution in satisfying whatever is freely desired by others in the open marketplace of supply and demand; (4) to each according to his need; and (5) similar treatment for similar cases [6].

In his extensive and thoughtful article on this subject, Outka [6] questions the possibility of justifying morally the goal of assuring comprehensive health services for every person irrespective of income or geographic location. After an analysis of each of the above approaches to social justice, he concludes that a reasonable and morally defensible case can, in fact, be made for equal access to health care delivery on the basis of the principle of (5) (above): similar treatment for similar cases.

This formulation has the added advantage of permitting one to deal with realities of the situation in which resources are finite and demands seem apparently insatiable. Outka writes:

Illness is the proper ground for the receipt of medical care . . . . So (1) if we accept the case for equal access, but (2) if we simply cannot, physically cannot, treat all who are in need, it seems more just to discriminate by virtue of categories of illness, for example, rather than between the rich ill and the poor ill . . . OR with Ramsey . . . we may urge a policy of random patient selection when one must decide between claimants for medical treatment unavailable to all . . . [6, p. 197].

## DIFFICULT DECISIONS

The principle of similar treatment for similar cases clearly permits *no* treatment for similar cases. The prospect of excluding from treatment whole categories of patients with certain conditions that are not amenable to generally successful treatment or ultimate rehabilitation — for example, cardiac shock, certain rare noncommunicable diseases, many forms of cancer — is less than an attractive one. Randomly selecting patients for limited treatment capacity is similarly troublesome given the potential for "fixing the lottery," either intentionally or through continued influence of long-standing societal inequities.

Unless, however, the public decides that all other social priorities are to be subjugated to health services delivery, difficult decisions will continue to be made, either consciously or by default. We all ought to be involved in considering, advancing, and debating the most just and ethical approaches to such decisions.

## REFERENCES

1. Dyke, A. Lectures, Kennedy Institute, Center for Bioethics, Washington, D.C., 1974.
2. Fried, C. Rights and health care — beyond equity and efficiency. *N. Engl. J. Med.* 293:241, 1975.

3. Fuchs, V. R. *Who Shall Live? Health Economics and Social Choice.* New York: Basic Books, 1974.
4. Funkhouser, G. R. Quality of care, part one. *Nursing 76* 6:22, 1976.
5. Illich, I. *Medical Nemesis: The Expropriation of Health.* New York: Pantheon, 1976.
6. Outka, G. Social justice and equal access to health care. *Perspect. Biol. Med.* p. 185, Winter 1975.
7. Ramsey, P. *The Patient as Person.* New Haven: Yale University Press, 1970. P. 240.

# Index

# Index